Liquid Crystal Research and Novel Applications in the 21st Century

Liquid Crystal Research and Novel Applications in the 21st Century

Guest Editor

Ingo Dierking

MDPI

Basel • Beijing • Wuhan • Barcelona • Belgrade • Novi Sad • Cluj • Manchester

Guest Editor
Ingo Dierking
Department of Physics and
Astronomy
The University of Manchester
Manchester
UK

Editorial Office
MDPI AG
Grosspeteranlage 5
4052 Basel, Switzerland

This is a reprint of the Special Issue, published open access by the journal *Crystals* (ISSN 2073-4352), freely accessible at: https://www.mdpi.com/journal/crystals/special_issues/MSNM4E5AGY.

For citation purposes, cite each article independently as indicated on the article page online and as indicated below:

Lastname, A.A.; Lastname, B.B. Article Title. *Journal Name* **Year**, *Volume Number*, Page Range.

ISBN 978-3-7258-3873-8 (Hbk)
ISBN 978-3-7258-3874-5 (PDF)
https://doi.org/10.3390/books978-3-7258-3874-5

Cover image courtesy of Ingo Dierking

Contents

About the Editor

Ingo Dierking

Ingo Dierking received his PhD in 1995 from the University of Clausthal in Germany. After an employment at the IBM TJ Watson Research Center in the US, working on electronic paper, he joined Chalmers University in Gothenburg, Sweden, as a Humboldt fellow. There he was appointed as docent, before joining the University of Darmstadt in Germany as lecturer for several years, and eventually moving to the Department of Physics and Astronomy at the University of Manchester in 2002. Ingo Dierking has published more than 175 scientific papers, as well as several books on topics of liquid crystal research. He is the 2009 awardee of the Hilsum medal of the British Liquid Crystal Society (BLCS), the 2016 winner of the Samsung Mid-Career Award for Research Excellence of the International Liquid Crystal Society (ILCS), the 2021 awardee of the prestigious Gray medal and in 2023 received the Luckhurst-Samulski award. Ingo Dierking is the editor of Liquid Crystals Today, a past President of the ILCS and a former Chair of the BLCS. His current research interests are broadly focused on soft matter systems with an emphasis on liquid crystals and LC-based composites with polymers and nanoparticles and the use of machine learning in liquid crystal research.

Preface

The idea for this Special Issue arose after several conferences on liquid crystals, when we realized that our research field had substantially departed from the original aspects of liquid crystal research: the synthesis and characterization of novel materials, and the drive to optimize their properties and develop new concepts to improve the LC displays that had taken the world by storm. In the last 15 years, this area has changed significantly. Of course, new materials are still being developed and displays are being improved, but there has been a shift in and widening of the scope of liquid crystal research. Colloidal liquid crystals have been developed, and active LC systems, biological liquid crystals, and polymer modification have grabbed the interest of researchers. The field of applications has widened significantly to lasers, biological sensors, (bio)-photonics, drug delivery, micro-cargo transport, smart glass, and much more.

It is an exciting and enlightening time for liquid crystal research, and I would like to thank all the authors for their efforts to contribute articles that reflect this broadening of liquid crystal-related research and its applications as we have moved into the 21st century. I would also like to thank the staff at MDPI's journal *Crystals*, and especially Lea Li, for all their work in organizing potential author lists, finding reviewers, and providing general editorial support.

And lastly, I hope that you, the reader, will find this Special Issue as interesting and thought-provoking as I did.

Ingo Dierking
Guest Editor

crystals

MDPI

rial

quid Crystal Research and Novel Applications in the st Century

Dierking

Department of Physics and Astronomy, University of Manchester, Oxford Road, Manchester M139PL, UK; ingo.dierking@manchester.ac.uk

Liquid crystals (LCs) are largely known as the materials used in flat panel displays, from simple pocket calculator displays to laptop screens, all the way to large TVs [1]. The technology is, in principle, still based largely on the original work from half a century ago, albeit of course with numerous advances and improvements [2] in materials, addressing, electrode patterns, alignment, and processing. While excellent displays with great viewing angle characteristics and video capability have been available since the turn of the millennium, fundamental display research has slowed down and progressed towards 3D vision, holographic displays, augmented and virtual reality devices (AV-VR) [3], and flexible displays.

In fundamental research of liquid crystals, interest has somewhat shifted from structure–property relationships like ferroelectric and antiferroelectric LCs (FLC and AFLC) [4–7], Twist Grain Boundary Phase (TGB) [8–10], or bent-shaped liquid crystals [11–14] towards a broader variety of questions, which are more closely related to other fields of research, like fundamental physics, optics and photonics, topology, biology, and nanotechnology. One can also observe an increasing overlap of topics, which indicates a growing synergy and multi-disciplinarity. An example would be the combination of liquid crystals and polymers to LC elastomers, which show promise in the application for soft robotics. Yet this should not take away from the interesting developments in the synthesis of new materials showing novel phases, particularly those of the nematic state, for example, the twist-bend phases, which by now have been studied extensively through materials [15,16] and theory [17]. Probably even more of interest in the future will be the recent discovery of ferroelectric nematic liquid crystals [18,19], the long sought "holy grail" of LC materials.

A topic that has largely raised interest from the fundamental and application point of view are rubber-like materials, elastomers, with liquid crystalline order [20]. These can be exploited as multifunctional materials, especially when prepared as composites [21]. Applications are mainly proposed for their use as smart, soft actuators [22], for example, in the field of soft, small-scale robotics [23], but also other applications as we shall see below. Into a similar category falls the technique of fibre drawing from liquid crystals, often from lyotropic phases. Such fibres have been known for quite some time in the form of Kevlar® or Nomex®, but more recently additional functionalities besides high tensile strength have been incorporated through the use of 2D materials such as graphene oxide [24,25] or MXenes [26]. These functionalities can comprise conductivity, so that one can weave high-tensile fabrics with the possibility to generate power.

Another trend that has set off an increasing activity in liquid crystal-related work is the development of advanced functional materials, using the self-organisation and self-assembly of liquid crystals together with properties that are electric, magnetic, mechanical, or optic in nature [27]. Liquid crystals used can be thermotropic or lyotropic, and the

ed: 24 March 2025
ed: 27 March 2025
ed: 28 March 2025

n: Dierking, I. Liquid Crystal ch and Novel Applications in Century. *Crystals* **2025**, *15*, tps://doi.org/10.3390/ 040321 functionality is often added via dispersion of colloids. The interactions between l crystal and colloidal particles lead to deformations of the director field and the c vation of point or line defects. Furthermore, these defects may interact to form tw three-dimension colloidal crystals [28]. Colloids of different shape and form, like p knots, or helices, embedded in a liquid crystal matrix can produce artificial materi great complexity [29,30]. Other examples can be found by dispersing low-dimens carbon allotropes in thermotropic liquid crystals [31], or the formation of lyotropic l crystal phases by dispersing shape-anisotropic colloids in isotropic solvents [32]. T materials open a whole new field of tunable and switchable devices in liquid crystal- nanotechnology and nanoscience [33].

Colloidal liquid crystals are closely related to active liquid crystals [34], where n bacteria, viruses, phages, or spindle-shaped cells, self-propelled, and often biolo colloids are dispersed in liquid crystals [35]. Again, the latter can be of the thermotro the lyotropic type but are often lyotropic to assure a longer survival time. Similarly, colloidal materials in an isotropic solvent, (often water), can also lead to the formati liquid crystal phases, depending on concentration. Active liquid crystals or living li crystals are gaining increasing interest as topological defects, for example, allow the ste of viruses or bacteria. Due to their monodispersion, they can also be exploited as m systems in soft matter physics [36].

Liquid crystals also increasingly draw on self-organisation mechanisms observed ture, bio-inspired liquid crystals and materials. Structures formed are often chiral [37 helical cholesteric structures found in certain beetles, which demonstrate quite astoni optical properties, such as the reflection of circular polarised light, but which could be used as active media in bio-lasers. Crosslinking of dispersed bifunctional mono can lead to bicontinuous films with widened selective reflection, or tunable refle wavelength, for example, in the use of smart glass, privacy windows, or scattering dis which can also be produced with flexible substrates. But also, Blue Phases and lyot systems can be achieved from nature-inspired materials. Examples of the latter are c colloidal type, for instance, liquid crystal phases of DNA or from cellulose nanocry (CNC), which can also be dried into solid films with selective reflection properties One cannot only observe the overlap with a range of photonic properties, but also i production of multifunctional materials as they were discussed above, for examp actuators and sensors in soft bio-inspired robotics [38] and generally as nature-ins liquid crystal-actuator materials [39]. Such actuators can be driven by electric or mag fields, light, temperature, or pressure [40], ideal for soft robots. Cellulose nanocrystals attracted much attention in recent years as colloidal liquid crystals on the basis of biol structures [41]. Their inherent chirality leads to lyotropic cholesteric liquid crysta which the pitch can be varied by a large variety of parameters, such as concentra aspect ratio, crystallite surface treatment, or the addition of salt. This varies the rhe and optical properties of the liquid crystal. Additionally, CNCs may be combined polymers or 2D materials in composites, varying mechanical properties.

Much related to bio- or nature-inspired liquid crystals is the use of these materia biomedical applications [42], due to their inherent biocompatibility and responsive This includes all areas of biomedicine, not only technical aspects such as bioimaging biosensing, also through wearable technologies, but also implants and tissue engine and of course drug delivery. For the encapsulation, transport, and delivery of d cubosomes have proven to be of substantial potential [43–45]. These are liquid cryst structures formed from the cubic phases of lipid molecules. Choice of lipid and amphip liquid crystalline phase, alky-chain length, stabilisers, and many more parameters al large variety of tuning mechanisms, which is beneficial for the development of the ideal structure for transport, delivery, and release of specific drugs.

One of the general topics of mathematical physics that has found its way into liquid crystal research over the last decade or so, and which is attracting increasing interest, is that of topology [46]. Also a topic in the wider field of soft matter, topology and knot theory, as well as defects and solitons, can be found in descriptions of polymers, DNA, proteins, living matter, or colloids. Examples involving liquid crystals are the following: (1) linear chains or 2D and 3D aggregates of colloids being held together by knots [28,47,48]; (2) motile bacteria in lyotropic nematic liquid crystals being steered by topological defects; or (3) the topology of liquid crystal director fields confined to droplets or shells. Closely related systems are topological solitons observed in liquid crystals [49,50]. The latter field has evolved quite rapidly over recent years with the experimental observation and description of skyrmions and skyrmion bags, hopfions, directrons and heliknotons, transformations between these solitons, interactions, and their dynamics and collective behaviour.

From a more applicational point of view, the alignment of liquid crystals on surfaces seems to have experienced a renaissance with the development of increasingly sophisticated techniques of producing patterned substrates, for example, through photo-patterned substrates for the alignment of liquid crystals for photonic applications [51]. But also, bio-inspired surfaces [52] can be used to tune the wettability of liquid crystals on surfaces to produce semi-droplets or droplets for micro-photonics; substrates can be patterned for tunable meta-materials or for the use in liquid crystal-based sensors for gases, solvents, or biomolecules.

Non-display applications of liquid crystals have taken hold of a considerable amount of efforts, now that large-area, high-performance screens are readily available at a decent price. We have already mentioned elastomers and their potential for robotics, as well as polymer stabilised- and polymer-dispersed liquid crystals for use in privacy windows, smart glass, and as environmental building materials for heat regulation. Other applications of liquid crystal polymers have been proposed as membranes in fuel cells [53]. Different liquid crystal phases can be used in self-assembled, tunable, soft photonics [54], or as light-driven materials in photochromic, photo-stimulated, or photo-modulated applications [55]. Of interest is certainly also the development of tunable lasers of liquid crystal-based materials [56,57], which comprise band-edge lasers of cholesteric and Blue Phases, and random lasers of nematics, polymer-dispersed liquid crystals (PDLC), quantum dot-doped LCs, and other nanoparticles dispersed in the liquid crystal. An interesting development can also be seen in micro-lasers, with whispering gallery modes observed from liquid crystal droplets [58]. A further field of LC applications that has generated significant impact is that of chemical and biochemical sensors on the basis of texture transitions of liquid crystals, either from homeotropic to planar orientation or vice versa [59–63]. These offer the opportunity to mass produce sensor devices at a reasonable price, which are easy to use and transportable, offering a potential self-test opportunity for the next pandemic.

At last, one also needs to mention that the field of computer modelling and simulation of liquid crystals has made some significant steps forward in recent years [64]. Of course, research can still not perform full atomic simulations of thermodynamic ensembles, neither for thermotropic liquid crystals and certainly not for lyotropic ones, which face the additional difficulty of the large number of solvent molecules present. But nevertheless, bent-core phases, the newly discovered nematic phases—like twist-bend nematics—are obtained, as are lyotropic systems and chromonics, as coarse-graining techniques have evolved and improved.

We have listed several aspects of liquid crystal research, together with a range of corresponding recent review articles, which I feel have held a preeminent position of interest in the last decade and which promise to continue to do so in the short- and me term future. This list is obviously somewhat biassed, and I am sure that readers may other topics that they feel deserve mentioning as well. In any case, the breadth of res around liquid crystals has clearly increased over recent years, which is also reflected l articles submitted to this Special Issue on liquid crystal research and applications 21st century. These can be arranged into five general areas, which are often interconn reflecting the multi-disciplinarity of the topics, also in relation to aspects mentioned a

A: Liquid crystal elastomers and polymer-modified liquid crystals, covering th of elastomers in biological applications [65] and multi-mode shape-morphing [66].

B: Photonics and sensors, summarising Tamm plasmons in liquid crystal device density functional theory calculations to interpret experimental IR spectra of chiral gens [68] and photo-physical properties of liquid crystals [69].

C: Topology and solitons in liquid crystals, reviewing chiral, topological, and kr colloids in liquid crystals [48], and studying topological defect annihilation of colloid-c LCs by machine learning [70].

D: Interactions with substrates and alignment, giving accounts of polarisation cou between ferroelectric liquid crystals and solid ferroelectrics [71], a detailed investigat the cholesteric lying helix state [72], and effects of photo-patterning conditions on anch strength [73].

E: Phase structure of thermotropic and lyotropic liquid crystals, providing a stru study of the nematic phase [74], and lyotropic structures from sphere-rod amphi compounds in a solvent [75].

I hope that this Special Issue of Crystals conveys the large range of research t covered by liquid crystal-based systems, from chemistry, physics, and mathemat biology and engineering. I hope that it also conveys the excitement of timely res topics available, ranging from topology and solitons to defect dynamics, from elastc and photonics to biomaterials, and phase structures to subtle interactions with surfa would like to thank all contributors for their time and engagement with this issue on crystal research and novel applications in the 21st century.

Conflicts of Interest: The author declares no conflict of interest.

References

1. Uchida, T. 40 years research and development on liquid crystal displays. *Jpn. J. Appl. Phys.* **2014**, *53*, 03CA02.
2. Kim, K.-H.; Song, J.-K. Technical evolution of liquid crystal displays. *NPG Asia Mat.* **2009**, *1*, 29.
3. Yin, K.; Hsiang, E.-L.; Zou, J.; Li, Y.; Yang, Z.; Yang, Q.; Lai, P.-C.; Lin, C.-L.; Wu, S.-T. Advanced liquid crystal devic augmented reality and virtual reality displays: Principles and applications. *Light Sci. Appl.* **2022**, *11*, 161.
4. Guo, Q.; Yan, K.; Chigrinov, V.; Zhao, H.; Tribelsky, M. Ferroelectric Liquid Crystals: Physics and Applications. *Crystals* **2** 470. [CrossRef]
5. Yoshizawa, A. Ferroelectric Smectic Liquid Crystals. *Crystals* **2024**, *14*, 350. [CrossRef]
6. Hird, M. Ferroelectricity in liquid crystals—Materials, properties and applications. *Liq. Cryst.* **2011**, *38*, 1467.
7. Lagerwall, S.T. *Ferroelectric and Antiferroelectric Liquid Crystals*; Wiley-VCH: Weinheim, Germany, 1999.
8. Goodby, J.W. Twist grain boundary and frustrated liquid crystal phases. *Curr. Opin. Colloid Interface Sci.* **2002**, *7*, 326. [Cro
9. Dhar, R. Twisted-grain-boundary (TGB) phases: Nanostructured liquid-crystal analogue of Abrikosov vortex lattices. *Transit.* **2006**, *79*, 175.
10. Selinger, J.V. Director Deformations, Geometric Frustration, and Modulated Phases in Liquid Crystals. *Annu. Rev. Condens. Phys.* **2022**, *13*, 49.
11. Takezoe, H.; Takanishi, Y. Bent-Core Liquid Crystals: Their Mysterious and Attractive World. *Jpn. J. Appl. Phys.* **2006**, *45*,
12. Reddy, R.A.; Tschierske, C. Bent-core liquid crystals: Polar order, superstructural chirality and spontaneous desymmetrisa soft matter systems. *J. Mater. Chem.* **2006**, *16*, 907. [CrossRef]
13. Eremin, A.; Jákli, A. Polar bent-shape liquid crystals—From molecular bend to layer splay and chirality. *Soft Matter* **2013**, [CrossRef]

Jesus Etxebarria, M. Blanca Ros, Bent-core liquid crystals in the route to functional materials. *J. Mater. Chem.* **2008**, *18*, 2919. [CrossRef]

Mandle, R.J. Designing Liquid-Crystalline Oligomers to Exhibit Twist-Bend Modulated Nematic Phases. *Chem. Rec.* **2018**, *18*, 1341. [CrossRef]

Mandle, R.J. A Ten-Year Perspective on Twist-Bend Nematic Materials. *Molecules* **2022**, *27*, 2689. [CrossRef]

Szmigielski, M. Theoretical models of modulated nematic phases. *Soft Matter* **2023**, *19*, 2675. [CrossRef]

Cruickshank, E. The Emergence of a Polar Nematic Phase: A Chemist's Insight into the Ferroelectric Nematic Phase. *ChemPlusChem* **2024**, *89*, e202300726. [CrossRef] [PubMed]

Osipov, M.A. On the origin of the ferroelectric ordering in nematic liquid crystals and the electrostatic properties of ferroelectric nematic materials. *Liq. Cryst. Rev.* **2024**, *12*, 14. [CrossRef]

Herbert, K.M.; Fowler, H.E.; McCracken, J.M.; Schlafmann, K.R.; Koch, J.A.; White, T.J. Synthesis and alignment of liquid crystalline elastomers. *Nat. Rev. Mater.* **2022**, *7*, 25. [CrossRef]

Wang, Y.; Liu, J.; Yang, S. Multi-functional liquid crystal elastomer composites. *Appl. Phys. Rev.* **2022**, *9*, 011301. [CrossRef]

Apsite, I.; Salehi, S.; Ionov, L. Materials for Smart Soft Actuator Systems. *Chem. Rev.* **2022**, *122*, 1349. [PubMed]

Hines, L.; Petersen, K.H.; Lum, G.Z.; Sitti, M. Soft Actuators for Small-Scale Robotics. *Adv. Mater.* **2017**, *29*, 1603483. [CrossRef]

Xu, Z.; Gao, C. Graphene in Macroscopic Order: Liquid Crystals and Wet-Spun Fibers. *Acc. Chem. Res.* **2014**, *47*, 1267. [CrossRef]

Yin, F.; Hu, J.; Hong, Z.; Wang, H.; Liu, G.; Shen, J.; Wang, H.-L.; Zhang, K.-Q. A review on strategies for the fabrication of graphene fibres with graphene oxide. *RSC Adv.* **2020**, *10*, 5722. [CrossRef]

Usman, K.A.; Qin, S.; Henderson, L.C.; Zhang, J.; Hegh, D.Y.; Razal, J.M. $Ti_3C_2T_x$ MXene: From dispersions to multifunctional architectures for diverse applications. *Mater. Horiz.* **2021**, *8*, 2886. [PubMed]

Uchida, J.; Soberats, B.; Gupta, M.; Kato, T. Advanced Functional Liquid Crystals. *Adv. Mater.* **2022**, *34*, 2109063. [CrossRef] [PubMed]

Musevic, I. Interactions, topology and photonic properties of liquid crystal colloids and dispersions. *Eur. Phys. J. Spec. Top.* **2019**, 227, 2455.

Smalyukh, I.I. Liquid Crystal Colloids. *Annu. Rev. Condens. Matter Phys.* **2018**, *9*, 207. [CrossRef]

Muševic, I. Nematic Liquid-Crystal Colloids. *Materials* **2018**, *11*, 24.

Draude, A.P.; Dierking, I. Thermotropic liquid crystals with low-dimensional carbon allotropes. *Nano Express* **2021**, *2*, 012002.

Dierking, I.; Al-Zangana, S. Lyotropic Liquid Crystal Phases from Anisotropic Nanomaterials. *Nanomaterials* **2017**, *7*, 305. [CrossRef]

Shen, Y.; Dierking, I. Perspectives in Liquid-Crystal-Aided Nanotechnology and Nanoscience. *Appl. Sci.* **2019**, *9*, 2512. [CrossRef]

Zhang, R.; Mozaffari, A.; de Pablo, J.J. Autonomous materials systems from active liquid crystals. *Nat. Rev. Mater.* **2021**, *6*, 437.

Zhao, J.; Gulan, U.; Horie, T.; Ohmura, N.; Han, J.; Yang, C.; Kong, J.; Wang, S.; Bin Xu, B. Advances in Biological Liquid Crystals. *Small* **2019**, *15*, 1900019.

Dogic, Z. Filamentous Phages as a Model System in Soft Matter Physics. *Front. Microbiol.* **2016**, *7*, 1013.

Wang, L.; Urbas, A.M.; Li, Q. Nature-Inspired Emerging Chiral Liquid Crystal Nanostructures: From Molecular Self-Assembly to DNA Mesophase and Nanocolloids. *Adv. Mater.* **2020**, *32*, 1801335.

Ilami, M.; Bagheri, H.; Ahmed, R.; Skowronek, E.O.; Marvi, H. Materials, Actuators, and Sensors for Soft Bioinspired Robots. *Adv. Mater.* **2021**, *33*, 2003139.

Shang, Y.; Wang, J.; Ikeda, T.; Jiang, L. Bio-inspired liquid crystal actuator materials. *J. Mater. Chem. C* **2019**, *7*, 3413.

Ahmed, F.; Waqas, M.; Jawed, B.; Soomro, A.M.; Kumar, S.; Hina, A.; Khan, U.; Kim, K.H.; Choi, K.H. Decade of bio-inspired soft robots: A review. *Smart Mater. Struct.* **2022**, *31*, 073002.

Casado, U.; Mucci, V.L.; Aranguren, M.I. Aranguren, Cellulose nanocrystals suspensions: Liquid crystal anisotropy, rheology and films iridescence. *Carbohydr. Polym.* **2021**, *261*, 117848.

Zhang, Z.; Yang, X.; Zhao, Y.; Ye, F.; Shang, L. Liquid Crystal Materials for Biomedical Applications. *Adv. Mater.* **2023**, *35*, 2300220. [CrossRef]

Barriga, H.M.G.; Holme, M.N.; Stevens, M.M. Cubosomes: The Next Generation of Smart Lipid Nanoparticles? *Angew. Chem. Int. Ed.* **2019**, *58*, 2958. [CrossRef] [PubMed]

Karami, Z.; Hamidi, M. Cubosomes: Remarkable drug delivery potential. *Drug Discov. Today* **2016**, *21*, 1359. [CrossRef] [PubMed]

Yaghmur, A.; Mu, H. Recent advances in drug delivery applications of cubosomes, hexosomes, and solid lipid nanoparticles. *Acta Pharm. Sin. B* **2021**, *11*, 871. [CrossRef]

Tubiana, L.; Alexander, G.P.; Barbensi, A.; Buck, D.; Cartwright, J.H.; Chwastyk, M.; Cieplak, M.; Coluzza, I.; Čopar, S.; Craik, D.J.; et al. Topology in soft and biological matter. *Phys. Rep.* **2024**, *1075*, 1.

Smalyukh, I.I. Review: Knots and other new topological effects in liquid crystals and colloids. *Rep. Prog. Phys.* **2020**, *83*, 106601. [CrossRef] [PubMed]

Yuan, Y.; Smalyukh, I.I. Chiral, Topological, and Knotted Colloids in Liquid Crystals. *Crystals* **2024**, *14*, 885. [CrossRef]

49. Tai, J.-S.B. Topological solitons in chiral liquid crystals. *Liq. Cryst. Today* **2023**, *32*, 45. [CrossRef]
50. Shen, Y.; Dierking, I. Recent Progresses on Experimental Investigations of Topological and Dissipative Solitons in Liquid Cr *Crystals* **2022**, *12*, 94. [CrossRef]
51. Kudreyko, A.; Chigrinov, V.; Hegde, G.; Chausov, D. Photoaligned Liquid Crystalline Structures for Photonic Applica *Crystals* **2023**, *13*, 965. [CrossRef]
52. Su, B.; Tian, Y.; Jiang, L. Bioinspired Interfaces with Superwettability: From Materials to Chemistry. *J. Am. Chem. Soc.* **201** 1727.
53. Guardià, J.; Reina, J.A.; Giamberini, M.; Montané, X. An Up-to-Date Overview of Liquid Crystals and Liquid Crystal Pol for Different Applications: A Review. *Polymers* **2024**, *16*, 2293. [CrossRef] [PubMed]
54. Ma, L.-L.; Li, C.-Y.; Pan, J.-T.; Ji, Y.-E.; Jiang, C.; Zheng, R.; Wang, Z.-Y.; Wang, Y.; Li, B.-X.; Lu, Y.-Q. Self-assembled liquid c architectures for soft matter photonics. *Light Sci. Appl.* **2022**, *11*, 270.
55. Bisoyi, H.K.; Li, Q. Light-Driven Liquid Crystalline Materials: From Photo-Induced Phase Transitions and Property Modul to Applications. *Chem. Rev.* **2016**, *116*, 15089. [PubMed]
56. Mysliwiec, J.; Szukalska, A.; Szukalski, A.; Sznitko, L. Liquid crystal lasers: The last decade and the future. *Nanophotonics* *10*, 2309.
57. Aljohani, O.; Dierking, I. Modern developments in lasing with liquid crystals. *Front. Mater.* **2024**, *11*, 1453744.
58. Muševič, I. Liquid-crystal micro-photonics. *Liq. Cryst. Rev.* **2016**, *4*, 1.
59. Tang, J.; Li, Z.; Xie, M.; Luo, Y.; Yu, J.; Chen, G.; Chen, Z. Liquid Crystal Based Label-Free Optical Sensors for Bioche Application. *Photonic Sens.* **2024**, *14*, 240203.
60. Zhan, X.; Liu, Y.; Yang, K.-L.; Luo, D. State-of-the-Art Development in Liquid Crystal Biochemical Sensors. *Biosensors* **20** 577. [CrossRef]
61. Luan, C.; Luan, H.; Luo, D. Application and Technique of Liquid Crystal-Based Biosensors. *Micromachines* **2020**, *11*, 176. [Cro
62. Oladepo, S.A. Development and Application of Liquid Crystals as Stimuli-Responsive Sensors. *Molecules* **2022**, *27*, 1453. [Cro [PubMed]
63. Carlton, R.J.; Hunter, J.T.; Miller, D.S.; Abbasi, R.; Mushenheim, P.C.; Tan, L.N.; Abbott, N.L. Chemical and biological se using liquid crystals. *Liq. Cryst. Rev.* **2013**, *1*, 29. [PubMed]
64. Wilson, M.R.; Yu, G.; Potter, T.D.; Walker, M.; Gray, S.J.; Li, J.; Boyd, N.J. Molecular Simulation Approaches to the Stu Thermotropic and Lyotropic Liquid Crystals. *Crystals* **2022**, *12*, 685. [CrossRef]
65. Shiralipour, F.; Akhtar, Y.N.; Gilmor, A.; Pegorin, G.; Valerio-Aguilar, A.; Hegmann, E. The Role of Liquid Crystal Elastom Pioneering Biological Applications. *Crystals* **2024**, *14*, 859. [CrossRef]
66. Zeng, X.; Zhou, T.; Li, L.; Song, J.; Duan, R.; Xiao, X.; Xu, B.; Wu, G.; Guo, Y. Reconfigurable Liquid Crystal Elastomer Di Patterns for Multi-Mode Shape Morphing. *Crystals* **2024**, *14*, 357. [CrossRef]
67. Reshetnyak, V.Y.; Pinkevych, I.P.; McConney, M.E.; Bunning, T.J.; Evans, D.R. Tamm Plasmons: Properties, Application Tuning with Help of Liquid Crystals. *Crystals* **2025**, *15*, 138. [CrossRef]
68. Deptuch, A.; Górska, N.; Murzyniec, M.; Srebro-Hooper, M.; Hooper, J.; Dziurka, M.; Urbańska, M. Density Functional T Calculations for Interpretation of Infra-Red Spectra of Liquid Crystalline Chiral Compound. *Crystals* **2024**, *14*, 645. [Cross
69. Hobbs, J.; Mattsson, J.; Nagaraj, M. Analysing the Photo-Physical Properties of Liquid Crystals. *Crystals* **2024**, *14*, 362. [Cro
70. Dierking, I.; Moyle, A.; Cepparulo, G.M.; Skingle, K.; Hernández, L.; Raidal, J. Machine Learning Analysis of Umbilic I Annihilation in Nematic Liquid Crystals in the Presence of Nanoparticles. *Crystals* **2025**, *15*, 214. [CrossRef]
71. Marni, S.; Barboza, R.; Oluwajoba, A.S.; Zamboni, R.; Lucchetti, L. Polarization Coupling between Ferroelectric Liquid Ferroelectric Solids: Effects of the Fringing Field Profile. *Crystals* **2024**, *14*, 425. [CrossRef]
72. Simdyankin, I.V.; Geivandov, A.R.; Kasyanova, I.V.; Palto, S.P. Local Orientation Transitions to a Lying Helix State in Ne Dielectric Anisotropy Cholesteric Liquid Crystal. *Crystals* **2024**, *14*, 891. [CrossRef]
73. Haputhanthrige, N.P.; Rajabi, M.; Lavrentovich, O.D. Effects of Photopatterning Conditions on Azimuthal Surface Anch Strength. *Crystals* **2024**, *14*, 1058. [CrossRef]
74. Deptuch, A.; Sęk, B.; Lalik, S.; Zając, W.; Ossowska-Chruściel, M.D.; Chruściel, J.; Marzec, M. Structural Study of Nematc Compound 5OS5. *Crystals* **2024**, *14*, 367. [CrossRef]
75. Haputhanthrige, N.P.; Zhou, Y.; Wei, J.; Gao, M.; Liu, T.; Lavrentovich, O.D. Liquid Crystalline Structures Formed by Spher Amphiphilic Molecules in Solvents. *Crystals* **2025**, *15*, 177. [CrossRef]

crystals

w

iral, Topological, and Knotted Colloids in Liquid Crystals

ian [1,*] and Ivan I. Smalyukh [1,2,3,4,*]

1 International Institute for Sustainability with Knotted Chiral Meta Matter (WPI-SKCM2), Hiroshima University, Higashi-Hiroshima, Hiroshima 739-8526, Japan
2 Department of Physics, University of Colorado, Boulder, CO 80309, USA
3 Department of Electrical, Computer and Energy Engineering, Materials Science and Engineering Program and Soft Materials Research Center, University of Colorado, Boulder, CO 80309, USA
4 Renewable and Sustainable Energy Institute, National Renewable Energy Laboratory and University of Colorado, Boulder, CO 80309, USA
* Correspondence: yeyuan@hiroshima-u.ac.jp (Y.Y.); ivan.smalyukh@colorado.edu (I.I.S.)

Abstract: The geometric shape, symmetry, and topology of colloidal particles often allow for controlling colloidal phase behavior and physical properties of these soft matter systems. In liquid crystalline dispersions, colloidal particles with low symmetry and nontrivial topology of surface confinement are of particular interest, including surfaces shaped as handlebodies, spirals, knots, multi-component links, and so on. These types of colloidal surfaces induce topologically nontrivial three-dimensional director field configurations and topological defects. Director switching by electric fields, laser tweezing of defects, and local photo-thermal melting of the liquid crystal host medium promote transformations among many stable and metastable particle-induced director configurations that can be revealed by means of direct label-free three-dimensional nonlinear optical imaging. The interplay between topologies of colloidal surfaces, director fields, and defects is found to show a number of unexpected features, such as knotting and linking of line defects, often uniquely arising from the nonpolar nature of the nematic director field. This review article highlights fascinating examples of new physical behavior arising from the interplay of nematic molecular order and both chiral symmetry and topology of colloidal inclusions within the nematic host. Furthermore, the article concludes with a brief discussion of how these findings may lay the groundwork for new types of topology-dictated self-assembly in soft condensed matter leading to novel mesostructured composite materials, as well as for experimental insights into the pure-math aspects of low-dimensional topology.

Keywords: liquid crystal; colloids; topology; chirality

n: Yuan, Y.; Smalyukh, I.I. Topological, and Knotted s in Liquid Crystals. *Crystals* 4, 885. https://doi.org/ /cryst14100885

nic Editor: Samo Kralj

d: 23 August 2024
l: 3 October 2024
d: 8 October 2024
ed: 11 October 2024

1. Introduction

Liquid crystals (LCs) are characterized by the long-range orientational order of constituting mesogens with anisotropic shapes, with the simplest form being a nematic LC consisting of rodlike molecules [1–3] (Figure 1a). Despite the crystal-like anisotropic properties stemming from the orientational ordering, the weak intermolecular interactions between LC mesogens still allow for uninhibited translational motions of the building blocks and thus liquid-like flow, earning the name "liquid crystal" of the medium that combines properties of crystals and liquids. Different from isotropic fluids such as water, LCs feature large anisotropic dielectric and optical properties as a result of the molecular orientation field. The combination of such anisotropy and the facile responsiveness to external stimuli enables far-reaching technological applications, notably the flat-panel liquid crystal displays. Such properties of LCs also provide a unique anisotropic environment for colloidal particles, giving rise to rich phenomena not observed in conventional isotropic solvents, opening avenues for addressing fundamental physics questions as well as exploring pre-designed development of novel functional materials [4–8]. Of particular interest is how the geometric shape, symmetry, and topology of colloidal particles interact with the

LC molecular orientation field to enrich this behavior, leading to unusual topologica distributions, colloidal interactions, and self-assembly processes.

Colloids are mixtures where particles of 1 nm to 10 μm are stably suspende solvent [9–11]. Particles of this size range are susceptible to Brownian motions which e the equilibrium suspension of colloidal systems, and are also responsible to other u colloidal behaviors including light scattering, depletion interaction, capillary assemb Common everyday objects including milk, fog, and clay are all colloids in nature, artificial colloidal systems have found applications in cosmetics, energy harvesting, materials, etc., as well as in understanding fundamental science questions ranging mechanisms of phase transitions to emergent properties of self-assembled materials [When introduced to a LC medium, colloidal particles force the surrounding LC molecu reorient in order to accommodate the volume occupied and interfacial boundary cond imposed by the foreign objects. Such accommodation manifests as the redistributi the LC orientation field including both continuous deformations and singular de which can spatially propagate far beyond the physical extent of the colloidal pa The equilibrium state of the LC colloidal system is determined by the configuratio minimizes the total energy cost associated with the field distortions, which also le long-range colloidal interactions mediated by the LC anisotropic elasticity. With the a to define how the LC orientation field is disturbed, the size, shape, and surface prop of colloidal inclusions become important contributing factors determining the equilil states and dynamic processes of the composite LC colloidal systems.

Early studies of LC colloids were usually carried out on highly symmetric partic trivial topological characteristics, such as spheres, rods, or discs [15,16]. The technolo advancement in colloidal fabrication [4,5,17,18] and microscopic imaging [19] in r years has allowed researchers to obtain colloidal particles of much more complex mo logical characteristics, as well as resolve fine details on LC field distributions induc these particles. Examples of fabrication techniques include two-photon photopol ization (2PP) and new chemical synthesis protocols to obtain chiral or knotted col particles in nano and micrometer size ranges and made of a large range of constituent rials ranging from polymerizable resin to noble metals, which are reviewed elsewher Imaging and optical manipulation techniques include three-photon excitation fluores polarizing microscopy (3PEF-PM) and holographic laser tweezer. Concurrently, opments in computational modeling have enabled accurate simulations of comple colloidal systems, providing insights into the underlying physics and guiding experin design [4,5,20].

In this review, the authors intend to provide a concise account of nematic LC co containing particles of complex geometries such as chiral, topological, knotted, and l features. These types of colloidal particles impart their symmetry-breaking and top cally nontrivial geometries into the LC orientation field in the forms of chirality-dic interactions, topological constraints on singular defects, as well as knotted field co rations, which is of both fundamental science and application significance. The re is structured as follows: after these introductory statements and remarks, the follo section will lay out the physical underpinnings of colloidal particles in LCs, overvie the relevant theories of LC elastic energy and topological defects, which will be foll by three main sections discussing colloidal particles of various symmetry and top characteristics; in the last section, we discuss further the context and implication colloids and potential future directions of development for this field.

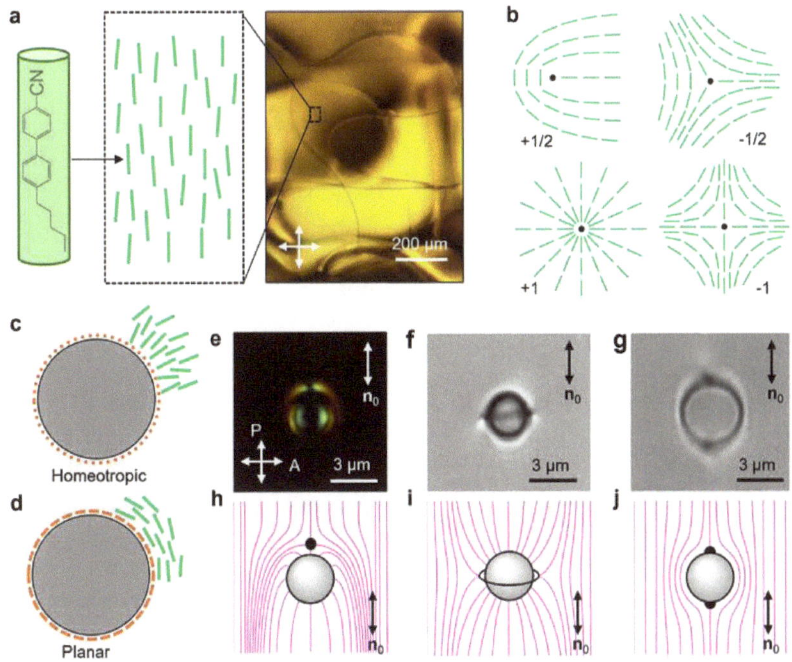

Figure 1. Colloids in liquid crystals (LCs). (**a**) Microscopic structure of a nematic LC with rod-like mesogens, i.e., pentylcyanobiphenyl (5CB). The micrograph (right image) shows the texture of a 5CB droplet observed under a microscope with crossed polarizers, polarization direction marked with white double arrows. Inset shows the chemical structure of 5CB molecules and their collective alignment within a small volume. (**b**) Topological defects in LCs of different winding numbers; green rods represent LC molecules. (**c**,**d**) Homeotropic and planar surface anchoring where LC molecules align perpendicular and parallel to the surface of colloidal inclusions. Orange dots and dashes represent surface functioning agents such as polymer grafting that impose the anchoring direction. (**e**,**f**) Micrograhs showing microspheres with homeotropic surface anchoring inducing "hedgehog" point defect and "Saturn ring" line defect; white double arrows indicate the crossed polarizers. (**g**) Micrographs showing microsphere with planar surface anchoring inducing "boojum" surface defects at the polar points of the sphere. (**h**–**j**) Corresponding schematics illustrating LC director field configurations around the colloidal spheres. The black dots and line represent the hedgehog defect, the Saturn ring loop, and the surface boojums, respectively. Schematics are not drawn to scale. The far-field director is shown by the double arrow marked with $\mathbf{n}_0$. Adapted from Ref. [21].

2. Basics of Physical Behavior of Colloidal Particles in LCs

The LC molecular ordering can be described using the so-called director field $\mathbf{n}(\mathbf{r})$ representing the local average of the molecular alignment, which can vary as a function of coordinates $\mathbf{r}$ on scales much larger than molecular dimensions. In contrast to a vector field, the constraint $\mathbf{n}(\mathbf{r}) = -\mathbf{n}(\mathbf{r})$ is imposed to comply with the non-polar nature of the LC molecular orientations [1]. In the undisturbed state, molecules of a nematic LC can orient along the same direction with the help of proper surface confinement, typically setting surface boundary conditions corresponding to a uniform distribution of the director $\mathbf{n}(\mathbf{r}) = \mathbf{n}_0$, which also corresponds to the energy-minimizing state of a uniaxial nematic LC. Deviating from such a "ground state" uniform alignment via producing spatial gradients of the director is energetically costly, which can be quantitatively described using the Frank-Oseen free energy density functional f_e [1]:

$$f_e = \frac{1}{2}K_{11}(\nabla \cdot \boldsymbol{n})^2 + \frac{1}{2}K_{22}(\boldsymbol{n} \cdot \nabla \times \boldsymbol{n})^2 + \frac{1}{2}K_{33}(\boldsymbol{n} \times \nabla \times \boldsymbol{n})^2 - K_{24}\nabla\cdot[\boldsymbol{n}(\nabla\cdot\boldsymbol{n}) + \boldsymbol{n} \times (\nabla \times \boldsymbol{n})],$$

where K_{11}, K_{22}, K_{33}, and K_{24} are the so-called elastic constants characterizing the ene costs of splay, twist, bend and saddle-splay director deformations. The typical val elastic constants is a few pN, for example for cyanopentylbiphenyl (5CB), a very con LC that is in the nematic phase at room temperature (Figure 1a). In colloidal systems, v the characteristic length scale of colloidal inclusions can be $L \sim 1$ μm, the ratio bet the elastic and the thermal energy is $KL/k_BT \sim 10^3$ ($K \sim 1$–10 pN is a representative e constant of LCs, k_B is the Boltzmann constant and T = 300 K is the room tempera Therefore, the elastic energy is sufficient to overcome thermal fluctuations and le elasticity-mediated colloidal interactions and self-assembled structures in LCs.

Another form of energetic cost is related to the surface of the colloidal par Boundary conditions imposed by grafted polymer or other functional groups can define a preferred direction (often called "easy axis") for the LC molecules to ali which can be normal, tangential, or sometimes conical to the surface [15,16,21]. Th most common cases are homoetropic (normal) and planar (tangential) surface anch (Figure 1c,d). The free energy density functional characterizing the energy cost c director deviation from the easy axes takes the Rapini-Papoular form [1]:

$$f_s = -\frac{1}{2}W(\boldsymbol{n}\cdot\boldsymbol{e})^2$$

where $\boldsymbol{e}$ is a unit vector along the easy axis and W is the surface anchoring strength a typical value in the range 10^{-6}–10^{-4} J/m^2; integrating this free energy density the area of LC interfaces with particles or confining substrates gives the overall er cost of LC-surface interaction. A parameter called "extrapolation length" can be de as $\xi_e = K/W$, yielding typical values in the range ~ 10^{-2}–1 μm. When particle's s much smaller than ξ_e, it cannot disturb the director field significantly enough and the bulk elastic energy contribution can be ignored; conversely, in the opposite re where the boundary conditions defined by the easy axis orientation at the interface c assumed infinitely strong, the energy cost of inserting the particle into the LC consi only that from elastic deformations and singular defects around the particle. Overall, most general case, the equilibrium distribution of the director field around the coll inclusions are determined by minimizing the total free energy of the system, which inc integrating over the bulk volume of the system V and the enclosing surface S for the and surface energies, respectively:

$$F_{\text{total}} = \int_V f_e\, dV + \int_S f_s\, dS.$$

When multiple particles are present, however, they can move relative to each otl minimize the total energy, which effectively determine the elasticity-mediated colloic teraction patterns in LCs. Under the one-constant approximation where all elastic cons are equal in value, the director field with small perturbations far away from the coll particles $\mathbf{n} = (n_x, n_y, 1)$ follows the Laplace-like equation [22]:

$$\Delta n_\mu = 0\ (\mu = x,\ y)$$

obtained by minimizing the Frank-Oseen free energy. Like that in electromagnetic the this equation can be solved using multipole expansion expressed in terms of sph harmonics, and the colloidal interactions mediated by the LC elasticity can thus be preted in the multipole paradigm [22–25]. However, it is worth mentioning that i regime when colloidal particles are too close to each other, such that the small-perturb approximation may no longer hold true, contributions from higher-order multipole nonlinear field effects may appear, which goes beyond the multipole expansion analysis. While the first experimental demonstrations of elastic interactions resembling that between electrostatic multipoles were made in 1997 when Poulin et al. discovered dipole–dipole-like interactions between water droplets with homeotropic anchoring in a nematic LC [15], Brochard and de Gennes in 1970 had considered such electrostatic analogy theoretically [22]. Thus far, elastic multipoles of zeroth till fourth order (monopole to hexadecapole) have been experimentally demonstrated using various type of LC colloidal systems [15,21,26,27] (Figure 1d), paving the way for establishing LC colloidal systems with structures and emergent behaviors mimicking that of atomic systems [28].

In addition to continuous deformations, singularities where the local director cannot be well-defined may also be induced by colloidal inclusions [3]. These singularities can exist in the bulk as defect points (also called hedgehogs) and defect lines (disclinations), as well as on the surface of the colloidal particles, where the surface point singularities are called boojums. In fact, the name "nematic" originated from the Greek word meaning thread, refers to the thread-like disclination lines observed in a nematic LC (Figure 1a). The defects can be classified by computing the winding number in 2D for surface boojums and cross section of disclinations, or topological charge in 3D for hedgehogs and compact closed loops of disclinations [29,30]. The winding number s counts the number of times as the director rotates by 2π following a loop circumnavigating the defect core once (Figure 1b); the sign indicates whether the rotation direction of the director is the same or opposite to that of the circumnavigation. Similarly, the topological hedgehog charge m counts the number of times the unit sphere is wrapped by the director on a surface enclosing the defect core [30]:

$$m = \frac{1}{4\pi}\int_S (\boldsymbol{n}\cdot\partial_1\boldsymbol{n} \times \partial_2\boldsymbol{n})\, dx_1 dx_2. \tag{5}$$

Several examples of singular defects with winding number of $\pm 1/2$ and ± 1 are illustrated in Figure 1b. In addition to representing the director distribution of surface boojums, they also serve as the cross-sectional schematic of disclination lines, with the defect core extending out of the viewing plane.

The topology of colloidal particles imposes constraints on the generation of singular defects. Following the Gauss-Bonnet theorem and assuming uniform far-field director $\mathbf{n}(|\mathbf{r}| >> L) = \mathbf{n}_0$, one finds that the total topological charge induced by a colloidal particle compensates that due to its boundary conditions and is equal to half of particle surface's Euler characteristic [31]

$$\textstyle\sum_i m_i = \pm\chi/2. \tag{6}$$

The Euler characteristic χ is a topological invariant that can be computed, for example, based on the number of vertices, edges, and faces of the surface, denoted by V, E, and F, respectively

$$\chi = V - E + F. \tag{7}$$

For surface boojums in cases of tangential degenerate surface boundary conditions, the mathematical theorems impose the requirement on the total winding number of surface defects [32]:

$$\textstyle\sum_i s_i = \chi. \tag{8}$$

However, it is important to note that the topological constraints alone cannot determine the exact distribution of topological defects. The constraint limits the net topological charge or winding number, while it is the minimization of the total free energy that determines the director field deformation and defect distribution that satisfy this topological constraint. For example, the elastic energy per unit length of a disclination line scales with the winding number as $\propto s^2$ [3], indicating that disclinations of higher strength involve significant field distortions and, thus, are energetically unfavorable, tending to split into lower-winding-number ones. For colloidal spheres with a homoetropic surface anchoring, either a hedgehog defect or a disclination loop named "Saturn ring" can be found accompanying the particle (Figure 1e,f,h,i). While both configurations satisfy the

constraint in Equation (6) since spherical surfaces have an Euler characteristic of
Saturn ring is the energetically preferred one when the sphere is small, i.e., <1 μm [2
contrast, the constraint in Equation (8) is fulfilled by two +1 surface boojums at both
for a colloidal sphere with planar surface anchoring (Figure 1g,j). More discussions c
interplay between topological defects and colloidal particles of non-trivial topology
Section 4.

In addition to the literature referenced above, interested readers can refer to in-depth reviews and textbooks [3–5].

3. Chiral Colloids

Chirality manifests itself in many different scales, ranging from subatomic eleme
particles to molecular enantiomers with opposite optical activity and to biological and
cosmological objects [33]. Chiral colloidal particles suspended in a LC may transf
broken symmetry manifestations into the surrounding medium, for example, in the
of low-symmetry director field distributions both in the close vicinity of the particl
in the far field, leading to topological defect distributions and novel colloidal intera
mediated by the LC elasticity. This is in direct contrast to colloidal spheres in ne
LCs inducing director distortions resembling symmetric distributions correspondi
dipoles, quadruples, or higher-order multipoles [5,16]. Martinez et al. used 2PP-base
microprinting to fabricate custom-designed chiral microparticles bound to substrates
cells (Figure 2a,b) [34]. Nonlinear optical imaging reveals that these surface-attached
cles impose a twist on the LC director field that propagates across the cell and gene
low-symmetry director distortions. A colloidal sphere inducing elastic dipole mome
found to be attracted to the surface-bound particles following monopole–dipole inter
scaling (Figure 2c). By engaging surface structures with chiral features, this findin
vides a new way to control LC molecular alignment and the ensuing elastic interac
demonstrating possibilities of directing the self-assembly of colloidal particles th
engineered surfaces.

The fundamental role of particle chirality on colloidal behaviors is further rev
with free-standing colloidal structures with chiral symmetry. Yuan et al. fabricate
loidal springs and helices of both handedness and re-suspended them in a nemat
(Figure 2d–i) [35]. Despite their complex shape, these chiral structures are topolog
equivalent to spheres, with an Euler characteristic $\chi = 2$. Given their planar surfa
choring, the winding numbers of the surface boojums should sum up to two follo
Equation (8). For the colloidal springs, two stable (corresponding to the lowest e
states) and metastable (local energy minima) orientations are observed experime
with their helical axes tilting slightly away from being perfectly orthogonal and para
$\mathbf{n}_0$, respectively (Figure 2d,e). In the stable state, the topological constraint is satisfi
two $s = 1$ boojums located at the edges of the spring's two end faces, while the meta
state has additional pairs of self-compensating boojums at positions where the surfac
mals are parallel to $\mathbf{n}_0$. Additionally, dipole-like pair interactions arise due to the di
field deformations with chiral symmetry induced by the microsprings: the interact
dependent on the angle that the center-to-center separation vector makes with $\mathbf{n}_0$ ar
interaction potential scales as d^{-3}, as expected for dipole–dipole interactions (Figure
The interesting observation is that the equivalent dipole direction depends on the re
handedness: same-handed microsprings interact as if they possess the same directi
dipolar moments, while the interaction direction flips when one of the springs chang
handedness (Figure 2f,g). These experimental observations are further supported b
numerical modeling based on the minimization of the Landau-de Gennes free energy
shape anisotropy of such colloidal particles also brings in anisotropic diffusive beh
coupled with the LC director field [36]. Overall, both reports demonstrate how chi
can be used as an important parameter to control and engineer the colloidal interactio
LCs. The findings bridge the microscopic chirality of individual particles with meso
self-assemblies of colloidal structures, which may enable novel optical, photonic, and

functional materials. While chirality transfer from guest molecules to the host LC medium is widely known and technologically utilized, e.g., in cholesteric LC displays and electro-optic devices, the fact that colloidal inclusions with dimensions three orders of magnitude larger than the host LC's molecules can also transfer chiral symmetry breaking into the otherwise achiral nematic host is a fundamentally important finding with a significant technological potential from the standpoint of view of the self-assembly of mesostructured functional materials.

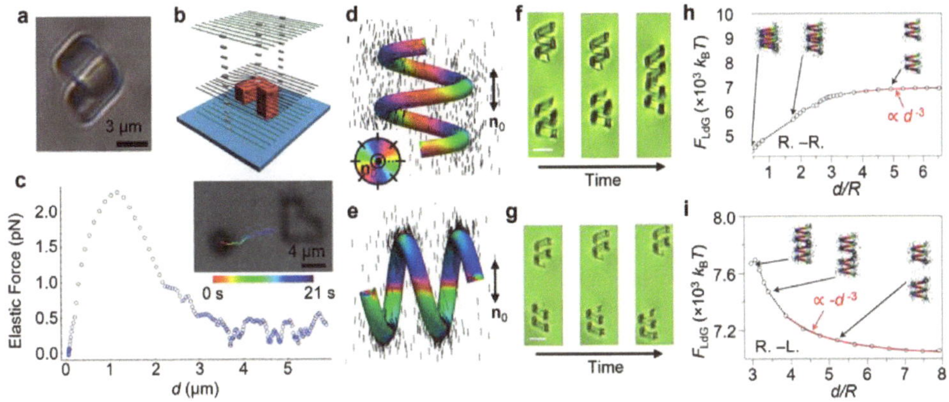

Figure 2. Chiral colloids in LCs. (**a**) Micrograph of a chiral microstructure obtained by 3D micro-printing. (**b**) Director field distortions around such a particle with planar surface anchoring bound to the confining substrate. The lines in the middle layer and the cylinders show that twist deformation is induced in the director field over the particle despite uniform far-field alignment. (**c**) Interaction forces vs. distance between the surface-bound chiral structure and a free-floating colloidal sphere. The inset is a micrograph of the interacting objects and the interaction trajectory is color-coded with time. (**d**,**e**) Director distributions around right-handed microsprings with a planar surface anchoring at energy-minimized positions. The double arrows indicate the far-field director $\mathbf{n}_0$; the color on the particles represents the orientation of the surface director projected to the plane orthogonal to $\mathbf{n}_0$; color scheme is shown as the inset of (**d**). (**f**,**g**) Snapshots of elasticity-mediated interactions between like- (**f**) and opposite- (**g**) handed microsprings, exhibiting attraction and repulsion, respectively, over the time of 10–100 s. Scale bars are 5 µm. (**h**,**i**) Numerically calculated Landau-de Gennes free energy vs. particle distances between like- (**h**) and opposite- (**i**) handed microsprings. When particles are far away from each other, the free energy scales as d^{-3} like that of dipole–dipole interactions. The distance between particles d is normalized by the particle radius R; the free energy F_{LdG} is normalized by the thermal energy $k_{\mathrm{B}}T$, where k_{B} is the Boltzmann constant and T is the room temperature. Adapted from Refs. [34,35].

4. Topological Colloids

Although not explicitly emphasized, colloidal particles of non-trivial topology, i.e., non-homeomorphic to spheres (square particles with a hole at the center) were introduced into a LC as early as 2009 (Figure 3a,b) [37]. These square particles induced quadrupolar deformations in the director field, though the central hole did not play an important role in determining the colloidal interactions in this particular system, which was the focus of that study [37]. The importance of particle topology, however, lies in its ability to control the distribution of induced topological defects by defining the net topological invariant of the defects as described in Section 2. While the cases are simple for topologically trivial spherical particles, the most common ones in colloidal sciences, Senyuk et al. and Liu et al. fabricated colloidal particles of nontrivial topology, handlebodies of various genus (i.e., number of holes) (Figure 3c,d), and demonstrated that the total charge or winding number of topological defects induced by particles is consistent with topological surface characteristics of the colloidal inclusions, following the relations given by the Gausss–

Bonnet and Poincare–Hopf index theorems [31,32]. For the surface of a particle, its
characteristic is related to the genus g by $\chi = 2 - 2g$, where genus g can be intui
understood as the number of holes of the surface. Thus, the topological constrai
Equations (6) and (8) becomes [31]

$$\sum_i m_i = \pm(1-g),$$

and [32]

$$\sum_i s_i = 2-2g.$$

The relations were experimentally confirmed while g was varied from 1 to 5
responding to handlebodies with 1 to 5 holes. For a g = 1 handlebody particle
homeotropic surface anchoring, the most commonly observed orientation is the pa
plane being perpendicular to $\mathbf{n}_0$ to minimize the free energy cost. Curiously, althou
defects are topologically required to form as Equation (9) is 0 when g = 1, the constra
satisfied by a pair of self-compensating disclination loop or an exterior disclination
and a hedgehog defect at the particle center, totaling net zero on the sum of the topolo
charge (Figure 3c,d) [31]. The same type of handlebody particle with g = 1 and p
surface anchoring induces two pairs of opposite-sign boojums so that the net wir
number is 0 [32]. More interestingly, the topological constraints are still obeyed even
external stimuli cause significant redistribution of the director field. This emergent beh
shows how topological defects can nontrivially emerge while both satisfying pure
theorems and yielding energetic minima of the LC–colloidal soft matter system. Scen
demonstrated in the above reports [31,32] include explorations of inter-transformatio
tween point defects and disclination loops at the central regions of the handlebodies'
when locally melting the LC, or when particles are reoriented following the applicati
external field. In other words, the total charge is a conserved quantity that is dictated
by the topological characteristic of the defect-inducing particles, i.e., the Euler characte
of the 2D LC–colloidal interface.

It is an important result vividly demonstrating how topological theorems res
constraints on induced defects that are still "soft" in the sense of allowing the syste
choose defect configurations of a certain net topological charge that are also energet
favorable. This energy-dependent selection of topology-satisfying configurations dis
resembles other types of director field transformations driven by energy minimiz
such as the radial director configuration in a cylindrical homoetropic confinement esc
into the third dimension [38]. Moreover, variations in morphological features of coll
particles can significantly diversify the type of defects and director deformations ind
and modulated by the surface topology even within the same constraint. Below, we di
two examples both involving g = 1 particles, where we find that the ensuing defect
director distributions are drastically different because of differences in the details o
particle geometry. Senyuk et al. discovered quarter-strength defect lines pinned to the
edges of torus-like particles with a large cross section of the tube measuring 5 × 5
(compared to 1 × 1 μm^2 of the handlebodies in Refs. [31,32]) (Figure 3e) [39]. Whi
total topological charge is still consistent with Equation (9), the quarter-strength d
lines are observed to migrate between edges of the particles by transforming into
integer disclinations across the surface of the particle, causing the particle to tilt
energy-minimizing orientation (Figure 3e–g). The disclinations can also be stretche
repositioned using optical tweezers while conserving the total topological charge. A
extreme case of changing geometric features while preserving topology involves fr
particles with many defect-inducing corners and edges once introduced to a LC. Has
et al. fabricated Koch star-based fractal particles from the 0th to the 3rd fractal iter
order [40]. As particles of genus g = 1 with homeotropic surface anchoring, the
topological charge induced should be 0 even though fractal shapes at high iteration o
exhibit complex geometric details (Figure 3h). It is observed that this constraint is ful
by pairs of oppositely charged defect points and disclinations found at the corners o

fractal particles, with the number of defects growing exponentially following the increased fractal order. However, when the fractal feature size is comparable to the nematic correlation length characterizing the mesoscopic size of the nematic ordering, director regions of reduced order parameters start to overlap, effectively causing local nematic-to-isotropic phase transitions in the vicinity of the fractal corners. Self-similarities of the nematic response such as defect distributions following the fractal stimulus are observed across different fractal iteration orders, despite being limited by the nematic correlation length. This work demonstrates that colloidal particles with fractal morphological complexity, while following the constraints imposed by the topological theorems, exhibit emergent behavior stemming from their fractal nature and may be instrumental in probing and understanding nematic order interactions with confining surfaces at limiting length scales.

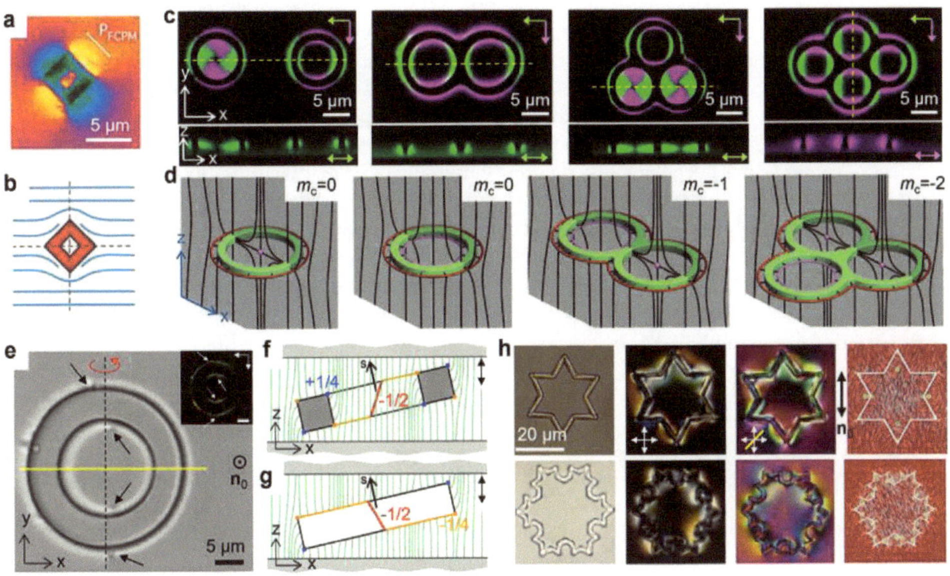

Figure 3. Topological colloids in LCs. (**a**,**b**) A square platelet with a central opening suspended in a LC. The image in (**a**) is taken with fluorescence confocal polarizing microscopy (FCPM); P_{FCPM} indicates the polarization direction of the excitation light. The schematic in (**b**) shows the director distribution around the platelet. (**c**,**d**) Colloidal handlebodies of various genera in LCs. Panels in (**c**) are micrographs obtained by overlapping fluorescence images with orthogonal excitation polarizations as indicated by the green and magenta arrows; insects below are cross-sectional images in the *xz* plane taken along the yellow dashed lines. The schematics in (**d**) represent the director (black lines) distortions and topological defects (red and purple lines; purple dots) induced by the handlebodies; the total topological charge is determined by the particle genus $m_c = 1 - g$. (**e**–**g**) A large torus-shaped colloidal particle with homeotropic surface anchoring suspended in a nematic LC, inducing $\frac{1}{2}$ and edge-pinned $\frac{1}{4}$ defect lines. The $\frac{1}{4}$ defect lines (blue lines in the schematic (**f**,**g**)) traverses along the edge of the particle and may jump between edges connected by $\frac{1}{2}$ defect lines (red lines in (**f**,**g**)). The black arrows in the micrograph in (**e**) indicate the location of bulk $\frac{1}{2}$ defect lines; tilting of the particle is indicated by the dashed line (rotation axis) and red curved arrow (tilt direction). Insets in (**e**) are obtained under crossed polarizers; polarization marked by white arrows. (**h**) Fractal colloidal particles with homeotropic surface anchoring in LCs. The first column is taken at elevated temperatures when the surrounding LC is in isotropic phase; the middle two columns are taken under crossed polarizers (white double arrows) and crossed polarizers with a retardation waveplate (yellow line). The last column is a computer-simulated director field distribution around such particles. Adapted from Refs. [31,37,39,40].

In addition to the static distributions of topological defects and director deforma dynamic processes involving topological colloids in LCs provide the possibility of exp the out-of-equilibrium interplay between particle topology and the topological feat the director field. One cannot change the topology of a particle's surface without bre it. However, the effective topology can be changed, for example, by smoothly op and closing holes in a 2D surface. Exploiting the anisotropic shape deformation elastomers upon phase transitions, Yuan et al. fabricated ring-shaped colloidal par and discovered that when heated, the aspect ratio of the rings changes and eventuall closed holes at the center of the particles open up after the temperature is above the ne isotropic phase transition point (Figure 4a) [41]. Moreover, 2D crystal-like lattices of LC elastomeric rings suspended in the unpolymerized nematic host show aniso collective actuations upon phase transitions. The experiments conducted so far, wi director fields of elastomeric particles and unpolymerized LC matching at the interfac not explore the induction and transformation of defects as the net topological charge director inside and outside the rings stays neutral. However, how such a change in pa topology may lead to the evolution of topological features of the director field for ca particles made of non-LC material and with well-defined boundary conditions at inte has been considered computationally. Numerical simulations of such systems by R et al. have provided insights: for a toroidal particle with two disclination loops, shri of the central hole results in the inner disclination looping closing into a point defec eventually disappears as the hole is fully closed (Figure 4c) [36]. We may expect that f developments in the area of shape-morphing colloids may allow for the direct observ of the interaction and dynamic interplay between the topological characteristics components.

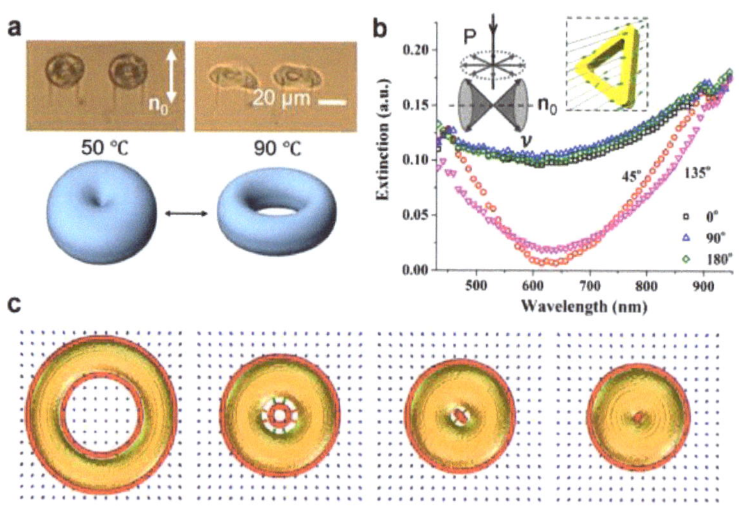

Figure 4. Stimuli-responsive topological colloids. (**a**) Ring-shaped microparticles made with crystal elastomers change shape upon temperature elevation over the nematic–isotropic transition point. The schematic in (**a**) shows the opening and closing of the central hole, effec changing the topology of the particle. (**b**) Alignment and polarization-dependent extinct plasmonic triangular nanoframes dispersed in a nematic LC. The schematic in (**a**) shows the d distortions caused by the triangular frame. The normal ν of the plane containing the nanofran the freedom to rotate in a cone shape with the far-field director $\mathbf{n}_0$ as the symmetry axis; **P** repr the polarization of the incident light. (**c**) Contraction of a ring-shaped particle with homeo anchoring and the induced disclination loops (indicated by red lines) in a nematic LC hos far-field director is perpendicular to the viewing plane. Adapted from Refs. [41–43].

Another interesting possibility recently considered experimentally is the mesoscale material behavior under conditions when the plasmonic nanoparticles dispersed in a nematic host have nontrivial topology. The complex geometry of triangular nanoframes made of gold and silver enables the generation of plasmonic effects as well as the cone-like orientation distribution of the particles' normal relative to the director when suspended in a nematic LC [43]. Differently from micron-scale dielectric particles with similar topology [44], polarization direction-dependent extinction spectra are observed upon incident light in the visible and infrared range (Figure 4b), consistent with the symmetry prescribed by the nanoframes' orientation distribution in the LC. Furthermore, this guest-host system enables the electric switching of the plasmonic extinction of the hybrid LC-nanocolloid material reminiscent of that of pure LCs in terms of timescales and threshold voltages but now allowing for tuning the surface plasmon resonance effects. This work complements existing works on plasmonic LC colloids with trivial topology at the time of publishing and may now be extended to other plasmonic particles of broken symmetry or nontrivial topology [45].

5. Knotted and Linked Colloids

Different from our everyday experience, a knot commonly studied in topology cannot be undone because the ends are connected to form a closed loop. The simplest form of a mathematical knot is a ring, a.k.a., an unknot, the behavior of which as a physical object with a tubing-like surrounding immersed in LCs is described in the previous section as genus $g = 1$ particles. More complex structures, i.e., non-trivial knots, in a LC director field induce knotted, linked, and other topologically non-trivial field configurations, allowing for experimental insights into predictions from knot theory and into the interplay of topologies of knotted surfaces, field deformation, and defects. Martinez et al. fabricated colloidal trefoil and pentafoil knots of tangential and homeotropic surface anchoring and demonstrated that a rich variety of director field and defect configurations may arise from the minimization of the total elastic energy of the system while still following the same topological constraints imposed by theorems from pure math (Figure 5a–e) [46]. For all cases studied of torus knots in the paper, the Euler characteristic is $\chi = 0$, the same as for rings or $g = 1$ handlebodies, which constrains the total topological charge or winding number to be zero. In a nematic LC, trefoil particle knots with either type of surface anchoring tend to orient with their corresponding torus plane orthogonal to $\mathbf{n}_0$ in their ground state (Figure 5a,b) where the total elastic energy is the lowest but are also found in metastable orientations such as those with the torus plane parallel to $\mathbf{n}_0$. Boojums induced by a trefoil knot with planar anchoring are usually located at the particle's surface where the surface normal is parallel to $\mathbf{n}_0$ (Figure 5a,c). A total number of 12 boojums can be found in such cases for trefoil-knot-shaped particles, with half of them having the winding number of +1 and the other half −1, adding to the net total charge of zero to satisfy the topological constraint. Similarly, a pentafoil knot with planar anchoring in the stable orientation state will induce 20 such self-compensating defects (Figure 5d), or in other words, four times the number of turns the knot string winding around the circular axis of the corresponding torus. For trefoil knots with homeotropic surface anchoring, however, the particles usually induce closed disclination loops (Figure 5e). As an example, for the case of the particle's torus plane perpendicular to $\mathbf{n}_0$, one observes two defect loops that are both trefoil knots tracing along the tube of the knotted trefoil-shaped particle and linked with each other and with the particle knot, forming a three-component link of knots (inset of Figure 5e). Knotted disclination loops can also be induced by multiple topologically simple spherical particles in twisted LCs where complex configurations of closed disclination loops are promoted by chiral symmetry breaking [47,48].

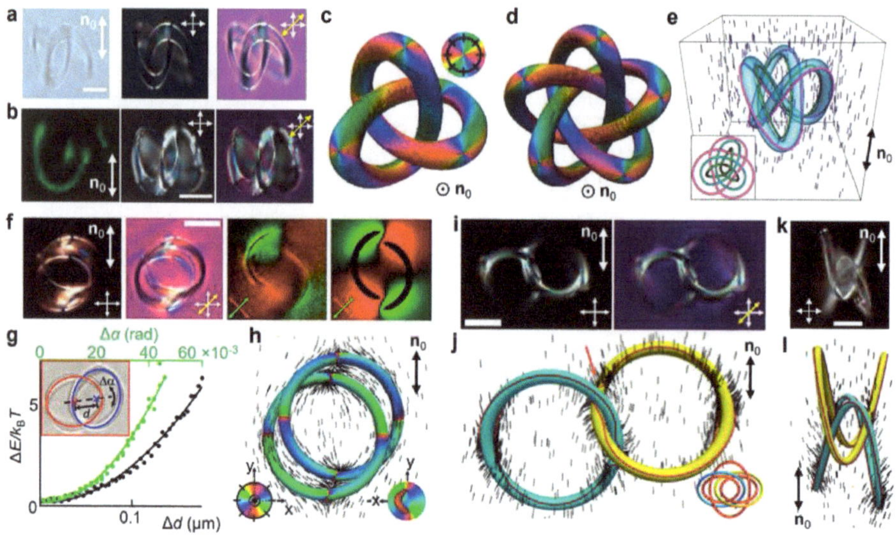

Figure 5. Knotted (**a–e**) and linked (**f–l**) colloids in LCs. (**a**,**b**) Optical micrographs of a trefoi with planar (**a**) and homeotropic (**b**) surface anchoring suspended in LCs. Crossed white d arrows indicate the direction of the polarizer and the analyzer and the yellow double arrow ind the direction of the slow axis of a 530 nm retardation plate. (**c**,**d**) Numerically simulated directo distributions on the surface of trefoil (**c**) and pentafoil (**d**) knots with planar surface anchoring color represents the orientation of the surface director projected to the plane orthogonal to **n** color scheme is shown as the inset of (**c**). The far-field director $\mathbf{n}_0$ is perpendicular to the sample as marked by the dot in a circle. (**e**) Schematic of defect lines (represented by green and magenta induced by and entwined with a trefoil knot of homeotropic anchoring obtained from num simulation. (**f**) Polarizing, fluorescence, and simulated micrographs of linked colloidal rings tangential surface anchoring suspended in LCs. The green and red double arrows represe excitation polarization for fluorescence imaging. (**g**) Elastic interaction energy vs. deviation the equilibrium center-to-center separation Δd (black symbols) and orientation $\Delta\alpha$ (green sym as defined in the inset. (**h**) Numerically simulated director field distributions on the surfa a Hopf link at the position in (**f**). The color represents the director orientation as defined i inset. (**i**, **k**) Polarizing micrographs of Hopf (**i**) and Salomon (**k**) links with homeotropic su anchoring suspended in LCs. (**j**,**l**) corresponding simulated director field distributions and d field (represented by the red lines) induced by the links. The red arrow in (**j**) points at the loc where the disclination line jumps from one colloidal loop to the other. The double arrows mark represent the far-field director. Adapted from Refs. [46,49].

Colloidal particles consisting of separate but linked components provide an way of generating complex topological defects, such as multi-component links. W suspended in LCs, these multi-component linked yet disconnected colloidal surfaces r in elastic coupling of the linked component through director deformations and topolo defects that they share in their tubular surroundings. Martinez et al. fabricated Hop Salomon links with the linking numbers Lk = ±1 and ±2, respectively (Figure 5f–l The linking number is a topological invariant representing the number of times that individual constituting loop winds around the other. The topological constraint requ net zero of the total winding number or hedgehog charge given that the Euler characte is $\chi = 0$ for all colloidal links studied. For Hopf links with planar surface anchoring most commonly observed director configuration contains eight boojums on the pa surface, with the two constituting colloidal unknot laying close to each other and planes crossing (Figure 5f,h). The minimization of elastic energies associated wit director distortion and topological defects holds the linked components in their equilibrium positions despite thermal fluctuations. The elastic coupling between the constituting loops can be characterized by tracking their positional and angular diffusion (Figure 5g) and is found to be Hookean-like with the interaction force in the pN range. For Hopf links with homeotropic surface anchoring, the elastic coupling manifests itself in configurations where the induced singular defect lines jump from one component to the other, serving as an elastic string holding the two components together (Figure 5i,j). In the more complex cases of Salomon links, twice as many surface boojums tend to be induced compared with their Hopf counterparts, while pairs of individual disclination loops can be seen following each of the linked components with homeotropic surface anchoring (Figure 5k,l). However, many metastable particle orientations and director field configurations can be observed for linked particles of any given linking number and surface anchoring. A non-exhaustive survey presented in Ref. [49] shows that colloidal linked rings and accompanying singular defect loops can form many topologically distinct multi-component systems even with the same number of unknots and linking numbers.

The colloidal systems of links and knots in LCs described above exemplify the emergent behaviors and complexities of topological particles and director distortions even though topological constraints for particles of Euler characteristic $\chi = 0$ do not require the presence of topological defects. Importantly, the disclination loops and knots would not be even topologically stable in a polar system, so the fact that such a large variety of multi-component links and knots of defects can be found demonstrates the complexity of the interplay between topologies of surfaces and nonpolar molecular alignment fields.

6. Conclusions and Future Perspectives

A central topic in colloidal science is to investigate the emergent behavior of constituting colloidal particles as they interact and self-assemble. They are ideal candidates for modeling the interaction and dynamics of atomic matter, such as self-assembly, crystallization, phase transition, etc., and even envisioning artificial forms of matter and metamaterials that go well beyond the complexity of natural systems [50–53]. Compared to the isotropic colloidal spheres and solvents commonly used in colloidal crystals, LC colloids provide new forms of long-range and anisotropic interactions mediated by the LC elasticity through director field deformation [5]. The examples described in this review have shown that varying the morphological characteristics of constituting particles such as symmetry and topology leads to a diversity of director configurations and colloidal behaviors. These colloidal "big atoms" with features of chiral symmetry, nontrivial topology, as well as knotted and linked components may readily contribute to the development of colloidal superstructures and exceed the diversity of atomic lattices. In addition, solitonic structures in cholesteric LCs involving topological defects and knotted and linked director configurations can also serve as building blocks of colloidal matter [6]. Although this review focused on nematic colloidal systems, other types of LC phases with additional partial positional degrees of ordering and chiral superstructures, such as cholesteric, columnar, and smectic LCs, may provide further symmetry-breaking director distributions and solvent-mediated interactions for LC colloids. These more complex LC field configurations add to the toolbox accessible for exploring fundamental mathematical and physical questions as well as developing novel functional materials with emergent properties and practical utility.

Historically, Lord Kelvin considered a model of atoms where different chemical elements in the periodic table would correspond to different knotted vortices [6]. Even though this model turned out to be incorrect, curiously, within the "colloidal atom" paradigm involving nematic colloids and vortices shaped as knots, researchers can now create modern-day analogs of Kelvin's topologically protected particles, thus opening doors to self-assembly of mesostructured materials with topology- and chirality-based design principles.

Future directions to be pursued as essential for the field of LC colloids may include the (i) large-scale self-assembly of low-symmetry and complex-topology colloidal particles;

(ii) design and demonstration of out-of-equilibrium processes including stimuli-respc dynamic rotational and translational behavior and the ensuing collective and eme dynamic patterns [44,54]; (iii) design of colloidal composite materials of novel viscoe electromagnetic, photonic, etc., properties; and (iv) use of ensuing novel composite n als to control transmission and directional scattering of light, as needed for applica in smart windows, displays, and electro–optic devices should be explored [55]. Like knotted disclination lines guide the formation of crystal-lattice-like structures of coll spheres in chiral LCs [4,6], colloidal knots or other chiral and topological particles as as LC solitonic structures may assemble into organization patterns not accessible to atomic counterparts. Such novel assemblies may open possibilities to chiral, topolc and knotted colloidal metamatter with pre-designed physical properties.

The topology- and geometry-guided colloidal self-assembly approach combine and hard materials at the mesoscale in order to develop solids and fluids that e: emergent physical behavior not encountered in conventional material systems, wi inverse design and pre-engineering capabilities. Future research will need to focus c derstanding and the control of material behavior arising from highly controlled mes interpenetration of ordered solid and liquid crystalline building blocks, with the comp exhibiting solid, liquid, or mesophase behaviors controlled by weak stimuli. The d and realization of desired physical properties in these composites can be based o following strategies: (1) templating the mesoscale organization of quantum and plas nanoparticles by mesoporous structures enabled by lyotropic liquid crystals to ha quantum mechanical and plasmon–exciton interactions; (2) coupling between topolc structures of continuous and singular field configurations in the solid and soft n mesoscale subsystems; (3) discovery of low-symmetry fluids in hybrid molecular-col systems enriched by exploiting particle's geometric shapes, surface charging and chi and (4) emergent effects in magnetic, colloidal, and topological solitonic spin ice sy arising in soft-hard matter systems. Electrical realignment of the LC component composites may in the future allow for the rearrangement and reorientation of aniso nanoparticles, leading to unprecedented control over self-assembled nanostructure to dramatic changes in the material emergent behavior and properties. The LC coll research transcends traditional disciplinary boundaries of physics, topology, chen engineering, and materials science and will advance our knowledge of mechanis the nanoscale self-organization phenomena of importance to achieving many scie and technological goals of renewable energy research. Since our abilities to harvest, efficiently use, and convert energy among different forms are highly dependent on able materials and their properties, this research may lead to cheaper and more eff renewable energy technologies, a new breed of energy-efficient information display electro–optic consumer devices, as well as a fertile ground for new basic science, transformative impacts.

Author Contributions: Writing—original draft preparation, Y.Y.; writing—review and editir and I.I.S. All authors have read and agreed to the published version of the manuscript.

Funding: Y.Y. acknowledges the financial support from the Japan Society for the Promotion of S (JSPS KAKENHI grant number JP24K23088) when writing the manuscript. I.I.S. acknowledg financial support of the fundamental research on colloids (the U.S. Department of Energy, of Basic Energy Sciences, Division of Materials Sciences and Engineering, under Award ER contract DE-SC0019293 with the University of Colorado in Boulder).

Acknowledgments: I.I.S. acknowledges the hospitality of the International Institute for Sustain with Knotted Chiral Meta Matter (WPI-SKCM^2) in Japan while on a sabbatical stay, as well hospitality of the Kavli Institute for Theoretical Physics in Santa Barbara while participating "Nanoparticle Assemblies" KITP program, when he was partially working on this review.

Conflicts of Interest: The authors declare no conflict of interest.

rences

de Gennes, P.G.; Prost, J. *The Physics of Liquid Crystals*, 2nd ed.; Oxford Science Publications: Oxford, UK, 1993.

Chaikin, P.M.; Lubensky, T.C. *Principles of Condensed Matter Physics*; Cambridge Univ. Press: Cambridge, UK, 2000.

Andrienko, D. Introduction to liquid crystals. *J. Mol. Liq.* **2018**, *267*, 520–541. [CrossRef]

Muševič, I. *Liquid Crystal Colloids*; Springer: Cham, Switzerland, 2017.

Smalyukh, I.I. Liquid crystal colloids. *Annu. Rev. Condens. Matter Phys.* **2018**, *9*, 207–226. [CrossRef]

Smalyukh, I.I. Review: Knots and other new topological effects in liquid crystals and colloids. *Rep. Prog. Phys.* **2020**, *83*, 106601. [CrossRef] [PubMed]

Uchida, J.; Soberats, B.; Gupta, M.; Kato, T. Advanced functional liquid crystals. *Adv. Mater.* **2022**, *34*, 2109063. [CrossRef] [PubMed]

Concellon, A. Liquid crystal emulsions: A versatile platform for photonics, sensing and active matter. *Angew. Chem. Int. Ed. Engl.* **2023**, *62*, e202308857. [CrossRef]

Lu, P.J.; Weitz, D.A. Colloidal particles: Crystals, glasses and gels. *Annu. Rev. Condens. Matter Phys.* **2013**, *4*, 217–233.

Li, Z.; Fan, Q.; Yin, Y. Colloidal self-assembly approaches to smart nanostructured materials. *Chem. Rev.* **2021**, *122*, 4976–5067. [CrossRef]

Everett, D.H. manual of symbols and terminology for physicochemical quantities and units, appendix ii: Definitions, terminology and symbols in colloid and surface chemistry. *Pure Appl. Chem.* **1972**, *31*, 577–638. [CrossRef]

Yamanaka, J.; Okuzono, T.; Toyatama, A. *Colloidal Self-Assembly*; Springer: Singapore, 2023.

Huang, Y.; Wu, C.; Chen, J.; Tang, J. Colloidal self-assembly: From passive to active systems. *Angew. Chem. Int. Ed. Engl.* **2024**, *63*, e202313885. [CrossRef]

Barrat, J.L.; Del Gado, E.; Egelhaaf, S.U.; Mao, X.; Dijkstra, M.; Pine, D.J.; Kumar, S.K.; Bishop, K.; Gang, O.; Obermeyer, A.; et al. Soft matter roadmap. *J. Phys. Mater.* **2024**, *7*, 012501. [CrossRef]

Poulin, P.; Stark, H.; Lubensky, T.C.; Weitz, D.A. Novel colloidal interactions in anisotropic fluids. *Science* **1997**, *275*, 1770–1773. [CrossRef] [PubMed]

Stark, H. Director field configurations around a spherical particle in a nematic liquid crystal. *Eur. Phys. J. B* **1999**, *10*, 311–321. [CrossRef]

Hueckel, T.; Hocky, G.M.; Sacanna, S. Total synthesis of colloidal matter. *Nat. Rev. Mater.* **2021**, *6*, 1053–1069. [CrossRef]

Senyuk, B.; Meng, C.; Smalyukh, I.I. Design and preparation of nematic colloidal particles. *Langmuir* **2022**, *38*, 0900–9118. [CrossRef] [PubMed]

Lee, T.; Smalyukh, I.I. Conventional and nonlinear optical microscopy of liquid crystal colloids. In *Liquid Crystals with Nano and Microparticles*; Lagerwall, J.P.F., Scalia, G., Eds.; World Scientific Publishing: Singapore, 2017; Volume 1, pp. 179–207.

Ravnik, M.; Žumer, S. Landau–de Gennes modelling of nematic liquid crystal colloids. *Liq. Cryst.* **2009**, *36*, 1201–1214. [CrossRef]

Senyuk, B.; Puls, O.; Tovkach, O.; Chernyshuk, S.; Smalyukh, I.I. Hexadecapolar nematic colloids. *Nat. Commun.* **2016**, *7*, 10659. [CrossRef]

Brochard, F.; de Gennes, P.G. Theory of magnetic suspensions in liquid crystals. *J. Phys.* **1970**, *31*, 691–708. [CrossRef]

Pergamenshchik, V.M.; Uzunova, V.A. Colloidal nematostatics. *Condens. Matter Phys.* **2010**, *13*, 1–29. [CrossRef]

Chernyshuk, S.B.; Tovkach, O.M.; Lev, B.I. Theory of elastic interaction between arbitrary colloidal particles in confined nematic liquid crystals. *Phys. Rev. E* **2014**, *89*, 032505. [CrossRef]

Jackson, J.D. *Classical Electrodynamics*, 3rd ed.; Wiley: New York, NY, USA, 1998.

Yuan, Y.; Liu, Q.; Senyuk, B.; Smalyukh, I.I. Elastic colloidal monopoles and reconfigurable self-assembly in liquid crystals. *Nature* **2019**, *570*, 214–218. [CrossRef]

Senyuk, B.; Mozaffari, A.; Crust, K.; Zhang, R.; de Pablo, J.J.; Smalyukh, I.I. Transformation between elastic dipoles, quadrupoles, octupoles, and hexadecapoles driven by surfactant self-assembly in nematic emulsion. *Sci. Adv.* **2021**, *7*, eabg0377. [CrossRef] [PubMed]

Yuan, Y.; Tasinkevych, M.; Smalyukh, I.I. Colloidal interactions and unusual crystallization versus de-mixing of elastic multipoles formed by gold mesoflowers. *Nat. Commun.* **2020**, *11*, 188. [CrossRef] [PubMed]

Lubensky, T.C.; Pettey, D.; Currier, N.; Stark, H. Topological defects and interactions in nematic emulsions. *Phys. Rev. E* **1998**, *57*, 610–625. [CrossRef]

Kleman, M.; Lavrentovich, O.D. Topological point defects in nematic liquid crystals. *Philos. Mag.* **2006**, *86*, 4117–4137. [CrossRef]

Senyuk, B.; Liu, Q.; He, S.; Kamien, R.D.; Kusner, R.B.; Lubensky, T.C.; Smalyukh, I.I. Topological colloids. *Nature* **2013**, *493*, 200–205. [CrossRef]

Liu, Q.; Senyuk, B.; Tasinkevych, M.; Smalyukh, I.I. Nematic liquid crystal boojums with handles on colloidal handlebodies. *Proc. Natl. Acad. Sci. USA* **2013**, *110*, 9231–9236. [CrossRef]

Wagnière, G.H. *On Chirality and the Universal Asymmetry*; John Wiley & Sons: Zurich, Switzerland, 2008.

Martinez, A.; Lee, T.; Asavei, T.; Rubinsztein-Dunlop, H.; Smalyukh, I.I. Three-dimensional complex-shaped photopolymerized microparticles at liquid crystal interfaces. *Soft Matter* **2012**, *8*, 2432–2437. [CrossRef]

Yuan, Y.; Martinez, A.; Senyuk, B.; Tasinkevych, M.; Smalyukh, I.I. Chiral liquid crystal colloids. *Nat. Mater.* **2018**, *17*, 71–79. [CrossRef] [PubMed]

Loudet, J.C.; Hanusse, P.; Poulin, P. Stokes drag on a sphere in a nematic liquid crystal. *Science* **2004**, *306*, 1525. [CrossRef]

37. Lapointe, C.P.; Mason, T.G.; Smalyukh, I.I. Shape-controlled colloidal interactions in nematic liquid crystals. *Science* **200** 1083–1086. [CrossRef]
38. Urbanski, M.; Reyes, C.G.; Noh, J.; Sharma, A.; Geng, Y.; Jampani, V.S.R.; Lagerwall, J.P.F. Liquid crystals in micron-scale dr shells, and fibers. *J. Phys. Condens. Matter* **2017**, *29*, 133003. [CrossRef] [PubMed]
39. Senyuk, B.; Liu, Q.; Yuan, Y.; Smalyukh, I.I. Edge pinning and transformation of defect lines induced by faceted colloidal ri nematic liquid crystals. *Phys. Rev. E* **2016**, *93*, 062704. [CrossRef] [PubMed]
40. Hashemi, S.M.; Jagodic, U.; Mozaffari, M.R.; Ejtehadi, M.R.; Musevic, I.; Ravnik, M. Fractal nematic colloids. *Nat. Commun* *8*, 14026. [CrossRef] [PubMed]
41. Yuan, Y.; Keller, P.; Smalyukh, I.I. Elastomeric nematic colloids, colloidal crystals and microstructures with complex topolog *Matter* **2021**, *17*, 3037–3046. [CrossRef]
42. Ravnik, M.; Čopar, S.; Žumer, S. Particles with changeable topology in nematic colloids. *J. Phys. Condens. Matter* **2015**, *27*, 3 [CrossRef] [PubMed]
43. Yuan, Y.; Smalyukh, I.I. Topological nanocolloids with facile electric switching of plasmonic properties. *Opts. Lett.* **20** 5630–5633. [CrossRef]
44. Lapointe, C.P.; Hopkins, S.; Mason, T.G.; Smalyukh, I.I. Electrically-driven multi-axis rotational dynamics of colloidal plate nematic liquid crystals. *Phys. Rev. Lett.* **2010**, *105*, 178301. [CrossRef]
45. Wang, P.; Krasavin, A.V.; Liu, L.; Jiang, Y.; Li, Z.; Guo, X.; Tong, L.; Zayats, A. Molecular plasmonics with metamaterials. *Rev.* **2022**, *122*, 15031–15081. [CrossRef]
46. Martinez, A.; Ravnik, M.; Lucero, B.; Visvanathan, R.; Žumer, S.; Smalyukh, I.I. Mutually tangled colloidal knots and in defect loops in nematic fields. *Nat. Mater.* **2014**, *13*, 258–263. [CrossRef]
47. Jampani, V.S.R.; Škarabot, M.; Ravnik, M.; Čopar, S.; Žumer, S.; Muševič, I. Colloidal entanglement in highly twisted nematic colloids: Twisted loops, Hopf links, and trefoil knots. *Phys. Rev. E* **2011**, *84*, 031703. [CrossRef]
48. Čopar, S.; Tkalec, U.; Muševič, I.; Žumer, S. Knot theory realizations in nematic colloids. *Proc. Natl. Acad. Sci. USA* **201** 1675–1680. [CrossRef]
49. Martinez, A.; Hermosillo, L.; Tasinkevych, M.; Smalyukh, I.I. Linked topological colloids in a nematic host. *Proc. Natl. Ac USA* **2015**, *112*, 4546–4551. [CrossRef]
50. Frenkel, D. Playing tricks with designer atoms. *Science* **2002**, *296*, 65–66. [CrossRef] [PubMed]
51. Poon, W. Colloids as big atoms. *Science* **2004**, *304*, 830–831. [CrossRef] [PubMed]
52. Li, B.; Zhou, D.; Han, Y. Assembly and phase transitions of colloidal crystals. *Nat. Mater. Rev.* **2016**, *1*, 15011. [CrossRef]
53. Kim, S.; Svetlizky, I.; Wetiz, D.A.; Spaepen, F. Work hardening in colloidal crystals. *Nature* **2024**, *630*, 648–653. [CrossRef]
54. Bishop, K.J.M.; Biswal, S.L.; Bhart, B. Active colloids as models, materials, and machines. *Annu. Rev. Chem. Biomol. Eng.* **20** 1–30. [CrossRef]
55. De La Cruz, J.; Liu, Q.; Frazier, A.W.; Senyuk, B.; Smalyukh, I.I. Cellulose-based photonic structures as optical filters and gain regulating films. *ACS Photonics* **2018**, *5*, 2468–2477. [CrossRef]

crystals

MDPI

w

e Role of Liquid Crystal Elastomers in Pioneering ological Applications

Shiralipour [1,2,3], Yeganeh Nik Akhtar [1,2,3,†], Ashley Gilmor [1,4,†], Gisele Pegorin [1,4,†], ham Valerio-Aguilar [1,4,†] and Elda Hegmann [1,2,3,4,5,6,*]

1 Advanced Materials and Liquid Crystal Institute, Kent State University (KSU), Kent, OH 44240, USA; fshirali@kent.edu (F.S.); ynikakht@kent.edu (Y.N.A.); asumme20@kent.edu (A.G.); gpegorin@kent.edu (G.P.); avaleri2@kent.edu (A.V.-A.)
2 Cell Biology and Molecular Genetics Graduate Program, Kent State University (KSU), Kent, OH 44240, USA
3 Department of Biological Sciences, Kent State University (KSU), Kent, OH 44240, USA
4 Materials Science Graduate Program, Kent State University (KSU), Kent, OH 44240, USA
5 Biomedical Sciences Program, Kent State University (KSU), Kent, OH 44240, USA
6 Brain Health Research Institute, Kent State University (KSU), Kent, OH 44240, USA
* Correspondence: ehegmann@kent.edu
† These authors contributed equally to this work.

Abstract: Liquid crystal elastomers have shown an attractive potential for various biological applications due to their unique combination of mechanical flexibility and responsiveness to external stimuli. In this review, we will focus on a few examples of LCEs used with specific applications for biological/biomedical/environmental systems. So far, areas of innovation have been concentrating on the integration of LCEs to enhance stability under physiological conditions, ensure precise integration with biological systems, and address challenges related to optical properties and spatial control of deformation. However, several challenges and limitations must still be addressed to fully realize their potential in biomedical and environmental fields, and future research should focus on continuing to improve biocompatibility, response to the environment and chemical cues, mechanical properties, ensuring long-term stability, and establishing cost-effective production processes. So far, 3D/4D printing appears as a great promise to develop materials of high complexity, almost any shape, and high production output. However, researchers need to find ways to reduce synthesis costs to ensure that LCEs are developed using cost-effective production methods at a scale necessary for their specific applications' needs.

Keywords: liquid crystals; liquid crystal elastomers; 3D printing; anisotropy; advanced manufacturing; additive manufacturing; orientational order

n: Shiralipour, F.; Nik Akhtar, or, A.; Pegorin, G.; Aguilar, A.; Hegmann, E. The Liquid Crystal Elastomers in ing Biological Applications. 2024, 14, 859. /doi.org/10.3390/ 100859

nic Editor: Alberta Ferrarini

d: 10 September 2024
: 27 September 2024
ed: 28 September 2024
ed: 29 September 2024

1. Introduction

1.1. Overview of Liquid Crystal Elastomers

At the time Otto Lehmann and Friedrich Reinitizer [1,2] encountered a double melting phenomenon in 1888 that will be described as a fourth state of matter (or liquid crystals, LCs), they did not realize the impact that their discovery would have on future technological applications. However, Lehmann visualized a technological application when, in 1909 [3], he suggested that some LCs can potentially act as artificial muscular driving motors. Subsequently, de Gennes in 1997 [4,5] supported this notion and a myriad of liquid crystal elastomer (LCE) applications have since been acknowledged. Most of these applications were related to properties combining orientational anisotropy of LCs with known elasticity of elastomers, and LCEs came to be known as materials capable of reversible actuation [6]. While the first discovery of LCs was related to the study of cholesteryl benzoate and related biological compounds, the main LC applications were heavily focused on liquid crystal displays (LCDs). LCDs were centered on the optical properties of LCs in the presence/absence of an electric field. Since the last two decades, most of the new applications have been back to their "roots" and target towards biological/biom applications, tissue engineering (TE), advanced manufacturing (AM), and soft robotics In this review, we aim to summarize most of the latest LCE's biological applications

1.2. LCE Significance in Biological Applications

LCs exist abundantly in living systems, and many of them are found as lyo liquid-crystalline (LLC) phases and are a base for studying biomimetic chemistry. LL predominantly found in cell membranes, mainly made of phospholipids [10]. Silk is an example of LCs; the organization of the proteins that form silk presents a liquid c phase [11,12]. DNA can also form liquid crystal phases [13]. During two-dimens (2D) cell culture experiments, LC behavior has been observed where cells form topol defects. These topological defects have shown certain biological effects that can indu extrusion and potentially cell death [14]. Tissue Engineering (TE) combines the know of both biology and engineering with the determination of finding ideal alternati restore or regenerate ailing or impaired tissue [15], and most of those alternatives on three-dimensional (3D) systems looking for better outcomes than those found systems [16,17]. Most of these 3D systems need to overcome most commonly f limitations, such as being non-toxic, providing effective support for cells and for tissue, promoting extracellular matrix, sustaining 3D cell growth, and having appro mechanical properties similar to those tissues of interest [18–22].

LCEs combine the orientational anisotropy of LCs with the intrinsic elasticity o tomers and have been recognized as ideal materials capable of reversible actuation [Biological/biomedical applications of LCEs as part of TE have become an importan of current academic research as responsive and anisotropic 3D cell scaffolds. With last decade, AM, or 3D printing, with its layer-by-layer fabrication process, has all the creation of complex 3D structures with high reproducibility in endogenous en ments. For 3D printing, the use of synthetic-based bio-inks has been used, showing cellular responses, biocompatibility, and capability of ECM formation [25,26]. Fig shows an example where LCE scaffolds are suitable for most types of somatic and imm cells. LCEs have been shown to be ideal substrates for cell proliferation, differenti and maturation, as well as serving as an ideal host for primary cell development ar in vitro studies of myelination. Figure 1 shows a study conducted in a 3D environ that promoted ECM formation and maturation processes of co-cultured neuroblast (SH-SY5Y) and human oligodendrocyte cells (MO3.13). Indicating that LCE scaffold impact cellular function, differentiation, and other cell processes, enabling the mimi an endogenous environment for long-term studies.

At the beginning, many LCEs included polycaprolactone (PCL) and polyeth glycol (PEG)-based inks. LCE-inks have been reported, for example, as nematic-l thermally responsive LCE ink [27], as direct-write printed into 3D structures, or as operating temperature direct ink writing (HOT-DIW) [28]. It is important to not processing can increase the likelihood of external stimulation of LCE shape actuatio Much has changed since the first LCE reports, where now LCE actuators are no longe thin film devices; they can now be complex 3D structures. LCE 3D printing formula are constantly improving, allowing the development of thermal- or photo-crosslinked and increasing a wide range of printing possibilities [30–34].

We are aware that each year we are faced with a fast-growing aging populatio inevitably brings age-related diseases [15]. On top of that, accidents occur involvi loss of mobility, organ, and limb function. To combat this, the medical community extensively on prosthetic and rehabilitation devices to bring back mobility to injured p Looking ahead, there has been a shift in focus onto soft robotics and medical roboti the purpose of finding better, individualized, more human-like alternatives, integ novel sensors, actuators, and materials into prosthetics to help increase life expec and improve the quality of life of affected individuals. The main objective is to s tute old technology that involved passive and at times skin-irritable materials [35,36] for flexible electronics and soft devices [37–39] that respond dynamically to external stimuli, including temperature, chemical cues, electricity, and light, promoting multifaceted applications [40]. Soft materials used in robots require different levels of biocompatibility, biomimicry, sensing, actuation, and computation/informatics [41]. Most importantly, there is a need for mechano-sensation electronics that can provide sensations of cold/hot, touch, pain, and comfort [42]. LCEs have been recently seen as soft responsive materials that can revolutionize the realm of soft robotics, leading to a new era of versatility and adaptability [43].

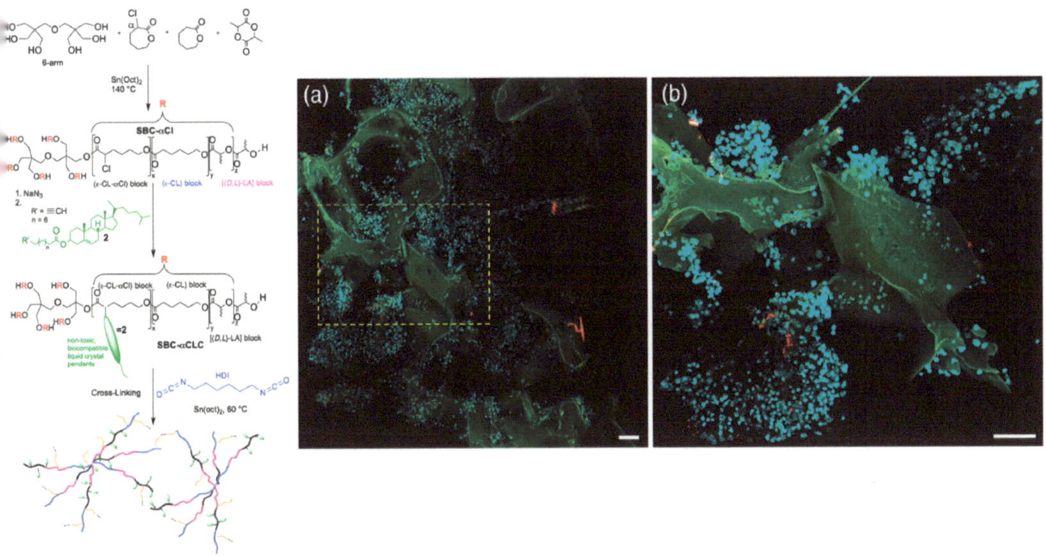

Figure 1. Example of liquid crystalline elastomers (LCEs). Scheme shows the synthetic pathway of a LCE star block-copolymer with cholesterol liquid crystal as pendant units and (**a**) confocal image of primary mouse brain cells grown on LCE scaffold and (**b**) enlarged image of selected area in yellow (cell nuclei stained in blue, scale is 60 µm). Reprinted from *J Appl Polym Sci* [25]. Copyright (2023) with permission from Wiley (Hoboken, NJ, USA).

2. Fundamentals of Liquid Crystal Elastomers

LCEs are unique materials that combine the properties of liquid crystals with elastomers, resulting in materials with distinct physical and chemical characteristics. LCEs consist of LC mesogens (rod-like molecules) that are incorporated into a crosslinked polymer network (see Figure 2) [44]. Mesogens can be attached to the polymer backbone (main-chain LCEs) or connected via flexible spacers (side-chain LCEs). LCEs can be either monodomain, where the mesogens are uniformly aligned throughout the material, or polydomain, where the alignment varies spatially. Monodomain LCEs exhibit more shape-changing abilities.

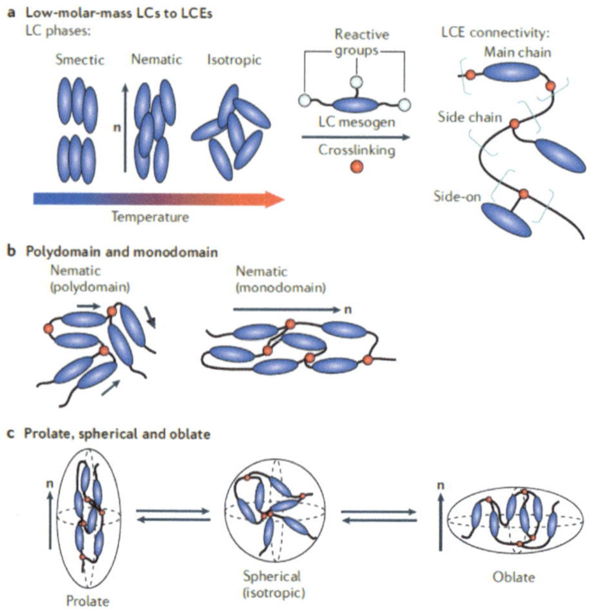

Figure 2. Types of liquid crystalline elastomers (LCEs). (**a**) The rod-like mesogens ar nected via reactive functional groups in the main chain, end-on side chains, or side-on side c (**b**) Polydomain LCEs have local domains of nematic order, and monodomain LCEs have a con director throughout the network. (**c**) In a monodomain LCE, the chain configuration can be p or oblate, with a spherical configuration in the isotropic state. Reprinted from *Nat. Rev. Ma* Copyright (2021) with permission from Springer Nature (Berlin/Heidelberg, Germany).

2.1. Chemical and Physical Properties

LCs are composed of anisotropic molecules called mesogens, which can be roc (calamitic), disc-like (discotic), or lath-like (sanidic) in shape. These mesogens exhibit p transitions based on temperature (thermotropic LCs) or concentration (lyotropic LCs), the most common phases being nematic, smectic, and cholesteric. Mesogens can various functional groups such as thiols, acrylates, or epoxides, which are crucial for polymerization and cross-linking into LCEs. The stability of these phases is influence the chemical structure of the mesogens and the polymer backbone. For instance, side- LCEs [23,46] can show stable nematic or smectic phases over a wide temperature r (see Figure 2). The elastomeric properties of LCEs arise from crosslinking the pol chains, either through covalent bonds or physical interactions. Crosslinking density significantly impact the mechanical properties and thermal behavior of the LCEs [47] chemical structure and functionalization of mesogens affect their alignment within which can be controlled using techniques such as mechanical stretching, surface align field-assisted alignment, and shear alignment [45,48].

LCEs exhibit anisotropic mechanical properties due to the alignment of liquid cr molecules within the elastomer matrix, allowing them to respond directionally to ext stimuli like light. For instance, certain LCEs are designed to respond to light sti particularly those containing azobenzene groups. Azobenzene can undergo a rever transformation between its trans and cis isomers when exposed to UV light, resulting change in the molecular alignment [45,49].

They display rubber-like elasticity and can undergo significant deformation, wit presence of polymer networks and liquid crystalline molecules imparting them witl and super-soft elasticity, making them suitable for applications like artificial muscle actuators. The thermo-elastic properties of LCEs enable them to change shape in respor temperature changes, particularly notable in side-chain liquid crystal elastomers (SC-LCEs), where different mesophases exhibit stability over a wide temperature range. Additionally, LCEs can remember and return to their original shape after deformation when exposed to certain stimuli, such as temperature changes or light, leveraging this shape-memory behavior in applications requiring precise and reversible actuation [45,49]. LCEs can be designed to respond to various chemical stimuli, which makes them useful in the development of sensors and responsive materials. This responsiveness is often achieved by incorporating photoresponsive or thermoresponsive groups within the polymer matrix [48].

2.2. Synthetic Strategies

There are different synthesis and alignment methods that lead to the production of LCEs with diverse molecular structures and properties. Several of these are displayed in Figure 3. The various approaches are described below and include the following:

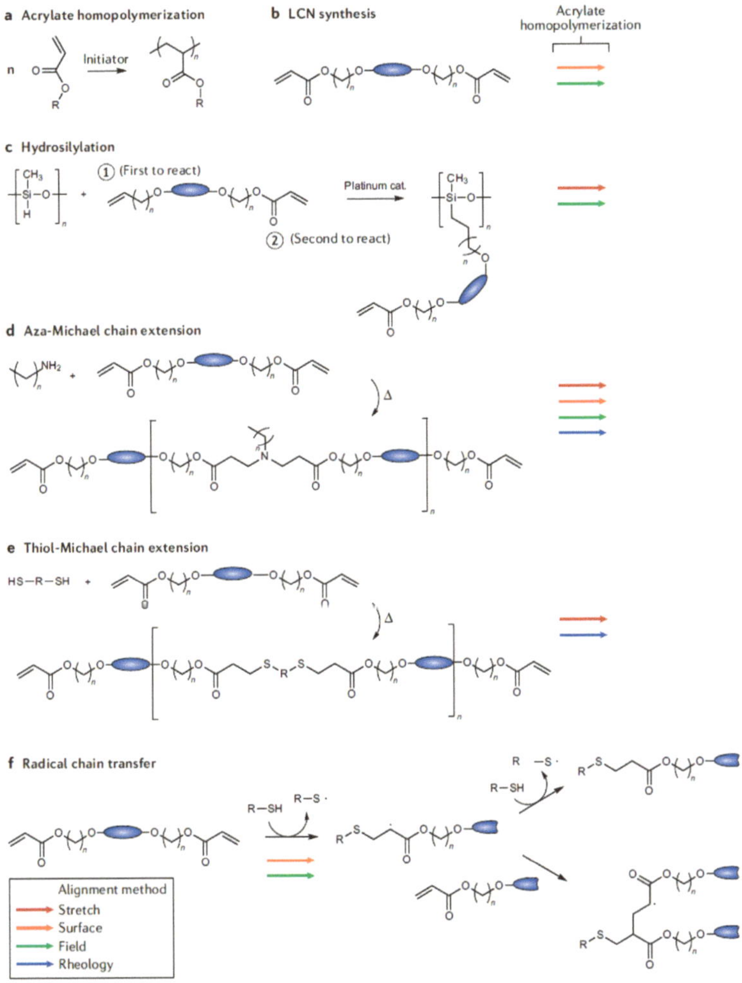

Figure 3. Synthetic strategies for crosslinking LC polymers and their corresponding alignment methods. Reprinted from *Nat. Rev. Mat* [45]. Copyright (2021) with permission from Springer Nature.

Mechanical alignment: This technique includes a two-phase reaction. Initially, the liquid crystal elastomer (LCE) mixtures undergo partial polymerization. They are then mechanically stretched to align the polymer chains. While maintaining this stre alignment, the polymerization process is completed, locking in the oriented structur

The field-assisted alignment can be divided into two primary categories: (1) Electri alignment uses electric fields to orient LC molecules by taking advantage of the mole uneven response to electric fields in different directions. (2) Magnetic field aligr utilizes magnetic fields to position liquid crystal molecules. It relies on the mole varying magnetic properties along different axes. Both techniques exploit the aniso nature of LC molecules to achieve alignment [50].

The 3D printing alignment technique uses the fluid dynamics inherent in 3D pri to orient LC molecules. As the material is pushed through a printer nozzle, it experi forces that stretch and shear the material. The mechanical stress suffered during extr naturally causes the LC molecules to line up (or orient) in a preferred direction [51].

Chain extension reactions include three subcategories: (1) Aza-Michael Addition, combines LC monomers containing two acrylate groups with primary amines. (2) Michael addition involves reacting extra diacrylate LC monomers with thiols. (3) Thic reactions involve a gradual polymer growth process, linking monomers with two groups to LC monomers featuring diallyl ether functionalities. These reactions all to lengthen molecular chains, but each uses different chemical pathways and func groups to achieve this goal [49,52].

Hydrosilylation is a two-step process where LC monomers and crosslinkers react linear polysiloxane in the presence of a platinum catalyst, followed by mechanical aligr and final polymerization [49].

The dynamic covalent chemistry method incorporates dynamic covalent bond can break and reform under stimuli. This technique allows alignment under loa subsequent retention of alignment after stimuli removal [53].

White et al. reported an extensive list of work related to LCEs, including finding to simplify the synthesis process and control the mechanical properties of the mat through thiol-acrylate and thiol-ene reactions [54–56].

3. LCEs in Biomedical Engineering

3.1. Drug Delivery Systems

The application of topical medications and remedies has been a method used thr out human history. Over the past several decades, transdermal drug delivery has b growing field, often via adhesive patches applied to intact skin. An increasing nu of medications are now available in a transdermal patch form, as this route can ad issues related to other routes of drug administration, such as pain from injections, inc tent serum medication levels, increased side effects, decreased bioavailability, and p noncompliance [57].

One issue that presents with this approach to medication administration is the l ability to control the drug delivery in an "on/off" manner. While some patches can d drugs slowly over a period of time, none can adjust the delivery rate or stop the drug r altogether while the patch is still in place on the skin. To address this concern, No et al. immobilized a liquid crystal (LC) monooxyethylene trimethylolpropane (M between sheets of a hydrophobic, porous Celgard membrane. Their goal was to utili thermos-responsive properties of the thermotropic LC to control drug permeation thr the membrane. Permeability of antipyretic and analgesic medications acetaminophe ethenzamide and non-steroidal anti-inflammatory drugs indomethacin and ketoprofe studied, both in vitro in diffusion cells and on excised rat skin and in vivo on the abdo of rabbits. Testing was performed for temperatures of 32–38 °C, and for all medica tested, drug permeation through the membrane was reported to have been depende the temperature. For all drugs except ketoprofen, the reported "on/off" temperature v the phase transition temperature of the liquid crystal, 38 °C; for ketoprofen, the perme increased proportionally with the temperature, which they report is likely related ethanol in the solution. The results of this work indicate that incorporating LC in a delivery system may be a way to control the drug release due to the transition between an ordered and isotropic system, where the pathways for the drug molecules to pass through the membrane are limited in the liquid crystalline state [58].

Along similar lines, Chen and colleagues embedded a liquid crystal cholesteryl oleyl carbonate (COC) into a nylon membrane to be tested for the penetration of the bronchodilator medication salbutamol sulfate. They found that the drug penetration correlated with the amount of LC in the membrane as well as with temperature change, as it was tested over the range of 10–25 °C. According to their results, the COC made the surface of the membrane hydrophobic, whereas salbutamol sulfate is hydrophilic, so it was not able to permeate through the hydrophobic membrane. However, at certain higher concentrations, drug penetration increased again, which they report could be due to aggregation of the COC on the surface, creating open channels through which the drug could travel. Therefore, drug penetration through their LC-embedded polymeric membrane was both rate-controlled and thermally responsive, further evidence that LC materials can be utilized to modulate the permeation of medications through a membrane. However, it is important to note that the phase transition temperature of the LC used in this study was not suitable for body temperature applications, and this would need to be addressed before it could be used for drug delivery [59].

This approach was further demonstrated by Inoue et al. with the development of thermoresponsive membranes containing a side-chain liquid crystal polymer (LCP) cast onto a polyvinylidene difluoride (PVDF) support membrane. It was found that the nematic-isotropic transition temperatures (T_{NI}) of the LCPs varied with the amount of mesogenic pendants added to the polymer chain, so that membranes could be designed to have T_{NI} at body temperature, addressing the issue of membrane performance at biologically relevant temperatures. In this work, permeation of the drugs indomethacin and vitamin B_{12} through the LCP/PVDF membranes was tested and found to be inhibited in the nematic phase but substantially increased in the isotropic phase (Figure 4) [60].

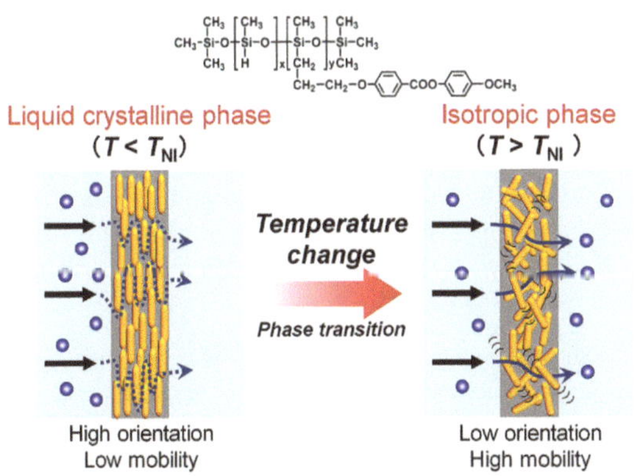

Figure 4. Mechanism for drug permeation through a thermoresponsive membrane composed of liquid crystal polymer (LCP) and polyvinylidene difluoride (PVDF). At temperatures below the nematic-isotropic (NI) transition temperature, the molecular arrangement of the nematic state prevents the drug from passing through the membrane. In the isotropic state above the transition temperature, the drug molecules may pass through. Reprinted from *J Membrane Science* [58]. Copyright (2019) with permission from Elsevier (Amsterdam, The Netherlands).

The aforementioned studies have utilized thermotropic LC materials within other polymers to act as physical barriers to the permeation of drug molecules rather than to encapsulate and release the drug itself. In contrast, Stepulane and collaborators developed a method of loading and delivering medications in a lyotropic liquid crystal elastomer (LLCE). LLCEs were composed of polydimethylsiloxane (PDMS), triblock copolymer diacrylated Pluronic F127 (DA-F127), and water in various ratios, resulting in four diff LCE samples with different mechanical properties. The samples were polymerized thin sheets for characterization and experimental procedures. Two drugs were test determine the ability of the LCEs to load and deliver a polar medication, Vancomycin a nonpolar medication, Ibuprofen. Both the drug loading and subsequent elution performed in solution using small pieces of each LCE that were punched out from polymerized sheets. Elution was monitored and measured via UV spectrometry at sp time intervals until no more drug release was noted (Figure 5). It is reported that tw the four LLCEs delivered a high amount of Vancomycin over a period of 6–7 days, related to their higher PDMS content that contributes to a strong network of crosslin the elastomer. The results were similar to the release of Ibuprofen, but it was noted th release of Ibuprofen had a steadier release rate, as it is found inside the micelles due hydrophobicity. In this work, the researchers were able to show that an LLCE system be utilized for the delivery of both hydrophobic and hydrophilic drugs, although the composition of the LCE for each drug may need to be determined [61].

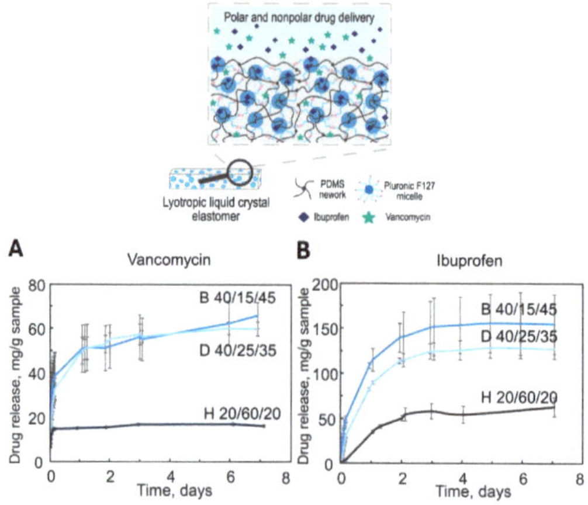

Figure 5. Drug release from lyotropic LCEs of (**A**) vancomycin in Milli-Q water and (**B**) ib fen in 1 wt% SDS buffer. Samples composed of polydimethylsiloxane (PDMS), triblock co mer diacrylated Pluronic F127 (DA-F127), and water in the following ratios: B (40–15–45 D (40–25–35 wt%), and H (20–60–20 wt%). Reprinted from *Colloid Surface B* [59]. Copyright with permission from Elsevier.

Thermotropic LC materials have been shown to make drug delivery systems tunable by the ability to control permeation, and lyotropic LCEs offer an attractive me of encapsulating and releasing drugs over a sustained period. Investigation into application of LCEs has been limited; however, the results of these previous works sh encourage more researchers to pursue this area further and develop practical metho using LCEs for real drug delivery systems.

3.2. Soft Robotics

Soft robotics often refers to Shape Memory Polymers (SMP) that are flexible, sens and possess good mechanical properties. Ideally, the material should be able to respo external stimuli such as bending, stretching, twisting, and compression. The search material that reacts and works as an organic muscle fiber is an ongoing field with m branches working for the same goal. Since 1969, with the prediction of Pierre deGenne the subsequent synthesis of the first perfect monodomain nematic elastomer by Küppe Heino Finkelmann in 1991 [62], LCEs have been sought to be a great solution for muscle material. They showed three important characteristics: orientational order in amorp soft materials, responsive molecular shape, and quenched topological constraints.

One of the biggest downsides for the utilization of LCEs in the soft robotic field is the low mechanical properties. For a natural muscle, the work capacity can reach 40 J kg^{-1} with a maximum strain over 40% [63]. A synthetic LCE shows potentially closer values but at times is not enough. Many efforts have been made in the field to increase the strain and work capacity exerted by the materials. One of the strategies is to dope the LCE with conducting materials. Rigid dopants like Carbon Nanotubes (CNTs) are embedded as sheets, forming a multilayer structure in combination with the LCE [64] and were reported to increase the strain and work capacity in response to an applied voltage. Another example is liquid dopants, which are getting attention since liquid dopants do not interfere with the mobility or flexibility of the LCE. The liquid dopant is mixed in the solution of the LCE as dispersed droplets of eutectic gallium indium alloy (EGaIn), which facilitates the synthesis process [65].

Applications of these kinds of materials have not often been considered in the field of biology. Only lately, the TE field has been looking at the implementation of soft robots made of LCEs as scaffolding for cell culture. A clear example of this surge of applications for the LCE soft robots are all the "bio-inspired" actuators. Among these examples, we can mention the "Janus" soft robot composed of two layers that can undergo opposite deformations simultaneously, emulating the grip of a hand or, depending on the complexity of the 3D shape, even mimic the movement of a starfish [66]. The very popular azo-dies do not escape from the TE field; many efforts have been directed to improve the photo-mechanical effect deriving from the change in morphology during the trans-cis isomerization of these molecules in efforts to have a viable artificial muscle-like material [67]. But the area of interest is much larger than just artificial muscles. In the last few years, we have seen light-driven soft robots that mimic caterpillars [68], with more complex modes of movement like rolling [69] with the ability of stirring and even swimming [70].

There are reports of thermotropic LCEs in combination with thermotropic dyes that not only change shape but also color [71] (Figure 6). These are called "chameleon-" or "octopus-like" biomimetic materials, developed with the goal of achieving some camouflage behavior in the soft robot. Finally, some LCE soft robots are created to heal or replace damaged tissue in the human body. Recent studies report a soft robot made of a LC membrane capable of secreting both polar and nonpolar liquids, similar to the human skin [72]. Generated digital models and 3D printed scaffolds of blood vessels give us a different perspective on how the tunability of physical characteristics in LCEs can mimic the human body and maybe be implanted and grafted for wound healing and tissue regeneration [34].

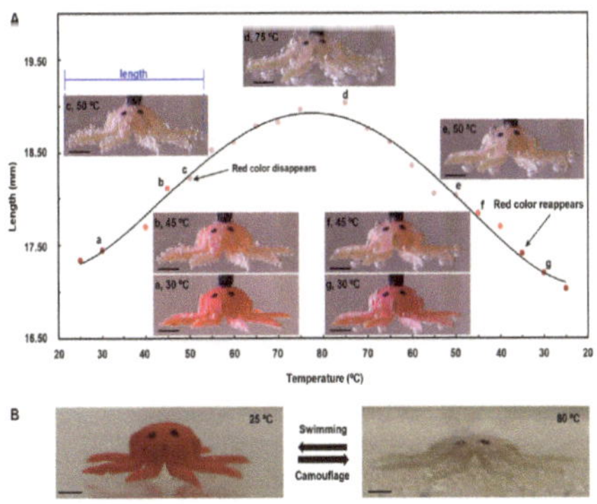

Figure 6. (**A**) Shape and color-changing thermochromic liquid crystal elastomer. Scale bar: 3 mm, (**B**) Octopus-like material mimicking the swimming/camouflage of the octopus in nature. Adapted with permission from Li et al. [69]. Permission conveyed through Copyright Clearance Center, Inc. (Danvers, MA, USA).

So far, advances in soft robotics research have focused on designing material are reactive, flexible, and robust. Research in the area of flexible electronics and sensing has progressed greatly; this has been thankfully due to efforts in bringing to advances in many areas of materials research. Multidisciplinarity has been the k advancing the research without sacrificing individual advances from the different Future success will depend on the combination of flexible and stretchable mech properties of materials combined with sensors and possibly the ability to self-healing path is to continue to improve materials design and properties for better bio-inspire robotics, biomedical applications, and prosthetics.

3.3. *Tissue Engineering*

3.3.1. Cell Culture Platforms

Cells in vivo experience a variety of environmental conditions, including extra-ce matrix (ECM) stiffness and elasticity, tensile forces, fluid viscosity and shear stress hydrostatic pressures. These features are a source of mechanical signals to the cells, are then translated into biochemical signals by cellular mechanotransduction [73]. of this process occurs through focal adhesions, which are large complexes of protein connect the cytoskeleton inside the cell with the ECM outside the cell. These pr undergo unfolding to expose binding sites, post-translational changes, or shuttling nucleus that trigger the activation of molecular pathways or mechanosensitive gene Studies have shown that cell behavior can be directed by altering the characteristics environment [75–77], such as patterning [78].

Babakhanova et al. developed a method to control the arrangement of micropa by employing an LCE coating that undergoes a trans-to-cis isomerization when irrac with UV light. To achieve this, LC cells were designed with unique patterns in me arrangements such that, upon isomerization, the spatial topography inside the cell transform, creating areas of lower potential energy at the desired locations. When sph particles of resin were placed onto the LCE coating, the microparticles would be direct gravity to these predesigned valleys. One main advantage of this method is the revers of the process: the LCE coating reverted completely back to its flat topography up to trans isomerization under visible light. Although this method had not been us biological applications, the authors noted that this type of device could prove use areas such as the control of cell behavior and targeted drug delivery [79]. Later on pursued an approach by making grooved substrates for cell cultures using smectic A coating over polyimide molecules of planar alignment. This geometry caused ar LC defects on the substrate, generating periodic grooves on the surface. For cell cul human dermal fibroblasts (hDFs) were used. Four substrates were tested: flat top with and without fibronectin (FN), and grooved topology with and without FN.

Cells were seeded and cultured for seven days or until confluent, after which an revealed that the cells seeded on flat substrates were multinucleated, elongated, an poor alignment (Figure 7). Whereas the grooved substrates with FN culture had a order parameter of 0.66, and the grooved one with FN had the highest amount of ce alignment with an order parameter of 0.81. The morphology of the cells on the gr substrate without FN was highly elongated with an aspect ratio of 6.4, but those wi were found with many actin filaments and filopodia and an aspect ratio of 3.2. From results, the authors concluded that depositing the FN was not necessary to have the adhere, align, or migrate during the culture. However, it is clear that the nanogr topology greatly affected the cell proliferation and alignment [80].

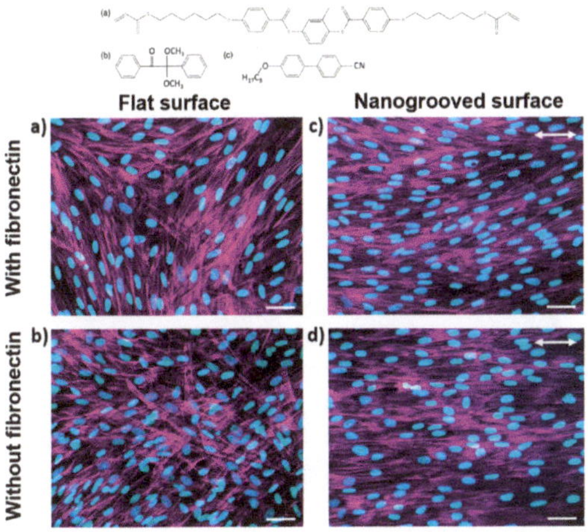

Figure 7. Fluorescent microscopy images of hDF grown on substrate (**a**) flat glass with fibronectin; (**b**) flat glass without fibronectin; (**c**) glass with nanogrooved SmA LCE coating and fibronectin; (**d**) glass with nanogrooved SmA LCE. Actin filaments stained with Alexa Fluor 488 phalloidin are presented in purple, while the cyan regions indicate the nuclei stained with DAPI. The arrows represent the direction of the nanogrooves. Scale bars 50 μm. Reprinted from *J Biomed Mater Res A* [78] Copyright (2020) with permission from Wiley.

Turiv and collaborators further studied the ability to control hDF behavior using an LCE substrate. In this case, plasmonic photoalignment was used to orient the molecules of the LCE precursor into the desired pattern, after which the photopolymerization reaction was used to secure this pattern. The specialized surface topography of the LCE formed when in contact with water, causing swelling of the LCE. In experiments, the authors showed that HDF cells grown on the LCE substrate aligned with the director, resulting in high scalar order parameters. Furthermore, the surface and number densities as well as phenotypes of the cells were affected. Positive topological charge defects were associated with an increased cell density and round shape, while negatively charged defects were found to have lower densities of cells and elongated shapes. Considering these results, the use of a photoaligned LCE may be a promising method of controlling cell proliferation, migration, and differentiation, as well as the possibility of developing environmentally responsive substrates that provide dynamic topographies for the cells [81].

3.3.2. Effects of Mechanical Properties of LCEs on Cell Behavior

LCEs' mechanical characteristics are extremely customizable, giving control over their stiffness, elasticity, and reaction to outside stimuli. Because of their ability to be customized, LCEs are especially appealing for use in dynamic cell culture conditions, where the mechanical characteristics of the cells are important factors influencing their behavior.

Agrawal et al. investigated the potential of monodomain LCEs for dynamic cell culture by making use of their reversible shape changes with heating cycles. Good adhesion and survivability were demonstrated by neonatal rat ventricular myocytes (NRVM) grown on LCE substrates. Because the LCEs could undergo cyclic uniaxial contraction and elongation, the cells were able to operate in a dynamic mechanical environment. Cell alignment, elongation, and differentiation are encouraged by this cyclic strain, which is like the physiological mechanical environment that cardiomyocytes encounter in vivo. It is important to note that the LCE surface was prepared for cell culture using methods like collagen coating, gold sputtering, and oxygen plasma cleaning. These alterations allowed the LCEs to function as dynamic substrates for cell culture while simultaneously enhancing cell adherence and preserving their mechanical characteristics [82].

In another study, Herrera-Posada et al. investigated magneto-responsive LCE nan posites enhanced with iron oxide nanoparticles. These LCEs could be remote-actuate to their anisotropic, directed nematic structures reacting to magnetic fields. The ner to-isotropic phase transition temperature was decreased using nanoparticles, ena actuation nearer physiological conditions. Higher nanoparticle concentrations enha the Young's modulus while decreasing the maximum thermomechanical deforma according to tensile testing. Reversible deformations under alternating magnetic were shown in magneto mechanical tests; the degree of deformation depended o concentration of particles. Under a 48 kA/m field, the maximum contraction for LCEs 0.5 and 0.7 wt% particle loadings was approximately 24%, indicating their potential fc in dynamic substrate applications. Furthermore, by functionalizing the surfaces of LCEs with collagen to facilitate NIH-3T3 fibroblast adhesion and proliferation, the investigated the biocompatibility of these LCEs. Staining with fluorescence and cor imaging verified strong cell adhesion and network development on the LCE subst According to their results, magneto-responsive LCEs with controlled deformation pr a viable means of establishing dynamic cell culture environments that replicate mec cal stimuli found in vivo, hence promoting tissue engineering and regenerative mec applications [83].

Prévôt et al. investigated the mechanical properties and cellular behavior of sn liquid crystal elastomers (LCEs), which are intended to be responsive scaffolds for t engineering. These LCE scaffolds offer mechanical properties that resemble different ti regulated degradability, and biocompatibility. Various LCE forms, such as films, foams microspheres, were investigated in this work to maximize pore size and shape for t growth and cell infiltration. Testing for mechanical properties demonstrated that LCEs have a Young's modulus (YM) that ranges from 2 to 4 MPa, depending on the linking density and structural arrangement. This means that these LCEs are approp for use in specific tissues, such as neurons. Studies on cell behavior concentrated o adhesion, growth, and orientation of different cell types, such as primary dermal fibro (hDF), SH-SY5Y neuroblastomas, and C2C12 myoblasts. High cell density and anisot growth were observed using fluorescence confocal microscopy on LCE scaffolds, w four-fold increase in cell proliferation when compared with standard porous films study showed that the liquid crystal moieties' mobility and alignment within the greatly affect cell behavior, promoting aligned cell proliferation and improving t engineering results [84].

LCEs' work by Shaha et al. in both polydomain and monodomain configura showed that the mesogen alignment and loading direction had a substantial impa mechanical properties including stiffness, elasticity, and damping capacity. In contr polydomain LCEs, which showed more isotropic qualities, mechanical property te revealed that monodomain LCEs showed increased stiffness and elasticity along the ali mesogen direction. Monodomain LCEs have a strong anisotropic behavior, which is u for applications that need directional mechanical features. This makes them partic effective at simulating the complex mechanical environment found in the intervert disc. The in vivo investigations comprised subcutaneous implantation of both solic porous LCEs in rats to evaluate tissue response and integration. The rats were moni over the 12 weeks, and the results showed that LCEs are biocompatible, facilitating encapsulation and ingrowth without causing detrimental effects on the surrounding ti Additionally, in vitro tests indicated that LCEs exhibit minimal swelling and maintain mechanical integrity under simulated physiological conditions [85].

Overall, better cell alignment, proliferation, and differentiation can be achiev LCEs by tailoring their mechanical properties and surface functions to resemble physi cal settings. These characteristics demonstrate how LCEs can further tissue engine and regenerative medicine.

3.4. Organ-on-a-Chip Applications

The traditional petri dish 2D system is by now over 100 years old [86]. The fast growth and easy experimental setup provided by these 2D systems is undeniable. Nonetheless, there is nothing that looks or behaves like a 2D system in any biological environment, so the results obtained from these systems may not translate well to the human body behavior of the same cells. Complexity is needed if one wants to study biological systems. Three-dimensional scaffolds help with this problem: the cells can grow in every direction, communicating properly with their neighbor cells, a closer behavior to the human body. Some drawbacks of this system are the inherent dark environment inside the 3D model, as it is sometimes difficult to observe what is happening to the cells inside the 3D scaffold, as well as the aggregates formed in 3D, creating a gradient of nutrient access and waste buildup [87].

Microfluidic organ-on-chip is a technology designed to overcome these difficulties. The device is designed to culture cells in a constantly moving environment, enclosed in a millimetric/micrometric chamber. This technology has the capacity of simulating the natural environment where the cells grow without the necessity of building the entire living organ [88]. The complexity of the microfluidic chip can increase as the necessity of the researcher needs, ranging from devices with a simple inlet and outlet to introduce flow [89] to as complex as introducing electric field gradients and producing dielectrophoresis effects inside the chip [90].

The addition of LC molecules into the organ-on-chip system clearly increases its potential uses. It has been shown that biological creatures like bacteria can be channeled and manipulated to follow specific paths using LC substrates [91], which can be considered a 2D system. Several studies dedicated to exploring the parameters of microfluidic channels as well as the anchoring condition of the LC molecule have concluded that by appropriately tuning these physical variables inside the chip, one can experimentally control and manipulate the profiles of flow velocity and director orientation across the channel [92]. It is still an open question whether biological entities like cells can be manipulated and sorted in these 3D microfluidic LC channels; nonetheless, the results and the direction of the latest experiments point out that those systems are close to being implemented.

4. Biosensing Applications

The use of liquid crystal-based biosensors has grown in recent years as a simple, direct, inexpensive, highly selective, and sensitive detection alternative. While there are only a few examples of LCEs as biosensors, we believe that it is important to mention early LC biosensors to encompass how these applications can potentially be the base for future LCE biosensor techniques. Gupta et al. described the construction of a cell composed of a gold thin film and a self-assembled monolayer in which target proteins, by binding to the surface of the monolayer, altered the roughness of the gold and consequently the orientation of the 4-cyano-4-pentylbiphenyl (5CB), allowing the amplification and transduction of signals in optical outputs [93]. Later, the group described the development of a system in which *"liquid crystals are used to amplify interfacial phenomena at fluid interfaces into optical images"* [94], inspired by earlier biomolecular investigations. 5CB is a widely used nematic liquid crystal whose orientational changes can easily be detected by polarized optical microscopy (POM). The apparatus consisted of treated glass surfaces as supports, copper grids to stabilize the nematic films (5CB), and an aqueous solution containing sodium dodecyl sulfate (SDS). The presence of SDS allows a reversible change from planar orientation (5CB-water interface) to homeotropic in the liquid crystalline bulk (Figure 8), which can be visualized under a polarized optical microscope (POM) [94]. This technique has opened the door to several other research projects that have revolutionized the field of biosensor technology.

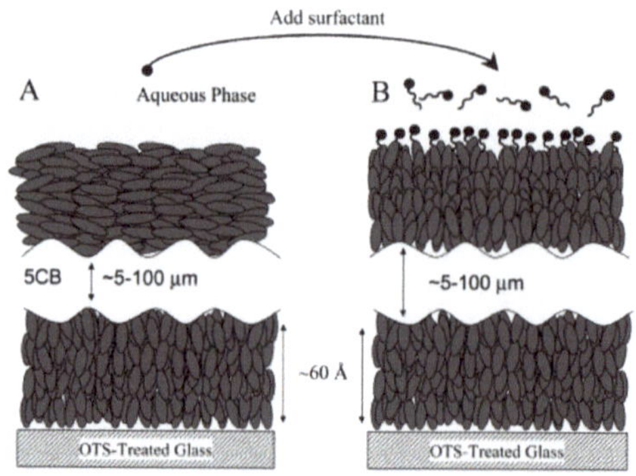

Figure 8. Representation of the 5CB-water interface: (**A**) in the absence of surfactant, 5CB has a orientation; (**B**) with the addition of surfactant, 5CB assumes a homeotropic configuration. Rep with permission from Brake and Abbott [92]. Copyright (2002) American Chemical Society.

With the aim of developing a simple and label-free tool for the diagnosis of path such as bacteria and fungi, Khan et al. proposed the construction of a TEM gri composed of a nematic liquid crystal blend (E7), a cationic surfactant (DTAB), single-stranded DNA probe (ssDNA). E7 is a eutectic LC mixture made of four c substituted polyphenyls prepared at a determined composition. E7 shows a single ne to isotropic (NI) transition at around T_{NI} = 61 °C [95,96]. This mixture presents a range of operating temperatures and has been widely used in the preparation of po dispersed liquid crystal (PDLC) for applications in several types of display devices device proposed by Kahn, then, in the presence of the target DNA, induces a char the orientation of the E7 molecules from homeotropic to planar (Figure 9), characte by the transition from black to bright colors due to the birefringence of the planar observed under POM. This device allows the detection of *Rhazictonia solani* and *Er carotovora* at concentrations ≥0.05 nM due to the signal amplification provided by the crystal molecules, as well as allowing the differentiation of even 2–3 bp mismatches the formation of different domains, guaranteeing high selectivity [97].

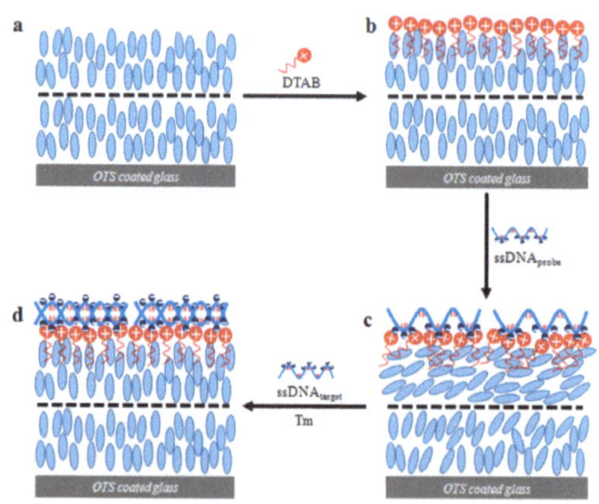

Figure 9. Illustration of the proposed LC cell for the detection of single-stranded DNA from *Rh nia solani* and *Erwinia carotovora*: (**a**) OTS-coated glass supporting TEM grid containing E7 mol (**b**) addition of surfactant (DTAB); (**c**) addition of single-stranded DNA probe causes orien change to planar orientation; (**d**) addition and hybridization of DNA target causes homeo orientation. Reprinted from *Sci Rep* [95]. Copyright © (2016) with permission from Springer N

In the race for diagnostic methods in the COVID-19 pandemic, liquid crystal-based biosensors emerged as a possible alternative. Researchers from The Ohio State University and Purdue University proposed using technology previously described in the literature to detect SARS-CoV-2 RNA from concentrations on the femtomolar (fM) scale; they developed a kit and smartphone app (Figure 10) that uses machine learning to classify image patterns and can provide a diagnosis quickly, easily, and affordably [98]. Figure 9 shows the LC device developed for the detection of SARS-CoV-2 RNA; in positive cases, the single-stranded RNA target hybridizes with the probe and promotes a change in orientation of E7 from planar to homeotropic in the aqueous interface, resulting in the disappearance of the birefringence and the formation of a dark pattern as observed on POM. The additional inset figures show the components of the device and the smartphone application developed for COVID diagnosis.

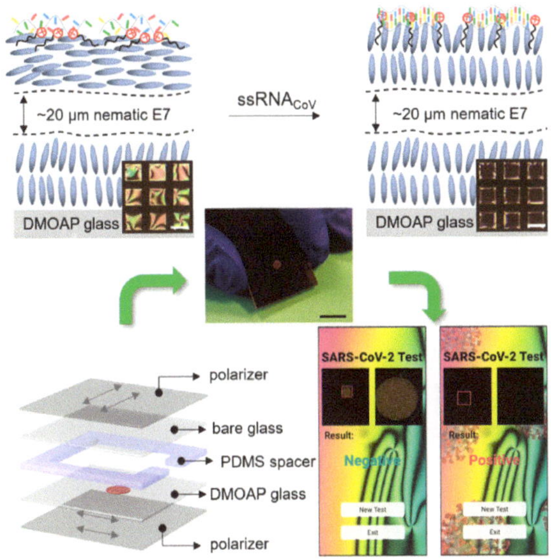

Figure 10. Liquid crystal cell developed for the detection of SARS-CoV-2 RNA; in positive cases, the single-stranded RNA target hybridizes with the probe and promotes a change in orientation, resulting in the formation of a dark pattern. Components of the device and smartphone application developed for COVID diagnosis. Reprinted from *Cell Rep Phys Sci* [96]. Copyright (2020) with permission from Elsevier.

The application of LC biosensors goes beyond what we can envision; they can become a new strategy for better detecting longtime known diseases. Insulin, for example, is one of the hormones that control the level of glucose in the blood [99] and is directly related to obesity, diabetes [100], and other illnesses. Liu et al. developed a sensor capable of measuring insulin between 0.1 and 10 nM within 5 min in diluted serum and urine samples. The apparatus consisted of a substrate containing a copper grid, immobilized dimethyloctadecyl[3-(trimethoxysilyl)propyl] ammonium chloride (DMOAP), and 5CB (Figure 11A), which has a homeotropic orientation in the presence of cetyltrimethylammonium bromide (CTAB) (Figure 11B). The G-quadruplex aptamer (IGA3) interacts electrostatically with the surfactant CTAB (Figure 11C), and upon binding to the insulin in the sample, IGA3 undergoes a conformational change to stretch, which consequently induces a planar orientation in the liquid crystal (Figure 11D) [101]. Figure 10 shows that when 5CB is in the homeotropic orientation (B–C), a dark image is formed, while in the presence of insulin, there is a change in orientation and the image obtained is bright.

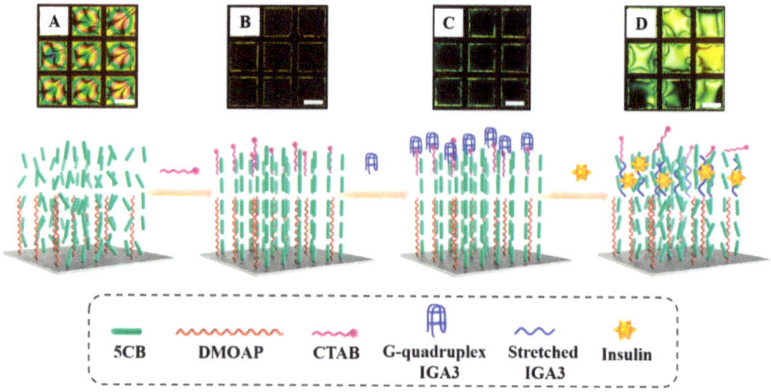

Figure 11. Optical patterns observed in POM: (**A**) Cell containing a glass substrate, coppe
immobilized DMOAP, and 5CB in PBS medium; (**B**) homeotropic orientation in the presence of
(**C**) IGA3 interacts electrostatically with CTAB; (**D**) in the presence of insulin, IGA3 is stretche
induces a planar orientation in 5CB. Reprinted from *J Colloid Interf Sci* [99]. Copyright (2022
permission from Elsevier.

Parkinson's disease is one of the most common disorders affecting the elderly in N
America and other parts of the world [102]. Previous studies by Zheng et al. identified
aptamers (F5R1 and F5R2) that efficiently bind to one of the potential Parkinson's di
biomarkers, the alpha-synuclein protein [103]. Based on these studies, Yang et al. prop
the use of the F5R1 structure to obtain a simple and inexpensive method for early diag
The device consists of a gold-coated glass and a self-assembled monolayer (C10SH + C
with the aptamer immobilized on the underside of the substrate. The cell is filled
5CB and has a homeotropic orientation in the absence of a biomarker, which is disru
by the folding of the DNA when it binds to the target protein, resulting in a change
dark to bright under the POM due to the planar alignment of the LC (Figure 12A).
showed a detection range of 50–400 nM and a low level of non-specific interactions
the controls used, despite lower sensitivity than ELISA tests [104]. Subsequently,
proposed another alternative to overcome some of the limitations found in the previ
described method, such as the cost and preparation of the substrate, detection time
sensitivity. This time, they used the gold grid on a DMOAP-coated substrate and C
at the LC-aqueous interface, with the same F5R1 aptamer used in the previous stu
resulting in a detection limit for alpha-synuclein of 10 pM (Figure 12B) [105].

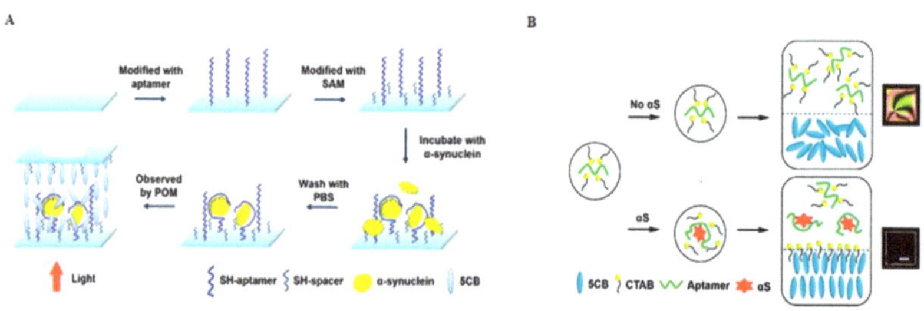

Figure 12. (**A**) Proposed model with a detection range of 50–400 nM of alpha-synuclein. Ad
with permission of [100] Copyright (2020) permission conveyed through Copyright Clearance C
Inc., (**B**) Optimizations performed on the biosensor resulted in a detection limit for alpha-syn
of 10 pM. Adapted with permission from Yang et al. [103], permission conveyed through Copy
Clearance Center, Inc.

Alzheimer's disease (AD) is another prevalent disease affecting the elderly. It was estimated that 6.7 million Americans over the age of 65 would be affected by AD in 2023, a number that is expected to nearly double by the year 2060 [106,107]. Thus, there is a desperate need for new tools for early diagnosis that are fast, cost-effective, and sensitive. In the search for new sensors, Kemiklioglu et al. developed a device that uses amyloid-beta-42 (Ab42) immunocomplex formation events as a strategy for detecting and diagnosing AD. The device consists of a glass coated with DMOAP filled with LCs (5CB) in which antibodies against Ab42 have been immobilized (Figure 13A–D). As a result, the addition of the Ab42 peptide was able to promote a change in the orientation of the LCs due to the formation of the antibody-antigen binding (Figure 13E); even at antigen concentrations as low as 1 pg/mL, Apolipoprotein E4 (ApoE4) was also added to evaluate the possible effects of its binding to the antigen in the presence of 5CB (Figure 13F); unfortunately, the effects observed were less pronounced than those of the antibody-antigen complex [108].

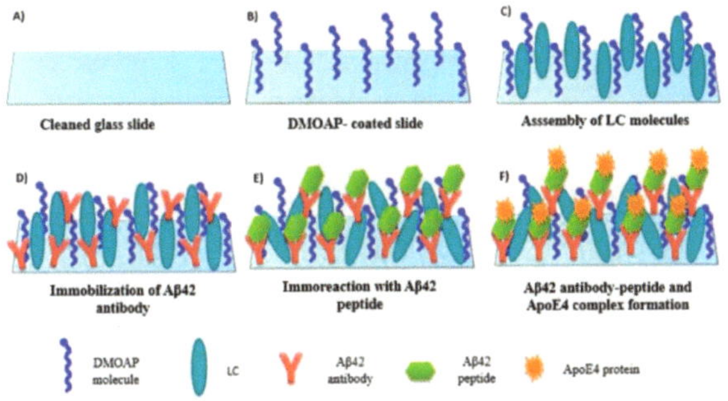

Figure 13. (**A**) Cleaned glass is used as a substrate; (**B**) DMOAP is used to coat the substrate; (**C**) the cell is filled with 5CB; (**D**) immobilization of antibody; (**E**) antibody-antigen binding process; (**F**) ApoE4 addition and complex formation. Reprinted from *J Biosci Bioeng* [106]. Copyright (2021) with permission from Elsevier.

Another example is acetylcholine (ACh), known as a neurotransmitter that is maintained at regulated levels for the transmission of nerve signals in a healthy organism. The enzyme acetylcholinesterase (AChE) is responsible for converting ACh to choline and acetic acid; however, due to certain factors, such as the pesticides malathion and fenobucarb, this activity is inhibited and can lead to continuous stimulation, which can cause fatal effects [109]. Inspired by the work of Wang et al. on the development of liquid crystal platforms for the evaluation of AChE [110,111] and enzymatic activity studies in microcapillaries [112,113], Nguyen et al. used OTS-coated microcapillaries containing 5CB droplets that form a monolayer in the presence of surfactant myristoylcholine chloride (Myr) and a four-petal pattern (Figure 14A–C). Once the enzyme AChE is present in the medium, it promotes the conversion of Myr to myristic acid and choline, preventing the formation of the monolayer and resulting in a bright double line pattern (Figure 14D). In the presence of malathion and fenobucarb, the enzymatic action is inhibited, and the double line pattern is not observed [114].

LCs can also be used as sensors to detect environmental pollutants that directly affect human health. It is known that mercury ions are highly toxic to living organisms and are unfortunately widely found as contaminants in water sources. One strategy for monitoring the contamination levels of this heavy metal uses octadecyltrichlorosilane (OTS)-coated glass with a TEM grid on top and a mixture of 5CB and potassium N-methyl-N-dodecyldithiocarbamate (MeDTC). MeDTC is used as a specific chelating agent for Hg^{2+} ions, which acts at the aqueous interface and promotes a change in the orientation of 5CB when the complex with mercury ions is formed (Figure 15). The detection limit of the described technique is 0.5 mM, and other common interferents present in the samples analyzed did not promote responses [115].

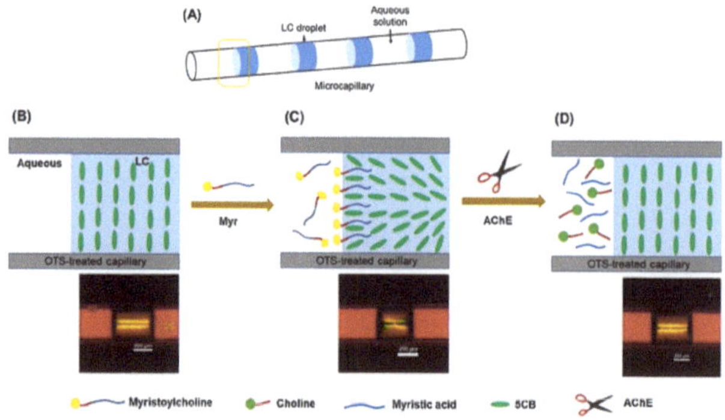

Figure 14. (**A**) OTS-coated microcapillaries containing 5CB droplets; (**B**) sensor in PBS me (**C**) Myr addition; (**D**) enzymatic cleavage of Myr to myristic acid and choline. The bottom the images corresponding to each POM event. Reprinted from *Colloids Surf B Biointerfaces* Copyright (2021) with permission from Elsevier.

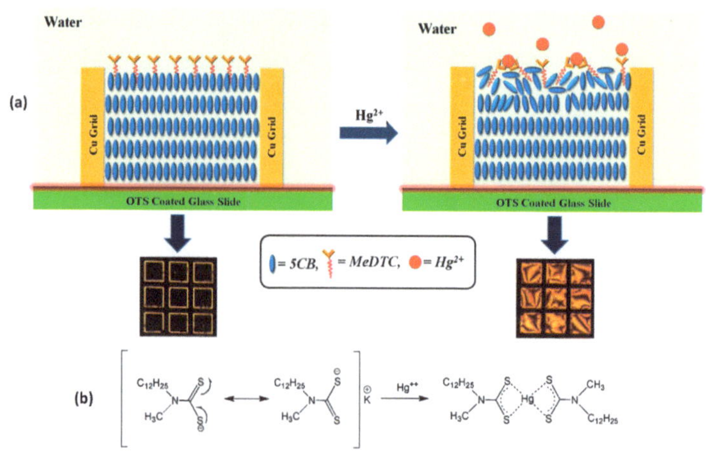

Figure 15. (**a**) Change in the 5CB orientation due to complexation MeDTC-Hg^{2+}, (**b**) Che of MeDTC-Hg^{2+} complexation. Reprinted from *Sensor Actuat B-Chem* [113] Copyright (2016 permission from Elsevier.

Firefighters and first responders are the most exposed to adverse environm situations where toxic gases such as chlorine (Cl_2) or phosgene ($COCl_2$) are often pr Protective equipment helps reduce the effects of these exposures, but they are still a when these protective measures are removed before they can predict if all toxic resulting during a fire have been eliminated. In most cases, one of the team mem is responsible for measuring the gases with a portable device, but often this me is not effective in reflecting the reality of each team member's exposure. Based o shortcoming, Prévôt et al. presented a zero-power sensor capable of measuring acut chronic exposure at ppm levels. Nanoparticles containing reactive ligands are used to print warning signs on a glass substrate. The cell is filled with nematic LCs that h homeotropic orientation in the absence of hazardous chemicals. When exposed to such as Cl_2, the transition to planar anchoring takes place and forms the unmista warning signal that can be observed under crossed polarizers (Figure 16) [116].

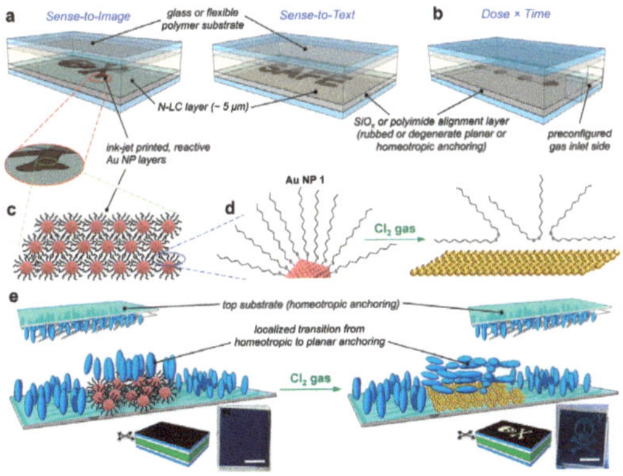

Figure 16. (**a**) Schematic representation of sensor models (figure and text); (**b**) representation of the sensor based on dose x exposure time; (**c**) Au NP pattern created by ink jet printer; (**d**) chemical reaction between Au NP and Cl_2; (**e**) change in orientation of N-LC in the presence of Cl_2. Reprinted from *Adv Mater Technol-Us* [114]. Copyright (2020) with permission from John Wiley and Sons.

Liquid crystals can also be combined with polymeric materials to create advanced sensors. Ailincai et al. prepared droplets (2–5 mm) using a microemulsion technique with different mass fractions between a cholesteric LC, cholesterol acetate (*L*-ChAc), and a polymeric matrix composed of polyvinyl alcohol boric acid (PVAB). The LC in the matrix exhibits planar anchoring, and changes in orientation were observed by different optical textures in the presence of sugars (*D*-glucose, *D*-galactose), amino acids (*D*-alanine, *L*-arginine), and DNA (salmon sperm), making the material promising in the field of biosensing [117]. The combination of LC and block copolymers for protein detection is another example found in the literature [118]. A TEM grid cell consisting of an interface between water and 5CB modified by the presence of PAA-b-LCP block copolymer (where PAA refers to poly(acrylic acid) and LCP to poly(4-cyanobiphenyl-4-oxy-undecylacrylate) was used as a method for the detection of albuminuria in urine samples, described by Khan et al. [119]. More recently, devices based on cholesteric LCE exhibit different colors when exposed to different mechanical stimuli, such as stretching or compression [120], under linear polarization and can be used as strain sensors or even in soft robots [121].

The use of LCEs in biological and medical applications will no doubt continue to increase as researchers develop more novel elastomers. Use of LC devices as sensors has been widely shown to be a successful application, and this function could easily translate into the medical setting in the form of diagnostic tools [122]. The intrinsic characteristics of LCEs, such as responsiveness to heat, light, and mechanical stress, may be harnessed in the development of diagnostic tools and imaging techniques. For example, various health conditions are associated with temperature variations in the body or in body parts, including cancer [123], blood clots [124], and infections [125].

For example, Li et al. designed LCE microspheres (LCEM) for use as pH sensors (Figure 17). Horseradish peroxidase (HRP) was immobilized on the microspheres (LCEM-HRP), which adopted two different configurations—radial and concentric. The configuration of the LCEM-HRP was dependent on the pH of the medium, and they were tested with various cells and found to be compatible. In addition, this detection method displayed single-cell resolution. The ability to determine the pH of individual cells could be applicable in the detection of cancer, since the pH of healthy cells is neutral while cancer cells exhibit a slightly acidic pH. These promising results indicate that novel LCE materials can be designed to aid in the diagnosis of medical conditions [126].

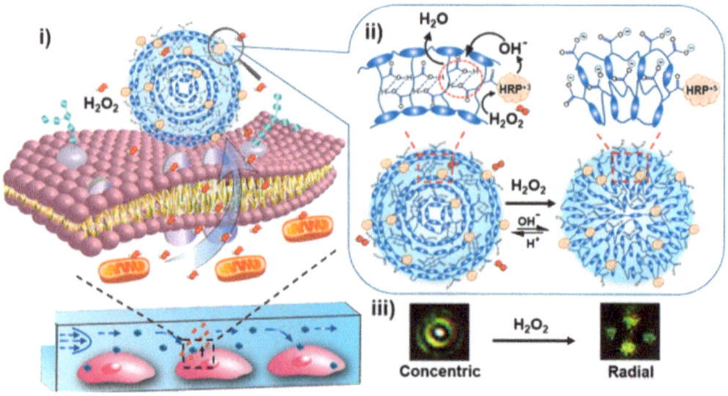

Figure 17. (**i**) Liquid crystal elastomer microspheres functionalized with horseradish pero (LCEM-HRP) immobilized on the cell membrane; (**ii**) the concentric and radial configurations LCEM-HRP, which transform in the presence of H_2O_2; and (**iii**) POM (polarized optical micro images of the LCEM-HRP in the two configurations. Reprinted from *Angewandte Chemie* Copyright (2020) with permission from Wiley.

Abadia et al. investigated biocatalytic LCE sensors for urea detection. In this urease is immobilized on the film, and the enzymatic response is transduced on a large due to the decrease in liquid crystalline order caused by the formation of ammoni Another use of LCE is in the detection of organophosphates such as malathion. I case, Abadia et al. proposed a film made of *N*,*N*-dimethylethylenediamine (DMEN 1,4-bis-[4-(6acryloyloxyhexyloxy)benzoyloxy-2-methylbenzene (C6M). The LCE i work was used to detect toxic pesticides by immobilizing hydrolytic pig liver est (PLE). In the presence of malathion, the PLE enzyme promotes catalytic hydroly form malathion monoacid, which in turn changes the pH of the medium and indu conformational change from a ribbon-like to a helical shape (see Figure 18). The s provides a quantitative relationship between the helical pitch change and a specific exp dose and is not affected by environmental variations and other chemical interferences

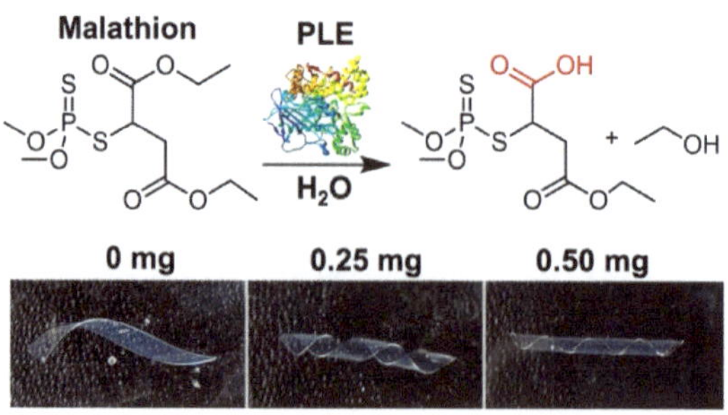

Figure 18. Malathion sensing using PLE-containing LCEs, showing the schematic of PLE hydr of malathion into malathion monoacid. Malathion monoacid lowers the pH of the solution, trig a response in the LCE. LCE ribbons in the presence of several doses of malathion at room tempe show the formation of a helix and the impact of the malathion dose on the helical pitch. Rep from *Sensors and Actuators B: Chemical* [125]. Copyright (2024) with permission from Elsevier.

5. 3D and 4D Printing of LCEs

5.1. 3D Printing

Digital Light Processing (DLP) stereolithography is effectively used for 3D pri LCEs. This technique employs a photocurable resin cured layer-by-layer to create cate 3D structures, incorporating LC moieties that promote cell alignment due to anisotropic ordering. A work by Prévôt et al. highlighted the successful use of a photocrosslinkable smectic-A (Sm-A) LCE for 3D printing, demonstrating its ability to promote cell proliferation and anisotropy. The study involved creating a digital 3D model of a brain structure, reproduced using a DLP printer. Synchrotron Small-Angle X-ray Diffraction (SAXD) data showed strong alignment of LCE layers, directing cell growth along anisotropic axes, essential for replicating organized native tissues and improving in vitro models [34].

Other 3D printing techniques have also been utilized in the production of LCEs, including extrusion printing, which leads to director alignment along the print line [27], as well as the ability to easily incorporate multiple materials into the same 3D printed object. This capability was shown by Jennifer Lewis, who reported the production of 3D-printed shape-memory LCEs that were not only actuated by temperature changes but also underwent network reconfigurations triggered by UV exposure by means of exchangeable dynamic bonds [128]. By exposing the LCE to UV while actuated, the programmed shape was "locked in". Multi-material printing was required to achieve this in 3D bulk scale, which was enabled by extrusion 3D printing of alternating layers of LCEs with and without the dynamic bonds.

In another display of multi-material printing, the same group developed a method of 3D printing "innervated" LCEs (iLCEs)—filaments containing a liquid metal (LM) core surrounded by a main chain LCE (Figure 19). They reported the design of a core-shell extrusion nozzle used to produce the iLCEs, which were printed in the nematic phase and UV-cured. Inclusion of the LM core allowed for actuation of the iLCE via joule heating and cooling induced by power input through the filament. Upon heating, the filaments contracted in the printing direction and were shown to lift over 200 times the LCE weight while doing so. The team also demonstrated the ability to program a desired shape for the iLCE, which was maintained in a closed-loop feedback system driven by sensation of the resistance of the iLCE fiber [129].

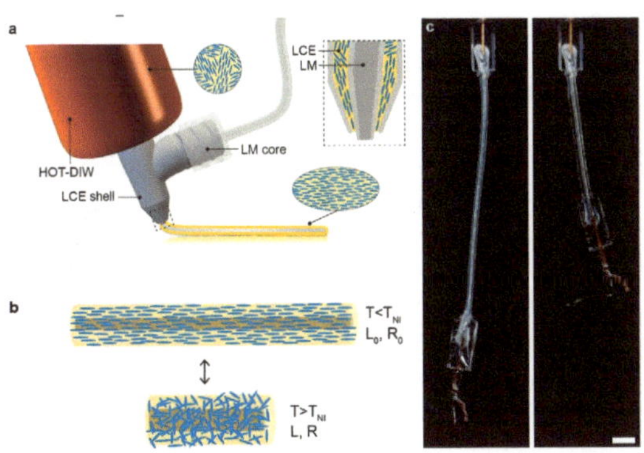

Figure 19. Innervated LM-LCE fibers. (**a**) Schematic illustration of core-shell 3D printing of iLCE fibers composed of a liquid metal (LM) core surrounded by a liquid crystal elastomer (LCE) shell, whose director is aligned along the print path. (**b**) Schematic illustration of iLCE actuation when cycled above and below the LCE nematic-to-isotropic transition temperature, TNI. (**c**) Images of an iLCE fiber before (**left**) and after (**right**) joule heating above TNI (scale bar = 5 mm). Reprinted from *Adv Mater* [127] Copyright (2021) with permission from Wiley.

5.2. 4D Printing

Significant progress has been achieved in 4D printing with LCEs pioneered by T. Ware using direct-write printing to create 3D structures capable of reversible shape changes by controlling the alignment of LC molecules during the printing process. With the use of this method, intricate, responsive structures capable of large deformations, alike volumetric contractions and snap-through transitions, could be fabricated and utilized in soft robotics and medical devices [130]. Proceeding from this basis, their work concentrated on m ing these LCEs at the molecular level to improve their actuation properties. By empl a two-stage thiol-acrylate reaction, they were able to fine-tune the material's net architecture, allowing for precise, programmable shape changes in response to th stimuli. This work demonstrated the potential for highly responsive and customi actuators in smart material applications [131]. Later on, Ware's team introduced a approach by combining liquid metal with LCEs, resulting in 4D-printed composites th only exhibited shape-changing properties but also enhanced electrical conductivity breakthrough opened new possibilities for multifunctional materials in soft electronic That same year, they explored the use of photo switchable actuators by incorpor azobenzene-based liquid crystals, enabling light-controlled, reversible shape changes innovation further demonstrated the versatility of 4D printing for creating adaptiv tems [132]. Consequently, the group pushed the boundaries of 4D printing by adding materials into their processes. By embedding bacteria into LCEs, they developed stru that could interact with their environment, paving the way for responsive, living sy with potential applications in healthcare and environmental monitoring [130].

McDougall et al. proposed greater efficiency and simplicity by eliminating ext supports through embedded printing, which promotes the extrusion of LCE (75/25 RM82:RM257) inside an easily removable gelatinous matrix (composed of wate laponite). Essential mechanical properties are attained with trifunctional acrylate, ar structure is cured with UV light, allowing reversible 3D shape changes upon heating, can be applied to biomedical and wearable devices [131]. The above examples highlig profound impact of 4D printing technologies on various fields, showcasing the pot for creating advanced, multifunctional materials.

6. Challenges and Limitations

While LCEs show potential for various biological applications, one of the key tations for utilizing LCEs in biological applications is biocompatibility. When thes terials contact biological tissues, they must not cause toxicity or unfavorable im responses [132]. To determine their safety, extensive in vitro and in vivo testing is r sary, which can be expensive and time-consuming [133].

Another significant limitation is the degradation and longevity of LCEs within b ical environments. They must maintain their structural integrity and functional prop over extended periods for long-term applications, such as implants or biosensors. For practical application, they must be resistant to the biochemical conditions of the hu body, which include temperature, pH, and different biochemicals [134].

The contact between the material and biological tissues must be carefully consi when integrating LCEs with biological systems. It is crucial to minimize inflamm reactions and adhere properly. The main challenge is creating an interface that focus biological interactions with the ordering transitions in the LC systems. Accurate ma lation of the mechanical attributes, like elasticity and strength, is also necessary to s the various demands of various biomedical uses. Partially crosslinked elastomers limited applicability in situations where precise mechanical behavior is essential, incl artificial muscles and responsive implants, due to their difficulty in spatially alterin direction of deformation [135].

The production processes for LCEs are often complex and require specialized e ment, posing practical limitations on their widespread adoption. Producing these mat at the scale required for clinical usage might be difficult due to the complexity and hig of their synthesis and manufacture. Another area that needs further work for sophisti optical applications in biological contexts is controlling optical qualities similar to a behavior and transparency under physiological conditions [132].

Some of these challenges have been addressed by recent developments. Proce methods include surface alignment, magnetic fields, mechanical straining, and 4D, have printing and have made it possible to create LCEs with a variety of mechanical c teristics and shape-changing behaviors. For instance, surface alignment techniques enable the creation of LCE films as substrates or coatings for flexible electronics, while 4D printing makes it possible to fabricate LCE actuators that can mimic the forces generated by human muscles. Furthermore, rheological research on biocompatible and biodegradable LCE inks has shed light on how to create and print new materials with desirable characteristics for specific uses [134].

7. Conclusions and Future Perspectives

7.1. Material Innovations

Recent advancements in the field of LCEs facilitate creative applications in a wide range of industries. The synthesis and orientation of LCEs have been one of the most key areas of development [28,42,43,52,63,135]. These developments have made it possible to produce materials that can fulfill the demanding specifications of applications in robotics, optics, and other areas [136].

Photothermal-driven LCEs are the result of researchers' efforts to improve the mechanical characteristics and responsiveness of LCEs to outside stimuli, such as heat and light. These materials have enormous potential for application in soft robotics, a field that depends heavily on controlled and exact movements. The photothermal effect in LCEs influences the ability of these materials to convert absorbed light into heat, leading to localized temperature changes. The LCEs undergo a phase transition because of this temperature difference, which modifies their molecular alignment and results in a reversible shape change. This effect's precision is important because it permits precisely controlled deformation that can be altered by varying the light's intensity, wavelength, and exposure time. For example, LCEs exposed to near-infrared (NIR) light can actuate quickly and significantly, which is important for applications that need a small amount of energy input and quick response times [136].

Another area of innovation is the integration of LCEs with other smart materials to create composites that exhibit enhanced functionalities. For example, combining LCEs with liquid metals made from 75% gallium and 25% indium results in soft, flexible composites that incorporate the beneficial aspects of both materials. Because of their exceptional mechanical strength and excellent conductivity, these composite materials are well suited for use in flexible electronics and wearable technology [137]. Furthermore, by adding magnetic nanoparticles to LCE matrices, magnetic composite films have been created, opening new possibilities, including remote actuation and enhanced magnetic field responsiveness [138].

Researchers introduced carbon fiber reinforcement to LCEs to improve their mechanical properties. Because of this integration, LCEs' tensile strength and endurance are significantly increased, making them appropriate for demanding applications in the aerospace and automotive industries. Due to their increased structural support and high strength-to-weight ratio, carbon fibers enhance the mechanical performance of LCE composites overall. For components that experience directional stress, such as airplane wings and automobile chassis, uniform load distribution is ensured by aligning carbon fibers within the LCE matrix. Moreover, carbon fibers' thermal conductivity aids in controlling the heat produced during LCE actuation, extending the operational lifespan of these composites [139].

A unique path of innovation in LCE applications was established by Mistry et al., who reported in 2018 the synthesis and characterization of acrylate-based LCEs with inherent auxeticity without porosity and suggested that these could be designed so as to tune the auxetic response [140]. This was later demonstrated in 2024 by Cooper et al. by the synthesis of the same acrylate nematic LCEs, whose optical and physical properties could be adjusted by their composition and synthetic conditions [141]. Auxetic materials present a negative Poisson's ratio: when the material is exposed to an applied extension along one axis, then the material will expand in the directions perpendicular to that axis. These unique materials are known to have impact resistance, shear resistance, and fracture toughness, among other desired properties [142]. The development of auxetic LCEs offers promising prospects for their use in various biological applications, such as cardiac stents [143], contact and optical lenses [141,144], and sports helmets [145].

7.2. Expanding Biological Applications

Liquid crystal elastomers (LCEs) are a remarkable new material with diverse a cations in medicine and biology. Their exceptional adaptability and ability to respo external stimuli make them ideal for a wide range of applications. LCEs enable pr environment-responsive medication release systems in drug delivery. In soft rob LCEs mimic natural muscle movements, advancing medical devices. For tissue engi ing, LCEs provide customizable scaffolds for improved cell growth. In biosensing, offer potential for advanced wearable and environmental sensors. LCEs' tunable erties, including shape memory and anisotropic responses, allow for creative desig biomedical engineering, including for load-bearing biomedical applications. How challenges remain in ensuring biocompatibility, maintaining long-term stability, and a ing cost-effective production. Future research should focus on improving mecha characteristics, reinforcing integration with biological systems, and developing innova like photothermal-driven LCEs and advanced composites. LCEs show great promis revolutionizing biological applications. Ongoing research and development will be cr to fully realizing their potential in advancing medical and technological fields.

Author Contributions: F.S., Y.N.A., A.G., G.P. and A.V.-A. contributed equally to this manus E.H. and F.S. conceived the idea of the manuscript. F.S., Y.N.A., A.G., G.P. and A.V.-A. pre the manuscript draft with contributions from all authors. E.H. directed, edited, and finalize manuscript with contributions of all authors. All authors have read and agreed to the publ version of the manuscript.

Funding: This research received no external funding.

Conflicts of Interest: The authors have no conflicts of interest to declare.

References

1. Lehmann, O. Über fliessende krystalle. *Z. Phys. Chem.* **1889**, *4*, 462–472. [CrossRef]
2. Reinitzer, F. Beiträge zur Kenntnis des Cholesterins. *Monatshefte Chem.* **1888**, *9*, 20. [CrossRef]
3. Lehmann, O. Les cristaux liquides. *J. Phys. Theor. Appl.* **1909**, *8*, 713–735. [CrossRef]
4. de Gennes, P.G. A semi-fast artificial muscle. *Comptes Rendus L'Acad. Sci. Ser. IIB Mech. Phys. Chem. Astron.* **1997**, *5*, 34 [CrossRef]
5. de Gennes, P.G.; Hebert, M.; Kant, R. Artificial muscles based on nematic gels. *Macromol. Symp.* **1997**, *113*, 39–49. [CrossRe
6. Warner, M.; Terentjev, E. *Liquid Crystal Elastomers*; Oxford University Press (OUP): Oxford, UK, 2003. [CrossRef]
7. Yakacki, C.M.; Saed, M.; Nair, D.P.; Gong, T.; Reed, S.M.; Bowman, C.N. Tailorable and programmable liquid-crystalline elast using a two-stage thiol-acrylate reaction. *RSC Adv.* **2015**, *5*, 18997–19001. [CrossRef]
8. Martella, D.; Paoli, P.; Pioner, J.M.; Sacconi, L.; Coppini, R.; Santini, L.; Lulli, M.; Cerbai, E.; Wiersma, D.S.; Poggesi, C. Liquid Crystalline Networks toward Regenerative Medicine and Tissue Repair. *Small* **2017**, *13*, 1702677. [CrossRef] [PubM
9. Ito, Y.; Chen, X.; Kang, I.-K. *Advances in Bioinspired and Biomedical Materials Volume 2*; ACS Symposium Series; American Che Society: Washington, DC, USA, 2017; Volume 1253.
10. Barón, M. Definitions of basic terms relating to low-molar-mass and polymer liquid crystals. *Pure Appl. Chem.* **2001**, *73*, 84 [CrossRef]
11. Kerkam, K.; Viney, C.; Kaplan, D.; Lombardi, S. Liquid crystallinity of natural silk secretions. *Nature* **1991**, *349*, 596–598. [Cro
12. Willcox, P.J.; Gido, S.P.; Muller, W.; Kaplan, D.L. Evidence of a Cholesteric Liquid Crystalline Phase in Natural Silk Spi Processes. *Macromolecules* **1996**, *29*, 5106–5110. [CrossRef]
13. Fontana, F.; Bellini, T.; Todisco, M. Liquid Crystal Ordering in DNA Double Helices with Backbone Discontinuities. *Macromol* **2022**, *55*, 5946–5953. [CrossRef]
14. Saw, T.B.; Doostmohammadi, A.; Nier, V.; Kocgozlu, L.; Thampi, S.; Toyama, Y.; Marcq, P.; Lim, C.T.; Yeomans, J.M.; Ladc Topological defects in epithelia govern cell death and extrusion. *Nature* **2017**, *544*, 212–216. [CrossRef] [PubMed]
15. Eggleson, K.K. Stem cell-based therapies: Promises, obstacles, discordance, and the agora. *Perspect. Biol. Med.* **2012**, *55*, [CrossRef] [PubMed]
16. Prévôt, M.; Bergquist, L.; Sharma, A.; Mori, T.; Gao, Y.; Bera, T.; Zhu, C.; Leslie, M.; Cukelj, R.; Korley, L.T.J.; et al. *New Develop in 3D Liquid Crystal Elastomers Scaffolds for Tissue Engineering: From Physical Template to Responsive Substrate*; SPIE: Philadelphi USA, 2017.

Prévôt, M.E.; Ustunel, S.; Bergquist, L.E.; Cukelj, R.; Gao, Y.; Mori, T.; Pauline, L.; Clements, R.J.; Hegmann, E. Synthesis of Biocompatible Liquid Crystal Elastomer Foams as Cell Scaffolds for 3D Spatial Cell Cultures. *J. Vis. Exp.* **2017**, *122*, e55452. [CrossRef]

Ustunel, S.; Prévôt, M.E.; Rohaley, G.A.R.; Webb, C.R.; Yavitt, B.; Freychet, G.; Zhernenkov, M.; Pindak, R.; Schaible, E.; Zhu, C.; et al. Mechanically tunable elastomer and cellulose nanocrystal composites as scaffolds for in vitro cell studies. *Mater. Adv.* **2021**, *2*, 464–476. [CrossRef]

Prévôt, M.; Hegmann, E. From Biomaterial, Biomimetic, and Polymer to Biodegradable and Biocompatible Liquid Crystal Elastomer Cell Scaffolds. In *Advances in Bioinspired and Biomedical Materials Volume 2*; ACS Symposium Series; American Chemical Society: Washington, DC, USA, 2017; Volume 1253, pp. 3–45.

Cheng, H.; Hill, P.S.; Siegwart, D.J.; Vacanti, N.; Lytton-Jean, A.K.R.; Cho, S.W.; Ye, A.; Langer, R.; Anderson, D.G. A Novel Family of Biodegradable Poly(ester amide) Elastomers. *Adv. Mater.* **2011**, *23*, H95–H100. [CrossRef]

Novak, U.; Kaye, A.H. Extracellular matrix and the brain: Components and function. *J. Clin. Neurosci.* **2000**, *7*, 280–290. [CrossRef] [PubMed]

Bartolo, P.; Domingos, M.; Gloria, A.; Ciurana, J. BioCell Printing: Integrated automated assembly system for tissue engineering constructs. *CIRP Ann.* **2011**, *60*, 271–274. [CrossRef]

Guo, H.; Saed, M.O.; Terentjev, E.M. Thiol–acrylate side-chain liquid crystal elastomers. *Soft Matter* **2022**, *18*, 4803–4809. [CrossRef]

Ustunel, S.; Prévôt, M.E.; Clements, R.J.; Hegmann, E. Cradle-to-cradle: Designing biomaterials to fit as truly biomimetic cell scaffolds—A review. *Liq. Cryst. Today* **2020**, *29*, 40–52. [CrossRef]

Ustunel, S.; Sternbach, S.; Prévôt, M.E.; Freeman, E.J.; McDonough, J.A.; Clements, R.J.; Hegmann, E. 3D Co-culturing of human neuroblastoma and human oligodendrocytes, emulating native tissue using 3D porous biodegradable liquid crystal elastomers. *J. Appl. Polym. Sci.* **2023**, *140*, e53883. [CrossRef]

Gungor-Ozkerim, P.S.; Inci, I.; Zhang, Y.S.; Khademhosseini, A.; Dokmeci, M.R. Bioinks for 3D bioprinting: An overview. *Biomater. Sci.* **2018**, *6*, 915–946. [CrossRef] [PubMed]

Ambulo, C.P.; Burroughs, J.J.; Boothby, J.M.; Kim, H.; Shankar, M.R.; Ware, T.H. Four-dimensional Printing of Liquid Crystal Elastomers. *ACS Appl. Mater. Interfaces* **2017**, *9*, 37332–37339. [CrossRef] [PubMed]

Kotikian, A.; Truby, R.L.; Boley, J.W.; White, T.J.; Lewis, J.A. 3D Printing of Liquid Crystal Elastomeric Actuators with Spatially Programed Nematic Order. *Adv. Mater.* **2018**, *30*, 1706164. [CrossRef] [PubMed]

Mistry, D.; Traugutt, N.A.; Yu, K.; Yakacki, C.M. Processing and reprocessing liquid crystal elastomer actuators. *J. Appl. Phys.* **2021**, *129*, 130901. [CrossRef]

Prévôt, M.E.; Ustunel, S.; Yavitt, B.M.; Freychet, G.; Webb, C.R.; Zhernenkov, M.; Hegmann, E.; Pindak, R. Synchrotron Microbeam Diffraction Studies on the Alignment within 3D-Printed Smectic-A Liquid Crystal Elastomer Filaments during Extrusion. *Crystals* **2021**, *11*, 523. [CrossRef]

Sharma, A.; Neshat, A.; Mahnen, C.J.; Nielsen, A.D.; Snyder, J.; Stankovich, T.L.; Daum, B.G.; LaSpina, E.M.; Beltrano, G.; Gao, Y.; et al. Biocompatible, Biodegradable and Porous Liquid Crystal Elastomer Scaffolds for Spatial Cell Cultures. *Macromol. Biosci.* **2015**, *15*, 200–214. [CrossRef]

Sharma, A.; Mori, T.; Mahnen, C.J.; Everson, H.R.; Leslie, M.T.; Nielsen, A.d.; Lussier, L.; Zhu, C.; Malcuit, C.; Hegmann, T.; et al. Effects of Structural Variations on the Cellular Response and Mechanical Properties of Biocompatible, Biodegradable, and Porous Smectic Liquid Crystal Elastomers. *Macromol. Biosci.* **2017**, *17*, 1600278. [CrossRef]

Prévôt, M.E.; Andro, H.; Alexander, S.L.M.; Ustunel, S.; Zhu, C.; Nikolov, Z.; Rafferty, S.T.; Brannum, M.T.; Kinsel, B.; Korley, L.T.J.; et al. Liquid crystal elastomer foams with elastic properties specifically engineered as biodegradable brain tissue scaffolds. *Soft Matter* **2018**, *14*, 354–360. [CrossRef]

Prevot, M.E.; Ustunel, S.; Freychet, G.; Webb, C.R.; Zhernenkov, M.; Pindak, R.; Clements, R.J.; Hegmann, E. Physical Models from Physical Templates Using Biocompatible Liquid Crystal Elastomers as Morphologically Programmable Inks For 3D Printing. *Macromol. Biosci.* **2023**, *23*, e2200343. [CrossRef]

Ring, M.E. The history of maxillofacial prosthetics. *Plast. Reconstr. Surg.* **1991**, *87*, 174–184. [CrossRef]

Heng, W.; Solomon, S.; Gao, W. Flexible Electronics and Devices as Human–Machine Interfaces for Medical Robotics. *Adv. Mater.* **2022**, *34*, 2107902. [CrossRef]

Cianchetti, M.; Laschi, C.; Menciassi, A.; Dario, P. Biomedical applications of soft robotics. *Nat. Rev. Mater.* **2018**, *3*, 143–153. [CrossRef]

Cianchetti, M.; Laschi, C. Pleasant to the Touch: By Emulating Nature, Scientists Hope to Find Innovative New Uses for Soft Robotics in Health-Care Technology. *IEEE Pulse* **2016**, *7*, 34–37. [CrossRef] [PubMed]

Yin, R.; Wang, D.; Zhao, S.; Lou, Z.; Shen, G. Wearable Sensors-Enabled Human–Machine Interaction Systems: From Design to Application. *Adv. Funct. Mater.* **2021**, *31*, 2008936. [CrossRef]

Yu, H.; Gold, J.I.; Wolter, T.J.; Bao, N.; Smith, E.; Zhang, H.A.; Twieg, R.J.; Mavrikakis, M.; Abbott, N.L. Actuating Liquid Crystals Rapidly and Reversibly by Using Chemical Catalysis. *Adv. Mater.* **2024**, *36*, 2309605. [CrossRef]

McEvoy, M.A.; Correll, N. Materials that couple sensing, actuation, computation, and communication. *Science* **2015**, *347*, 1261689. [CrossRef]

Hua, Q.; Sun, J.; Liu, H.; Bao, R.; Yu, R.; Zhai, J.; Pan, C.; Wang, Z.L. Skin-inspired highly stretchable and conformable matrix networks for multifunctional sensing. *Nat. Commun.* **2018**, *9*, 244. [CrossRef] [PubMed]

43. Schwartz, M.; Lagerwall, J.P.F. Embedding intelligence in materials for responsive built environment: A topical review on Crystal Elastomer actuators and sensors. *Build. Environ.* **2022**, *226*, 109714. [CrossRef]
44. Hussain, M.; Jull, E.I.L.; Mandle, R.J.; Raistrick, T.; Hine, P.J.; Gleeson, H.F. Liquid Crystal Elastomers for Biological Applic *Nanomaterials* **2021**, *11*, 813. [CrossRef]
45. Herbert, K.M.; Fowler, H.E.; McCracken, J.M.; Schlafmann, K.R.; Koch, J.A.; White, T.J. Synthesis and alignment of crystalline elastomers. *Nat. Rev. Mater.* **2021**, *7*, 23–38. [CrossRef]
46. Resetic, A.; Milavec, J.; Bubnov, A.; Pociecha, D.; Hamplova, V.; Gorecka, E.; Zalar, B.; Domenici, V. New Liquid Crys Elastomeric Films Containing a Smectic Crosslinker: Chemical and Physical Properties. *Crystals* **2023**, *13*, 96. [CrossRef]
47. Wang, Y.; Liu, J.; Yang, S. Multi-functional liquid crystal elastomer composites. *Appl. Phys. Rev.* **2022**, *9*, 011301. [CrossRef]
48. Brannum, M.T.; Steele, A.M.; Venetos, M.C.; Korley, L.T.J.; Wnek, G.E.; White, T.J. Light Control with Liquid Crystalline Elast *Adv. Opt. Mater.* **2019**, *7*, 1801683. [CrossRef]
49. Modes, C.; Warner, M. Shape-programmable materials. *Phys. Today* **2016**, *69*, 32–38. [CrossRef]
50. Babakhanova, G.; Turiv, T.; Guo, Y.; Hendrikx, M.; Wei, Q.H.; Schenning, A.; Broer, D.J.; Lavrentovich, O.D. Liquid c elastomer coatings with programmed response of surface profile. *Nat. Commun.* **2018**, *9*, 456. [CrossRef]
51. Pei, Z.; Yang, Y.; Chen, Q.; Terentjev, E.M.; Wei, Y.; Ji, Y. Mouldable liquid-crystalline elastomer actuators with exchan covalent bonds. *Nat. Mater.* **2014**, *13*, 36–41. [CrossRef]
52. Ware, T.H.; Perry, Z.P.; Middleton, C.M.; Iacono, S.T.; White, T.J. Programmable Liquid Crystal Elastomers Prepared by Thi Photopolymerization. *ACS Macro Lett.* **2015**, *4*, 942–946. [CrossRef]
53. Godman, N.P.; Kowalski, B.A.; Auguste, A.D.; Koerner, H.; White, T.J. Synthesis of Elastomeric Liquid Crystalline Po Networks via Chain Transfer. *ACS Macro Lett.* **2017**, *6*, 1290–1295. [CrossRef]
54. Fowler, H.E.; Pearl, H.M.; Hoang, J.D.; White, T.J. Liquid Crystal Elastomers Prepared by Thiol-Ene Photopolymeri Amenable to Surface-Enforced Alignment. *Macromolecules* **2024**, *57*, 2619–2627. [CrossRef]
55. Prausnitz, M.R.; Mitragotri, S.; Langer, R. Current status and future potential of transdermal drug delivery. *Nat. Rev. Drug* **2004**, *3*, 115–124. [CrossRef] [PubMed]
56. Nozawa, I.; Suzuki, Y.; Sato, S.; Sugibayashi, K.; Morimoto, Y. Preparation of thermo-responsive membranes. II. *J. Biomed. Res.* **1991**, *25*, 577–588. [CrossRef]
57. Chen, K.; Lin, Y.; Lin, S. Thermally on-off switching nylon membrane for controlling drug penetration. *Drug Deliv. Syst.* **19** 55–61. [CrossRef]
58. Inoue, Y.; Atsumi, Y.; Kawamura, A.; Miyata, T. Thermoresponsive liquid crystalline polymer membranes that undergo transition at body temperature. *J. Membr. Sci.* **2019**, *588*, 117213. [CrossRef]
59. Stepulane, A.; Ahlgren, K.; Rodriguez-Palomo, A.; Rajasekharan, A.K.; Andersson, M. Lyotropic liquid crystal elastom drug delivery. *Colloid. Surface B* **2023**, *226*, 113304. [CrossRef]
60. Küpfer, J.; Finkelmann, H. Nematic liquid single crystal elastomers. *Die Makromol. Chem. Rapid Commun.* **1991**, *12*, 71 [CrossRef]
61. Brochu, P.; Pei, Q. Advances in dielectric elastomers for actuators and artificial muscles. *Macromol. Rapid Commun.* **2010**, *31*, [CrossRef] [PubMed]
62. Kim, H.; Lee, J.A.; Ambulo, C.P.; Lee, H.B.; Kim, S.H.; Naik, V.V.; Haines, C.S.; Aliev, A.E.; Ovalle-Robles, R.; Baughman et al. Intelligently Actuating Liquid Crystal Elastomer-Carbon Nanotube Composites. *Adv. Funct. Mater.* **2019**, *29*, 19 [CrossRef]
63. Ambulo, C.P.; Ford, M.J.; Searles, K.; Majidi, C.; Ware, T.H. 4D-Printable Liquid Metal-Liquid Crystal Elastomer Composite *Appl. Mater. Interfaces* **2021**, *13*, 12805–12813. [CrossRef]
64. Xiao, Y.Y.; Jiang, Z.C.; Tong, X.; Zhao, Y. Biomimetic Locomotion of Electrically Powered "Janus" Soft Robots Using a L Crystal Polymer. *Adv. Mater.* **2019**, *31*, e1903452. [CrossRef] [PubMed]
65. Shimoga, G.; Choi, D.-S.; Kim, S.-Y. Bio-Inspired Soft Robotics: Tunable Photo-Actuation Behavior of Azo Chromo Containing Liquid Crystalline Elastomers. *Appl. Sci.* **2021**, *11*, 1233. [CrossRef]
66. Lu, X.; Zhang, H.; Fei, G.; Yu, B.; Tong, X.; Xia, H.; Zhao, Y. Liquid-Crystalline Dynamic Networks Doped with Gold Nan Showing Enhanced Photocontrol of Actuation. *Adv. Mater.* **2018**, *30*, e1706597. [CrossRef]
67. Cheng, Y.C.; Lu, H.C.; Lee, X.; Zeng, H.; Priimagi, A. Kirigami-Based Light-Induced Shape-Morphing and Locomotior *Mater.* **2020**, *32*, e1906233. [CrossRef]
68. Ma, S.; Li, X.; Huang, S.; Hu, J.; Yu, H. A Light-Activated Polymer Composite Enables On-Demand Photocontrolled M Transportation at the Liquid/Air Interface. *Angew. Chem. Int. Ed. Engl.* **2019**, *58*, 2655–2659. [CrossRef]
69. Li, Y.; Teixeira, Y.; Parlato, G.; Grace, J.; Wang, F.; Huey, B.D.; Wang, X. Three-dimensional thermochromic liquid crystal elas structures with reversible shape-morphing and color-changing capabilities for soft robotics. *Soft Matter* **2022**, *18*, 6857 [CrossRef]
70. Zhan, Y.; Broer, D.J.; Liu, D. Perspiring Soft Robotics Skin Constituted by Dynamic Polarity-Switching Porous Liquid C Membrane. *Adv. Mater.* **2023**, *35*, e2211143. [CrossRef]
71. *Biomaterials Science: An Introduction to Materials in Medicine*; Elsevier: Amsterdam, The Netherlands, 2020.
72. Martino, F.; Perestrelo, A.R.; Vinarsky, V.; Pagliari, S.; Forte, G. Cellular Mechanotransduction: From Tension to Function. *Physiol.* **2018**, *9*, 824. [CrossRef]

Nguyen, A.T.; Sathe, S.R.; Yim, E.K. From nano to micro: Topographical scale and its impact on cell adhesion, morphology and contact guidance. *J. Phys. Condens. Matter* **2016**, *28*, 183001. [CrossRef]
Ye, K.; Wang, X.; Cao, L.; Li, S.; Li, Z.; Yu, L.; Ding, J. Matrix Stiffness and Nanoscale Spatial Organization of Cell-Adhesive Ligands Direct Stem Cell Fate. *Nano Lett.* **2015**, *15*, 4720–4729. [CrossRef]
Tse, J.R.; Engler, A.J. Stiffness gradients mimicking in vivo tissue variation regulate mesenchymal stem cell fate. *PLoS ONE* **2011**, *6*, e15978. [CrossRef]
Abadia, A.V.; Herbert, K.M.; White, T.J.; Schwartz, D.K.; Kaar, J.L. Biocatalytic 3D Actuation in Liquid Crystal Elastomers via Enzyme Patterning. *ACS Appl. Mater. Interfaces* **2022**, *14*, 26480–26488. [CrossRef]
Babakhanova, G.; Yu, H.; Chaganava, I.; Wei, Q.H.; Shiller, P.; Lavrentovich, O.D. Controlled Placement of Microparticles at the Water-Liquid Crystal Elastomer Interface. *ACS Appl. Mater. Interfaces* **2019**, *11*, 15007–15013. [CrossRef]
Babakhanova, G.; Krieger, J.; Li, B.X.; Turiv, T.; Kim, M.H.; Lavrentovich, O.D. Cell alignment by smectic liquid crystal elastomer coatings with nanogrooves. *J. Biomed. Mater. Res. A* **2020**, *108*, 1223–1230. [CrossRef]
Turiv, T.; Krieger, J.; Babakhanova, G.; Yu, H.; Shiyanovskii, S.V.; Wei, Q.H.; Kim, M.H.; Lavrentovich, O.D. Topology control of human fibroblast cells monolayer by liquid crystal elastomer. *Sci. Adv.* **2020**, *6*, eaaz6485. [CrossRef]
Agrawal, A.; Adetiba, O.; Kim, H.; Chen, H.; Jacot, J.G.; Verduzco, R. Stimuli-responsive liquid crystal elastomers for dynamic cell culture. *J. Mater. Res.* **2015**, *30*, 453–462. [CrossRef]
Herrera-Posada, S.; Mora-Navarro, C.; Ortiz-Bermudez, P.; Torres-Lugo, M.; McElhinny, K.M.; Evans, P.G.; Calcagno, B.O.; Acevedo, A. Magneto-responsive liquid crystalline elastomer nanocomposites as potential candidates for dynamic cell culture substrates. *Mater. Sci. Eng. C Mater. Biol. Appl.* **2016**, *65*, 369–378. [CrossRef]
Khoo, I.C.; Clements, R.J.; McDonough, J.A.; Freeman, E.J.; Korley, L.T.; Cukelj, R.; Leslie, M.T.; Zhu, C.; Bera, T.; Gao, Y.; et al. New developments in 3D liquid crystal elastomers scaffolds for tissue engineering: From physical template to responsive substrate. In Proceedings of the Liquid Crystals XXI, San Diego, CA, USA, 25 August 2017.
Shaha, R.K.; Merkel, D.R.; Anderson, M.P.; Devereaux, E.J.; Patel, R.R.; Torbati, A.H.; Willett, N.; Yakacki, C.M.; Frick, C.P. Biocompatible liquid-crystal elastomers mimic the intervertebral disc. *J. Mech. Behav. Biomed. Mater.* **2020**, *107*, 103757. [CrossRef]
Harrison, R.G. The outgrowth of the nerve fiber as a mode of protoplasmic movement. *J. Exp. Zool.* **2005**, *9*, 787–846. [CrossRef]
Duval, K.; Grover, H.; Han, L.H.; Mou, Y.; Pegoraro, A.F.; Fredberg, J.; Chen, Z. Modeling Physiological Events in 2D vs. 3D Cell Culture. *Physiology* **2017**, *32*, 266–277. [CrossRef]
Bhatia, S.N.; Ingber, D.E. Microfluidic organs-on-chips. *Nat. Biotechnol.* **2014**, *32*, 760–772. [CrossRef]
Gong, Y.; Fan, N.; Yang, X.; Peng, B.; Jiang, H. New advances in microfluidic flow cytometry. *Electrophoresis* **2018**, *40*, 1212–1229. [CrossRef]
Zhang, H.; Chang, H.; Neuzil, P. DEP-on-a-Chip: Dielectrophoresis Applied to Microfluidic Platforms. *Micromachines* **2019**, *10*, 423. [CrossRef]
Genkin, M.M.; Sokolov, A.; Lavrentovich, O.D.; Aranson, I.S. Topological Defects in a Living Nematic Ensnare Swimming Bacteria. *Phys. Rev. X* **2017**, *7*, 011029. [CrossRef]
Sengupta, A. Topological microfluidics: Present and prospects. *Liquid Cryst. Today* **2015**, *24*, 70–80. [CrossRef]
Gupta, V.K.; Skaife, J.J.; Dubrovsky, T.B.; Abbott, N.L. Optical Amplification of Ligand-ReceptorBinding Using Liquid Crystals. *Science* **1998**, *279*, 2077–2080. [CrossRef] [PubMed]
Brake, J.M.; Abbott, N.L. An experimental system for imaging the reversible adsorption of amphiphiles at aqueous-liquid crystal interfaces. *Langmuir* **2002**, *18*, 6101–6109. [CrossRef]
Bedjaoui, L.; Gogibus, N.; Ewen, B.; Pakula, T.; Coqueret, X.; Benmouna, M.; Maschke, U. Preferential solvation of the eutectic mixture of liquid crystals E7 in a polysiloxane. *Polymer* **2004**, *45*, 6555–6560. [CrossRef]
Nolan, P.; Tillin, M.; Coates, D. Liquid Crystal Microdroplet Composition in a UV Cured PDLC Film. *Mol. Cryst. Liq. Cryst. Lett.* **1992**, *8*, 129–135. [CrossRef]
Khan, M.; Khan, A.R.; Shin, J.H.; Park, S.Y. A liquid-crystal-based DNA biosensor for pathogen detection. *Sci. Rep.* **2016**, *6*, 22676. [CrossRef]
Xu, Y.; Rather, A.M.; Song, S.; Fang, J.C.; Dupont, R.L.; Kara, U.I.; Chang, Y.; Paulson, J.A.; Qin, R.; Bao, X.; et al. Ultrasensitive and Selective Detection of SARS-CoV-2 Using Thermotropic Liquid Crystals and Image-Based Machine Learning. *Cell Rep. Phys. Sci.* **2020**, *1*, 100276. [CrossRef] [PubMed]
Rahman, M.S.; Hossain, K.S.; Das, S.; Kundu, S.; Adegoke, E.O.; Rahman, M.A.; Hannan, M.A.; Uddin, M.J.; Pang, M.-G. Role of Insulin in Health and Disease: An Update. *Int. J. Mol. Sci.* **2021**, *22*, 6403. [CrossRef]
Olefsky, J.M. The insulin receptor: Its role in insulin resistance of obesity and diabetes. *Diabetes* **1976**, *25*, 1154–1162. [CrossRef] [PubMed]
Chen, J.M.; Liu, Z.P.; Yang, R.Z.; Liu, M.J.; Feng, H.Q.; Li, N.; Jin, M.L.; Zhang, M.M.; Shui, L.L. A liquid crystal-based biosensor for detection of insulin driven by conformational change of an aptamer at aqueous-liquid crystal interface. *J. Colloid. Interf. Sci.* **2022**, *628*, 215–222. [CrossRef] [PubMed]
Willis, A.W.; Roberts, E.; Beck, J.C.; Fiske, B.; Ross, W.; Savica, R.; Van Den Eeden, S.K.; Tanner, C.M.; Marras, C.; Parkinson's Foundation, P.G. Incidence of Parkinson disease in North America. *NPJ Park. Dis.* **2022**, *8*, 170. [CrossRef]

101. Zheng, Y.; Qu, J.; Xue, F.; Zheng, Y.; Yang, B.; Chang, Y.; Yang, H.; Zhang, J. Novel DNA Aptamers for Parkinson's D Treatment Inhibit α-Synuclein Aggregation and Facilitate its Degradation. *Mol. Ther.-Nucleic Acids* **2018**, *11*, 228–242. [Cro [PubMed]
102. Yang, X.; Li, H.; Zhao, X.; Liao, W.; Zhang, C.X.; Yang, Z. A novel, label-free liquid crystal biosensor for Parkinson's d related alpha-synuclein. *Chem. Commun.* **2020**, *56*, 5441–5444. [CrossRef]
103. Yang, X.; Zhao, X.; Liu, F.; Li, H.; Zhang, C.X.; Yang, Z. Simple, rapid and sensitive detection of Parkinson's disease r alpha-synuclein using a DNA aptamer assisted liquid crystal biosensor. *Soft Matter* **2021**, *17*, 4842–4847. [CrossRef]
104. *2023 Alzheimer's Disease Facts and Figures*, Wiley: Hoboken, NJ, USA, 2023. [CrossRef]
105. Rajan, K.B.; Weuve, J.; Barnes, L.L.; McAninch, E.A.; Wilson, R.S.; Evans, D.A. Population estimate of people with cl Alzheimer's disease and mild cognitive impairment in the United States (2020–2060). *Alzheimers Dement.* **2021**, *17*, 1966 [CrossRef] [PubMed]
106. Kemiklioglu, E.; Tuncgovde, E.B.; Ozsarlak-Sozer, G. Development of liquid crystal biosensor for the detection of amyloid b levels associated with Alzheimer's disease. *J. Biosci. Bioeng.* **2021**, *132*, 88–94. [CrossRef]
107. Mileson, B.E.; Chambers, J.E.; Chen, W.L.; Dettbarn, W.; Ehrich, M.; Eldefrawi, A.T.; Gaylor, D.W.; Hamernik, K.; Hodgs Karczmar, A.G.; et al. Common Mechanism of Toxicity: A Case Study of Organophosphorus Pesticides. *Toxicol. Sci.* **19** 8–20. [CrossRef]
108. Wang, Y.; Hu, Q.Z.; Guo, Y.X.; Yu, L. A cationic surfactant-decorated liquid crystal sensing platform for simple and ser detection of acetylcholinesterase and its inhibitor. *Biosens. Bioelectron.* **2015**, *72*, 25–30. [CrossRef] [PubMed]
109. Wang, Y.; Hu, Q.Z.; Tian, T.T.; Yu, L. Simple and sensitive detection of pesticides using the liquid crystal droplet patterns pla *Sens. Actuat. B-Chem.* **2017**, *238*, 676–682. [CrossRef]
110. Kim, H.J.; Jang, C.H. Micro-capillary sensor for imaging trypsin activity using confined nematic liquid crystals. *J. Mol. Liq.* *222*, 596–600. [CrossRef]
111. Rim, J.; Jang, C.H. Detection of catalase activity with aldehyde-doped liquid crystals confined in microcapillaries. *Anal. Bi* **2018**, *560*, 19–23. [CrossRef]
112. Nguyen, D.K.; Jang, C.H. An acetylcholinesterase-based biosensor for the detection of pesticides using liquid crystals confi microcapillaries. *Colloids Surf. B Biointerfaces* **2021**, *200*, 111587. [CrossRef]
113. Singh, S.K.; Nandi, R.; Mishra, K.; Singh, H.K.; Singh, R.K.; Singh, B. Liquid crystal based sensor system for the real time det of mercuric ions in water using amphiphilic dithiocarbamate. *Sens. Actuat. B-Chem.* **2016**, *226*, 381–387. [CrossRef]
114. Prévôt, M.E.; Nemati, A.; Cull, T.R.; Hegmann, E.; Hegmann, T. A Zero-Power Optical, ppt- to ppm-Level Toxic Gas and Sensor with Image, Text, and Analytical Capabilities. *Adv. Mater. Technol.* **2020**, *5*, 2000058. [CrossRef]
115. Ailincai, D.; Pamfil, D.; Marin, L. Multiple bio-responsive polymer dispersed liquid crystal composites for sensing applicati *Mol. Liq.* **2018**, *272*, 572–582. [CrossRef]
116. Popov, P.; Mann, E.K.; Jakli, A. Thermotropic liquid crystal films for biosensors and beyond. *J. Mater. Chem. B* **2017**, *5*, 5061 [CrossRef]
117. Seo, J.M.; Khan, W.; Park, S.Y. Protein detection using aqueous/LC interfaces decorated with a novel polyacrylic acid block crystalline polymer. *Soft Matter* **2012**, *8*, 198–203. [CrossRef]
118. Li, L.; Bai, H.; Dong, X.; Jiang, Y.; Li, Q.; Wang, Q.; Yuan, N.; Ding, J. Flexible Capacitive Sensors Based on Liquid C Elastomer. *Langmuir* **2023**, *39*, 12412–12419. [CrossRef] [PubMed]
119. Park, H.; Lee, H.J.; Ahn, H.; Han, W.C.; Yun, H.S.; Choi, Y.S.; Kim, D.S.; Yoon, D.K. Mechanochromic Palettes of Chole Liquid Crystal Elastomers for Visual Signaling. *Adv. Opt. Mater.* **2024**, *12*, 2400266. [CrossRef]
120. Fallah-Darrehchi, M.; Zahedi, P.; Harirchi, P.; Abdouss, M. Performance of Liquid Crystalline Elastomers on Biologica Response: A Review. *ACS Appl. Polym. Mater.* **2023**, *5*, 1076–1091. [CrossRef]
121. Shimatani, A.; Hoshi, M.; Oebisu, N.; Iwai, T.; Takada, N.; Nakamura, H. Clinical significance of thermal detection of soft- tumors. *Int. J. Clin. Oncol.* **2020**, *25*, 1418–1424. [CrossRef]
122. Deng, F.; Tang, Q.; Zeng, G.; Wu, H.; Zhang, N.; Zhong, N. Effectiveness of digital infrared thermal imaging in detecting extremity deep venous thrombosis. *Med. Phys.* **2015**, *42*, 2242–2248. [CrossRef]
123. Fierheller, M.; Sibbald, R.G. A clinical investigation into the relationship between increased periwound skin temperatur local wound infection in patients with chronic leg ulcers. *Adv. Skin Wound Care* **2010**, *23*, 369–379; quiz 380–361. [CrossRe
124. Li, W.; Khan, M.; Lin, L.; Zhang, Q.; Feng, S.; Wu, Z.; Lin, J. Monitoring H_2O_2 on the Surface of Single Cells with Liquid C Elastomer Microspheres. *Angew. Chem.* **2020**, *132*, 9368–9373. [CrossRef]
125. Velasco-Abadia, A.; White, T.J.; Schwartz, D.K.; Kaar, J.L. 4D Cumulative Dose Sensing of Malathion Using a Biocatalytic L Crystal Elastomer with Chemical Memory. *Sens. Actuators B Chem.* **2024**, *400*, 134877. [CrossRef]
126. Davidson, E.C.; Kotikian, A.; Li, S.; Aizenberg, J.; Lewis, J.A. 3D Printable and Reconfigurable Liquid Crystal Elastomer Light-Induced Shape Memory via Dynamic Bond Exchange. *Adv. Mater.* **2020**, *32*, e1905682. [CrossRef]
127. Kotikian, A.; Morales, J.M.; Lu, A.; Mueller, J.; Davidson, Z.S.; Boley, J.W.; Lewis, J.A. Innervated, Self-Sensing Liquid C Elastomer Actuators with Closed Loop Control. *Adv. Mater.* **2021**, *33*, e2101814. [CrossRef] [PubMed]
128. Saed, M.O.; Ambulo, C.P.; Kim, H.; De, R.; Raval, V.; Searles, K.; Siddiqui, D.A.; Cue, J.M.O.; Stefan, M.C.; Shankar, M.R. Molecularly-Engineered, 4D-Printed Liquid Crystal Elastomer Actuators. *Adv. Funct. Mater.* **2018**, *29*, 1806412. [CrossRef]

Lu, X.; Ambulo, C.P.; Wang, S.; Rivera-Tarazona, L.K.; Kim, H.; Searles, K.; Ware, T.H. 4D-Printing of Photoswitchable Actuators. *Angew. Chem. Int. Ed. Engl.* **2021**, *60*, 5536–5543. [CrossRef] [PubMed]
Rivera-Tarazona, L.K.; Shukla, T.; Singh, K.A.; Gaharwar, A.K.; Campbell, Z.T.; Ware, T.H. 4D Printing of Engineered Living Materials. *Adv. Funct. Mater.* **2021**, *32*, 2106843. [CrossRef]
McDougall, L.; Herman, J.; Huntley, E.; Leguizamon, S.; Cook, A.; White, T.; Kaehr, B.; Roach, D.J. Free-Form Liquid Crystal Elastomers via Embedded 4D Printing. *ACS Appl. Mater. Interfaces* **2023**, *15*, 58897–58904. [CrossRef]
Ambulo, C.P.; Tasmim, S.; Wang, S.; Abdelrahman, M.K.; Zimmern, P.E.; Ware, T.H. Processing advances in liquid crystal elastomers provide a path to biomedical applications. *J. Appl. Phys.* **2020**, *128*, 140901. [CrossRef]
Zhang, Z.; Yang, X.; Zhao, Y.; Ye, F.; Shang, L. Liquid Crystal Materials for Biomedical Applications. *Adv. Mater.* **2023**, *35*, e2300220. [CrossRef]
Barnes, M.; Cetinkaya, S.; Ajnsztajn, A.; Verduzco, R. Understanding the effect of liquid crystal content on the phase behavior and mechanical properties of liquid crystal elastomers. *Soft Matter* **2022**, *18*, 5074–5081. [CrossRef]
Ustunel, S.; Pandya, H.; Prevot, M.E.; Pegorin, G.; Shiralipour, F.; Paul, R.; Clements, R.J.; Khabaz, F.; Hegmann, E. A Molecular Rheology Dynamics Study on 3D Printing of Liquid Crystal Elastomers. *Macromol. Rapid Commun.* **2024**, *45*, e2300717. [CrossRef]
Zhang, W.; Nan, Y.; Wu, Z.; Shen, Y.; Luo, D. Photothermal-Driven Liquid Crystal Elastomers: Materials, Alignment and Applications. *Molecules* **2022**, *27*, 4330. [CrossRef]
Ford, M.J.; Ambulo, C.P.; Kent, T.A.; Markvicka, E.J.; Pan, C.; Malen, J.; Ware, T.H.; Majidi, C. A multifunctional shape-morphing elastomer with liquid metal inclusions. *Proc. Natl. Acad. Sci. USA* **2019**, *116*, 21438–21444. [CrossRef]
Zhang, J.; Guo, Y.; Hu, W.; Soon, R.H.; Davidson, Z.S.; Sitti, M. Liquid Crystal Elastomer-Based Magnetic Composite Films for Reconfigurable Shape-Morphing Soft Miniature Machines. *Adv. Mater.* **2021**, *33*, e2006191. [CrossRef] [PubMed]
Xia, Y.; Mu, T.; Liu, Y.; Leng, J. Harnessing the power of carbon fiber reinforced liquid crystal elastomer composites for high-performance aerospace materials: A comprehensive investigation on reversible transformation and shape memory deformation. *Compos. Part A Appl. Sci. Manuf.* **2024**, *177*, 107943. [CrossRef]
Mistry, D.; Connell, S.D.; Mickthwaite, S.L.; Morgan, P.B.; Clamp, J.H.; Gleeson, H.F. Coincident molecular auxeticity and negative order parameter in a liquid crystal elastomer. *Nat. Commun.* **2018**, *9*, 5095. [CrossRef] [PubMed]
Cooper, E.J.; Reynolds, M.; Raistrick, T.; Berrow, S.R.; Jull, E.I.L.; Reshetnyak, V.; Mistry, D.; Gleeson, H.F. Controlling the optical properties of transparent auxetic liquid crystal elastomers. *Macromolecules* **2024**, *57*, 2030–2038. [CrossRef] [PubMed]
Mazaev, A.V.; Ajeneza, O.; Shitikova, M.V. Auxetic materials: Classification, mechanical properties and applications. *IOP Conf. Ser. Mater. Sci. Eng.* **2020**, *747*, 012008. [CrossRef]
Ebrahimi, M.S.; Noruzi, M.; Hamzehei, R.; Etemadi, E.; Hashemi, R. Revolutionary auxetic intravascular medical stents for angioplasty applications. *Mater. Des.* **2023**, *235*, 112393. [CrossRef]
Amigo-Melchior, A.; Finkelmann, H. A concept for bifocal contact- or intraocular lenses: Liquid single crystal hydrogels ("LSCH"). *Polym. Adv. Technol.* **2002**, *13*, 363–369. [CrossRef]
Foster, L.; Peketi, P.; Allen, T.; Senior, T.; Duncan, O.; Alderson, A. Application of auxetic foam in sports helmets. *Appl. Sci.* **2018**, *8*, 354. [CrossRef]

crystals

Article

Reconfigurable Liquid Crystal Elastomer Director Patterns fo Multi-Mode Shape Morphing

Xianbing Zeng [1], Tianfeng Zhou [1,2,*], Lei Li [1], Juncai Song [1], Ruijue Duan [1], Xiang Xiao [1], Baiqian Xu [1], Guanghao Wu [1] and Yubing Guo [1,*]

1 School of Medical Technology, Beijing Institute of Technology, Beijing 100081, China; xianb_zeng@bit.edu.cn (X.Z.); liooo@bit.edu.cn (L.L.); jcsong@bit.edu.cn (J.S.); Duanruijue@bit.edu.cn xiangxiao@bit.edu.cn (X.X.); xbq@bit.edu.cn (B.X.); bityswgh@bit.edu.cn (G.W.)
2 School of Mechanical Engineering, Beijing Institute of Technology, Beijing 100081, China
* Correspondence: zhoutf@bit.edu.cn (T.Z.); yguo@bit.edu.cn (Y.G.)

Abstract: Liquid crystal elastomers (LCEs) are a monolithic material with program three-dimensional (3D) morphing modes stemming from their designable non-uniform mol orientations (or director). However, the shape morphing mode is generally fixed when direct terns of LCEs are determined. Multi-mode shape morphing is difficult to achieve since di patterns cannot be reconfigured. Herein, we demonstrate the ability to reconfigure LCE di patterns and initial shapes—and thus shape morphing modes—by the manual assembly a assembly of LCE pixels. We measured the mechanical properties of LCEs with and without U and found their Young's moduli were 9.6 MPa and 11.6 MPa. We firstly fabricate LCE pixel designed director fields and then assemble 24 pixels with required director fields into an LC with a designed director pattern, which corresponds to a programmed shape morphing mod further exhibit that we can de-assemble the LCE film back into original pixels or new pixels wi ferent shapes and then re-assemble them into a new film with a different initial shape and di pattern, which corresponds to a second programmed shape morphing mode. Principally, v have a large amount of shape morphing modes if we have enough pixels. The demonstrated c ity of multi-mode shape morphing enhances functions of LCEs, which broadens their applic in soft robotics, programmable origami/kirigami, responsive surfaces, and so on.

Keywords: liquid crystal elastomers; multi-mode shape morphing; reconfigurable directo reconfigurable shapes; pixels assembly

Citation: Zeng, X.; Zhou, T.; Li, L.; Song, J.; Duan, R.; Xiao, X.; Xu, B.; Wu, G.; Guo, Y. Reconfigurable Liquid Crystal Elastomer Director Patterns for Multi-Mode Shape Morphing. *Crystals* **2024**, *14*, 357. https://doi.org/10.3390/cryst14040357

Academic Editor: Ingo Dierking

Received: 19 March 2024
Revised: 7 April 2024
Accepted: 8 April 2024
Published: 10 April 2024

1. Introduction

Liquid crystal elastomers (LCEs) exhibit up to 400% strain [1,2] under various ex stimuli, such as temperature [3], light [4], and magnetic field [5]. One advantage of compared to other stimuli-responsive materials, is their programmable three-dimens (3D) shape morphing from a monolithic material, which relies on the designable m ular orientations (or director) of LCEs introduced by mechanical stretching [2], mag field alignment [6–8], or surface alignment [9–11]. Due to their advanced shape mor capabilities, LCEs are an ideal candidate for artificial muscles [12,13], micro-mech actuators [14–16], and micro-robots [17].

Recent research efforts are focused on LCEs with programmable 3D initial shape 3D director patterns, to enable more freedom in designing shape morphing modes. F ample, inkjet printing can introduce 3D initial shapes of LCEs with director fields alo printing path [18,19]; photopatterned surfaces can command designable two-dimens (2D) LCE director patterns [20–22]; and optical 3D printing enables 3D LCE stru with printing path-independent director fields, which are determined by either ex magnetic fields [23] or commanding surfaces [10,24]. Recently, Sitti et al. realized structures with initial 3D shapes and 3D director patterns by assembling cubic pixels with a programmed uniform director field [25]. Generally, LCEs have fixed director patterns which correspond to a determined single shape morphing mode.

For real world applications, the capability of multi-mode shape morphing is in high demand for LCE structures to perform multi-functions. Currently, multi-mode shape morphing of LCEs can be realized through dynamic carbamate bonds [26], the synergistic use of photochemical and photothermal effects [27], or selective polymer chain decrosslinking [28]. Specifically, Ji et al. demonstrated programmed and reprogrammed shape morphing through heat-induced network rearrangement [26]; Priimagi et al. realized 6 different shapes from one actuator programmed with UV light and blue light [27]; Zhao et al. presented reprogrammable shape morphing and locomotion modes with the crosslinking and decrosslinking properties of polymer chains [28]. These studies demonstrate many appealing features of programmable shape morphing, while triggering the question of whether the director patterns of LCE can be reconfigured to enable completely different shape morphing modes.

In this paper, we propose a strategy to reconfigure LCE shape morphing modes by rearranging LCE director patterns, which are enabled by the manual assembly and de-assembly of LCE pixels. We firstly demonstrate the ability to fabricate LCE pixels with designed director fields and the ability to assemble these pixels into an LCE film with designed director patterns. We then show that the assembled LCE film can be de-assembled back into original pixels or new pixels with different shapes, which can be re-assembled into a new LCE film with distinct director patterns and thus a new shape morphing mode. As an example, we demonstrated two-times reconfigurations of LCEs director patterns (from -1 topological defect with 0 initial phase to -1 topological defect with $-\pi/4$ initial phase, and then to +1 topological defect with 0 initial phase) and the corresponding shape morphing modes. We further demonstrate that except for director patterns, LCE initial shapes can also be reconfigured by our strategy. We believe the demonstrated multi-mode shape morphing capability will enhance various applications of LCEs.

2. Design and Results

2.1. Concept of Reconfigurable LCE Director Patterns

We propose to realize reconfigurable LCE director patterns by re-arranging pixels with programmable director fields. We need to consider three critical points: (1) a strategy to fabricate LCE pixels with programmable director fields; (2) a strategy to assemble pixels together into a film with a designed director pattern; (3) a strategy to de-assemble the film back into the original pixels or completely different pixels. In this way, we can repeatedly assemble and de-assemble between pixels and LCE films with different designable director patterns.

The first critical point is enabled by the surface patterning of LCE film and laser cutting of the LCE film along designed orientations. As schematically shown in Figure 1a, we fabricated LCE films with planar alignment on the bottom surface and perpendicular alignment on the top surface, which corresponds to splay alignment. The chemical composition of LCE and the correspondingly weight ratio were presented in Figure S1. To verify the alignment of LCE, we checked polarized microscope images of uniform alignment and splay alignment, both of which were adopted in our research. Furthermore, the images indicated the good alignment quality of our samples (see Figure S6). Square-shaped LCE pixels with designable director fields can be achieved if we cut the film along corresponding orientations (α in Figure 1b). For example, we schematically present four pixels with different top-surface director fields (different α) in Figure 1c. Note that these pixels have splay alignment (Figure 1d). This kind of splay alignment results in a bending shape morphing towards the top surface along the planar alignment direction of each pixel (Figure 1e) [17,29]. Order change is a function with respect to temperature; in particular, at relatively low temperature, the order change might be much smaller than the one drawn in Figure 1e. We remark that we choose the splay alignment and square shapes of pixels for simplicity in pixel assembly and the design of shape morphing modes. In fact, w select various director fields and shapes of pixels, to increase the design freedom.

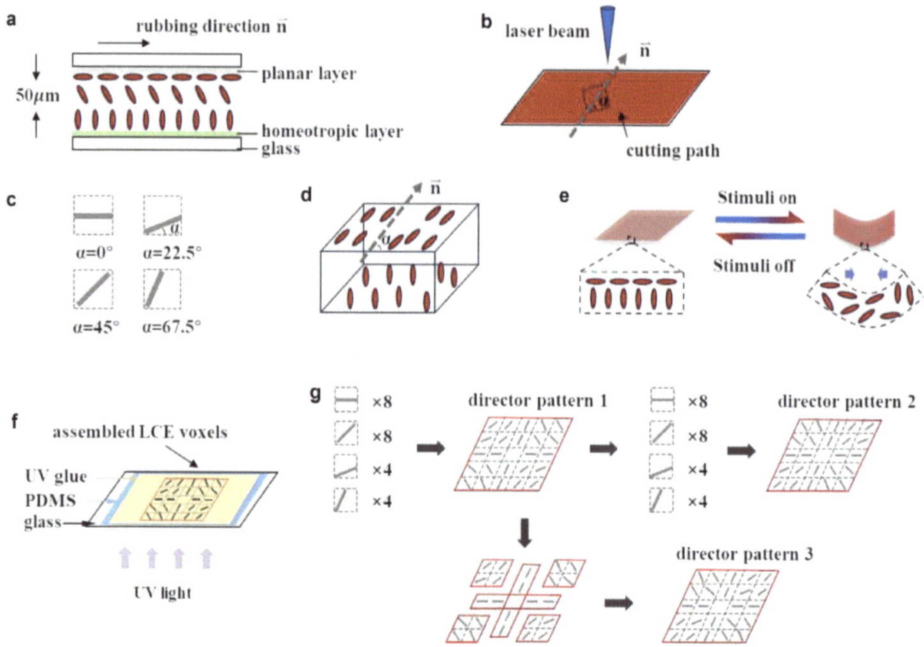

Figure 1. Schematic of reconfigurable LCE director patterns. (**a**) Schematic LCE director (**b**) LCE pixels with different director fields fabricated through laser cutting, where α is the between the planar alignment direction and one cutting edge. (**c**) A series of square pixels wi ferent director fields. (**d**) Schematic 3D LCE director field of a single pixel. (**e**) Schematic mor mode of an LCE pixel. (**f**) Schematic of LCE pixel assembly process. Pixels are assembled spin-coated UV glue and are sticked together after UV curing. (**g**) Schematic of LCE director p reconfiguring process.

The second critical point is realized by using UV glue to stick assembled pixels designed director patterns into an LCE film (Figure 1f). We can design the director pa of an LCE film and select pixels with proper director fields. These pixels are assemble top of a thin layer of spin-coated UV glue. After UV curing, these pixels will stick tog to form an LCE film with the designed director pattern.

The third critical point is fulfilled by cutting the assembled LCE film into ori pixels or completely different pixels, as schematically shown in Figure 1g. Note th can cut the LCE film partially or along different lines to form new pixels. These p either original pixels or new pixels, can be used to form a new LCE film with di designed director patterns.

2.2. *Effect of UV Glue and Pixel Geometry on Shape Morphing*

We proposed to use UV glue to assemble pixels into LCE films with designed sh and director patterns. It is important to explore the effect of UV glue film and pixel ge try on shape morphing. The most ideal situation is that neither the UV glue film nor geometry introduces obvious changes on shape morphing.

We firstly investigated the effect of UV glue film on the LCE shape morphing Figure S2 for more details). We mixed UV glue 9310 and 9300 in a 1:1 weight rati a proper adhesion performance, and we spin-coated the UV glue mixture onto a P (Polydimethylsiloxane) layer cured on a glass substrate with a rotation speed of 3000

An LCE film (6 mm in length and 3 mm in width) with splay alignment was placed on top of the spin-coated UV glue. After UV curing, the LCE film and UV glue film will stick together firmly, which will be peeled off from PDMS. Another LCE film without UV glue film is used for comparison. Then, these two samples are heated to 60 °C to compare their shape morphing behaviors (Figure 2a). We measured the bending angle θ of LCE films by adding the two bending angles (θ_1 and θ_2) of each end (Figure 2b). We present the time-dependent bending angles of these two samples in Figure 2c, which indicates similar stable bending angles (at 1.5 s). For better explanation of the experimental phenomenon, we conducted tensile tests on these two samples (see Figure S5). The results indicated that the Young's moduli of the LCE films without UV glue and with UV glue were 11.6 Mpa and 9.6 Mpa, respectively. To further illustrate the effect of UV glue on shape morphing, we also established a simplified physical model (see Figure S10 for more details). Undoubtfully, the thickness of UV glue film will affect the morphing performance. However, the influences introduced by UV glue are acceptable in our research due to the relatively small thickness of UV glue film (almost 25 µm, see Figure S4) and its low Young's modulus. Note that the LCE film with UV glue had a slower responding speed compared to the one without UV glue, which is reasonable since the presence of the UV glue film decreases the heat transfer speed. These experimental results prove the feasibility of UV-glue-assembled pixels.

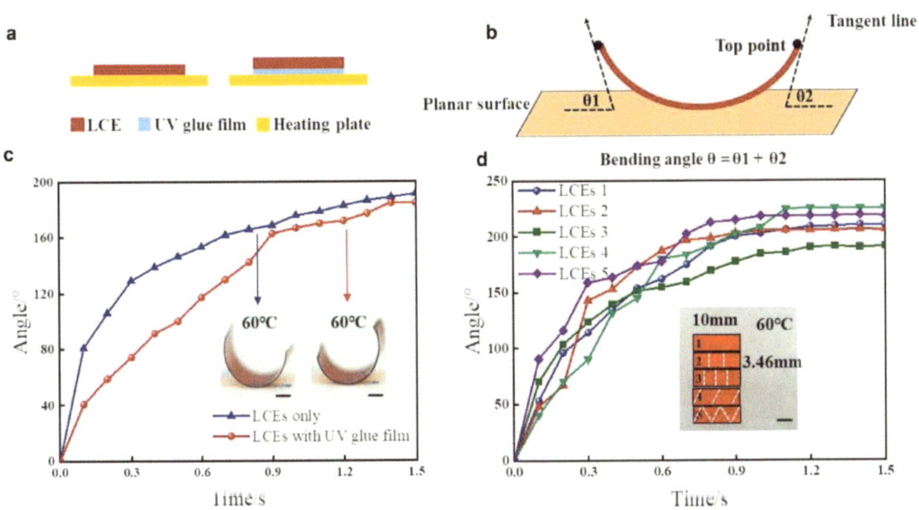

Figure 2. Effect of UV glue and pixel geometry on shape morphing of LCE. (**a**) Schematic heating setup of LCE with and without UV glue. (**b**) Method to measure bending angle for LCE films. (**c**) Bending angles of LCE film with and without UV glue as a function of time. Insets: photographs of LCE films at 1.5 s response time. (**d**) Bending angles of LCE films assembled with different pixels as a function of time. Insets: photographs of LCE films assembled with different pixels before shape morphing. All scale bars are 3 mm.

We then explored the effect of pixel geometry on LCE shape morphing. Both pixel size and pixel shape might be important in affecting the shape morphing of LCE films. Therefore, we assembled LCE films with pixels of different sizes (film 2 and film 3 in Figure 2d) and shapes (film 4 and film 5 in Figure 2d) and compared their shape morphing behavior with an LCE film that was directly fabricated (film 1 in Figure 2d). The time-dependent bending angles presented in Figure 2d indicate all films have similar stable bending angles and responding speeds, which is favorable for the further design of assembled LCEs with programmable director patterns. In other words, as long as the geometry and the director pattern are the same, the differences between directly fabricated LCEs and the assembled LCEs is acceptable for reconfiguring experiments. Therefore, we can use pixels with various shapes and sizes to create desired assembled LCEs, which further enh the versatility of our proposed strategy. We remark that the operating temperature LCE is between room temperature and 90 °C in this manuscript. However, the ne isotropic transition temperature of the LCE is as much as 200 °C [30], which allows operating temperature between room temperature and 200 °C.

2.3. Demonstration of Programmable Complex LCE Director Pattern from Pixels

We demonstrated the ability to fabricate LCE films with complex designable di patterns by assembling selected pixels with a proper director field. As schematically s in Figure 3a, we assembled 24 pixels into an LCE film with a +1 topological defect. that we only used the four types of pixels schematically shown in Figure 1c. Each pix be rotated by 90° to form a pixel with a different director field.

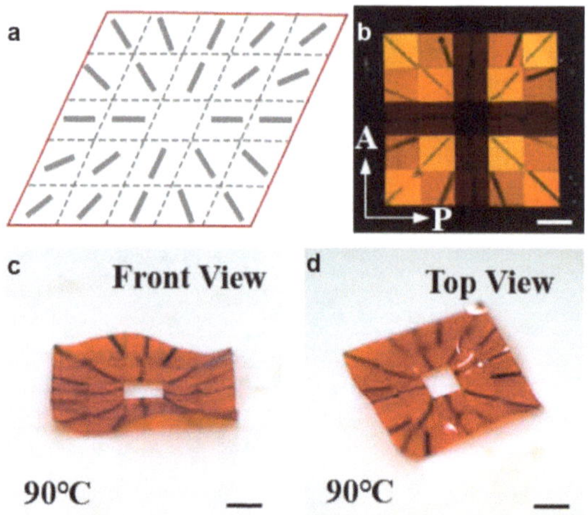

Figure 3. Optical characterization and shape morphing of LCE film assembled with square (**a**) Schematic of +1 topological defect director fields of an LCE film assembled from pixels. larized optical microscope image of the assembled LCE film, taken at room temperature. P represent polarizer axis and analyzer axis respectively. (**c**,**d**) Front (**c**) and top view (**d**) of LC when heated to 90 °C. Black lines in (**b**–**d**) represent director fields of each pixel. All scale b 3 mm.

The polarized optical microscope image (Figure 3b) shows a dark cross, whic resents a typical liquid crystal topological defect with charge 1. We observed pixels 3 types of light transmittance, corresponding to pixels with $\alpha = 0°$, $\alpha = 45°$, and $\alpha =$ or 67.5°. Here, we used pixels with planar alignment on both the top and bottom su (instead of splay alignment) for better characterization of the assembled LCE unde larized microscope. We then investigated the shape morphing behavior of the asser LCE film. Figure 3c,d present a front view and top view of the film at 90 °C, which s typical saddle shape. These results indicate that assembled LCE films with complex tor patterns show similar shape morphing behaviors as directly fabricated LCE film which is favorable for designing the shape morphing of assembled LCE films.

2.4. Reconfiguration of LCE Director Patterns

We then explored the ability to reconfigure LCE director patterns by using pixels splay alignment. We firstly designed an LCE film with a director pattern correspond a −1 topological defect with a 0 initial phase, as schematically shown in Figure 4a(i selected 24 pixels divided into 4 types, the same as those presented in Figure 1c. We heated the film to 30 °C, 60 °C, and 90 °C sequentially to investigate the shape mor behaviors. The polarized optical microscope images of the LCE at different temperatures implied different order parameters (see Figure S8 for more details) [31–33]. We present the side view (top row) and top view (bottom row) of the morphed shape in Figure 4a(iii). At 60 °C, we observed a four-fold symmetric shape. For better understanding, we schematically show the corresponding shape in Figure 4a(ii). At 90 °C, the shape morphing along one diagonal direction dominates due to the non-uniformity of heat.

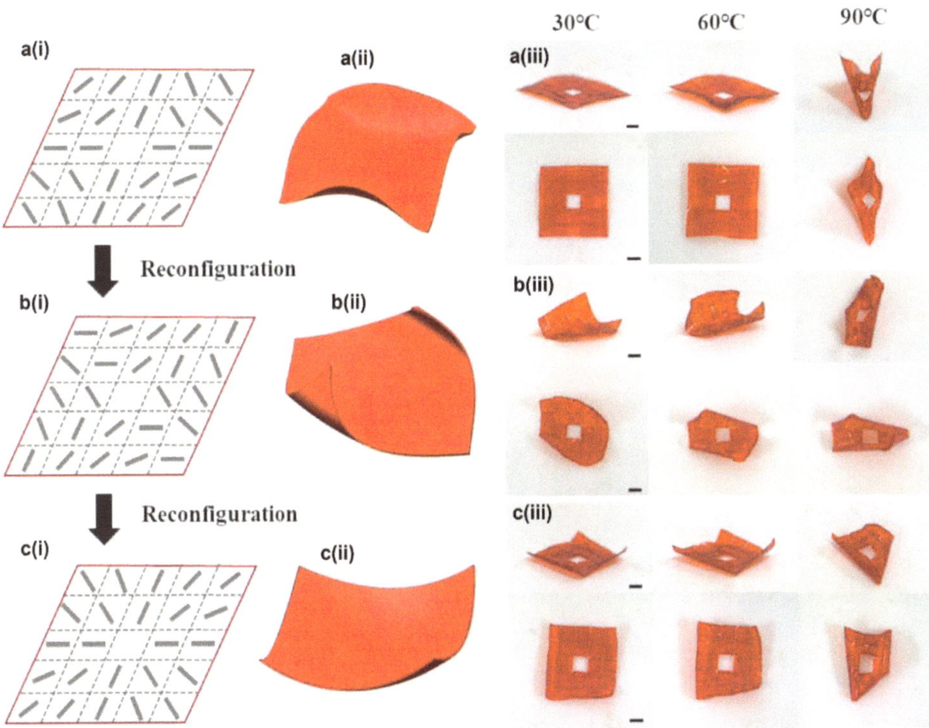

Figure 4. Reconfiguration of LCE director pattern. (**a(i–iii)**) Schematic director pattern (**a(i)**), schematic shape morphing (**a(ii)**), and experimental shape morphing (**a(iii)**) of LCE film with a director pattern of a −1 topological defect with 0 initial phase. (**b(i–iii)**) Schematic director pattern (**b(i)**), schematic shape morphing (**b(ii)**), and experimental shape morphing (**b(iii)**) of LCE film with a director pattern of a −1 topological defect with $-\pi/4$ initial phase. (**c(i–iii)**) Schematic director pattern (**c(i)**), schematic shape morphing (**c(ii)**), and experimental shape morphing (**c(iii)**) of LCE film with a director pattern of a +1 topological defect with 0 initial phase. All scale bars are 3 mm.

We then de-assembled the LCE film into original pixels (see Figure S3 for more details) and re-assembled these pixels into a new film with a director pattern corresponding to a −1 topological defect with $-\pi/4$ initial phase (Figure 4b(i)). We also performed experiments to prove the durability (see Figure S7) and reversibility of the LCE (see Figure S9). At 60 °C, we observed a bending shape morphing along one diagonal (Figure 4b(iii)). Finally, we de-assembled the LCE film into original pixels again and re-assembled these pixels into a new film with a director pattern corresponding to a +1 topological defect with 0 phase (Figure 4c(i)). At 60 °C, we observed a four-fold symmetric shape (Figure 4c(iii)) which differed from Figure 4a(iii).

We notice that in Figure 4a(i), the director fields are always perpendicular to two diagonals if we track the local director field along these diagonals; this causes expansion of the top surface along two diagonals upon actuation and makes the four corners bend downward. In contrast, director fields are always parallel to two diagonals in Figure 4c(i), which results in contraction of the top surface along two diagonals upon actuation and mak four corners bend upwards. In addition, director fields are always 45° to two diagon Figure 4b(i), and the shape morphing is apparently different from the previous two

In this section, we demonstrated the reconfiguration ability of LCE director pa by the assembly and de-assembly of LCE pixels with a selected director field. Thr two-times reconfiguration of the director pattern, we achieved three distinct shape m ing modes. We remark that more shape morphing modes can be realized by multi-reconfiguration.

2.5. Reconfiguration of Both LCE Shapes and Director Patterns

Reconfiguration of the director patterns already enables multiple shape mor modes. We further explored the ability to reconfigure LCE shapes and director pa simultaneously to allow for more shape morphing modes. We firstly assembled an square ring with circular alignment using 8 pixels (Figure 5a(i,ii)) and then reconfig both the shape and director pattern of the square ring to obtain N-shaped film wi axially symmetric director pattern (Figure 5b(i,ii)); this was finally reconfigured into shaped film with complex director pattern (Figure 5c(i,ii)). The shape morphing beha of these three films are completely different. Therefore, we can design shape mor modes through reconfiguring both the initial shapes and director patterns of LCE fil

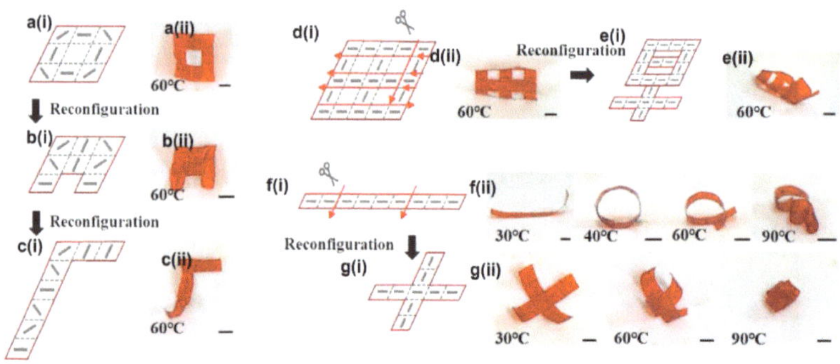

Figure 5. Reconfiguration of both LCE shapes and director patterns. (**a(i,ii)–c(i,ii)**) Multi-shape morphing realized with 8 pixels. (**d(i,ii),e(i,ii)**) LCE director pattern and shape reco ration between two Chinese characters. (**f(i,ii),g(i,ii)**) Reconfiguration between a helix and a g realized through reconfiguration of LCE director patterns and shapes. Red arrows in figures sent cutting tracks. All scale bars are 3 mm.

We then demonstrated the flexibility of our reconfiguring strategy to partial assemble the LCE films and introduce a new LCE film with programmable shap director pattern. For example, we firstly assembled an LCE film with the shape of th nese character "田" using 21 pixels with $\alpha = 0°$ (Figure 5d). We then selectively sepa the LCE film into 11 pixels via 8 cuttings (red arrows in Figure 5d). We finally assen these 11 pixels into another Chinese character "早" (Figure 5e). We obtained distinct s morphing modes from these two films.

We finally demonstrated multi-functions of LCE-based soft robots, enabled b reconfiguration of LCE films. We firstly assembled nine pixels (using an $\alpha = 0°$ tor field) into a stripe, which transformed into a circle at 40 °C and a helix above (Figure 5f). We then de-assembled the stripe into three pixels with two cuttings (r rows in Figure 5f) and assembled the three pixels into a cross shape, which could wc a gripper at high temperature (Figure 5g(ii)). The helix [5] and gripper [17] are widely in soft robotics for locomotion and grabbing, respectively. Reconfiguration between two shapes may allow for cargo transportation and more functions. We expect the r figuration of LCE shapes and director patterns will introduce more advanced functions of LCE-based soft robots.

3. Discussion

In this research, we demonstrated a strategy to reconfigure both LCE director patterns and LCE initial shapes for improving multi-mode shape morphing. Compared with previous researches [26–28], one obvious advantage of this work lies in the reconfiguration of complex LCE director patterns. For simplicity, we demonstrated the reconfiguration of LCE director patterns assembled from 24 square-shaped pixels (Figure 4). We remark that different numbers of pixels with various designable initial shapes can be applied to further improve the complexity of achievable LCE configurations and thus complex shape morphing modes.

In principle, we should be able to reconfigure LCE director patterns and LCE initial shapes for an unlimited number of times. Ideally, we need a reversible glue to transfer between adhesion and non-adhesion states. However, even though we tried our best, we could not find this kind of glue. The glue we used cannot be removed after de-assembly of our LCE films, and for each reconfiguration process, the new LCE film will add a new layer of UV glue. This significantly reduces the number of times we can reconfigure the LCE films. Furthermore, a smaller thickness of UV glue will minimize the influence on the shape morphing of assembled LCE structures. We expect a newly developed reversible glue can solve this problem and enable unlimited times reconfiguration of the LCE director patterns and initial shapes. Further improvement includes the application of a larger LCE pixel thickness and gluing the pixels on the sides.

We also use manual assembly in this research to assemble pixels into LCE films, which restricts the assembly efficiency; furthermore, it is time-consuming to assemble a large number of pixels and it is difficult to assemble pixels on the micrometer scale. Therefore, we believe self-assembly should be incorporated in our reconfiguration in the future, to significantly improve our strategy. In this research, we focused on assembling pixels into two-dimensional (2D) LCE films. In fact, the strategy demonstrated can be easily applied in reconfiguring three-dimensional (3D) LCE structures. In addition, as the LCE pixels were fabricated by laser beam, we could design arbitrary shapes and sizes of the LCE pixels, even down to the micro-scale. Considering the manual assembly process, the minimum controllable size of the pixel would be 50 μm [25].

To improve the versatility of the proposed strategy on the reconfiguration of the LCE director pattern, we proposed the following future studies: (1) self-assembly processes to gather pixels into desired shapes and director patterns with high efficiency; (2) a better glue to stick pixels together so that they can be de-assembled via a simple method; and (3) the 3D assembly of voxels. We believe an improved reconfiguration strategy of LCE director patterns will find enhanced applications in soft robotics, medical micro-robotics, reconfigurable origami/kirigami, and so on.

4. Conclusions

In conclusion, we demonstrated a strategy to reconfigure both LCE director patterns and initial shapes for multi-mode shape morphing. We firstly fabricated LCE pixels with designed director fields by using surface alignment and laser cutting, and we further demonstrated the reconfiguration of both LCE director patterns and LCE shapes by the assembly and de-assembly of pixels with designed director fields. In particular, we can de-assemble LCE films into pixels different from the original ones, improving the flexibility of our reconfiguration strategy. We finally showed that this strategy enabled multi-functions of LCE-based soft robotics. We believe these demonstrated capabilities will enable more advanced functions of LCE based actuators.

5. Materials and Methods

Materials: The LCE was made from a mixture containing 65 wt% of liquid c monomer Rm006, 32 wt% LC monomer Rm257, 2 wt% of light-responsive molec Ethyl-N-(2-hydroxyethyl)-4-(4-nitrophenylazo) aniline (Disperse Red 1, Sigma Aldri Louis, MO, USA)), and 1 wt% of photo-initiator (2-benzyl-2-dimethylamino-4-morph butyrophenone, MERYER), in which Rm006 and Rm257 served as acrylate-function LC monomers and LC crosslinker, respectively. Disperse Red 1 was added as red c color materials for convenience in operation and observation. The weight ratio of R and RM257 was selected as approximately 2:1, which benefits the alignment of LCE d the fabrication process. All materials were used as received. UV glue (Type 9310 (zhen Tegu New Material Co., Ltd., Shenzhen, China)) was used to assemble two g into a liquid crystal cell. PDMS was mixed in a 10:1 weight ratio and then spin-c on the glass. After curing at high temperature (e.g., 100 °C), the PDMS film serv the surface on which UV glue was spin-coated for better separation of the assembled films from the glass substrate. An UV glue mixture of Type 9310 and Type 9300 (She Tegu New Material Co., Ltd., Shenzhen, China) with a weight ratio of 1:1 was prepa balance the adhesion and Young's modulus for pixel assembly.

Liquid crystal cells: To fabricate liquid crystal cells, two glass substrates wer cleaned for 15 min using ultrasound and then for 40 min using UV ozone. After th glass substrates were either spin-coated with a 2 wt% water solution of polyvinyl al with a rotation speed of 4000 rpm for 30 s or spin-coated with PI1211 at a rotation of 1500 rpm for 30 s, respectively. The PVA-coated glass substrate and the PI1211 c substrate were heated at 120° for 10 min and at 180° for 30 min. After cooling, the coated glass slide was rubbed unidirectionally with a cloth for uniform planar aligr and subsequently blown with air blast to remove surface dust particles. Tiny dr UV glue were picked up by sharp needle and placed on the four corners of the PVA-c substrate. Then, the PI1211-coated substrate was assembled with the PVA-coated sub to form a cell. Note that spherical spacers of 50 μm diameter were mixed in the U to determine cell thickness. After that, an UV LED was used to cure the UV glue.

LCE films: To prepare LCE films, the monomer mixture was magnetically stir 150 °C for 5 min. Then, the mixture was filled into the cell by a capillary force at 150 ° cooled down to room temperature to reach the nematic phase for programmed align Note that the whole process needed to be performed in the absence of light. Then, we the UV LED to polymerize the LC mixture for 2 h. After curing via the UV LED, the L was opened from one side by blade. Then, we used the laser beam to cut LCEs into according to our design. In particular, the model of the laser equipment was AMT 20-100-W (Industrial Picosecond Lasers, Advanced Optowave Corporation, Ronkon NY, USA) and the wavelength of the laser beam was 1064 nm. Moreover, the relat rameters we applied were as follow: power: 500 w; repetition rate: 200 kHz; ve 400 mm/s; and cycle times: 300.

Pixel assembly: To assemble pixels into an LCE film, a glass substrate was thoro cleaned for 15 min using ultrasound and then spin-coated with PDMS mixture at a of 2500 rpm for 30 s. The coated glass was then cured on a heating plate at 100 °C for 1 to form a PDMS film. After that, the PDMS film was spin-coated with UV glue mixt a speed of 3000 rpm for 30 s. Then, pixels with designed director fields were assembl top of the UV glue mixture. After curing of the glue, the LCE film was peeled off fro PDMS substrate.

Supplementary Materials: The following supporting information can be downloaded at: //www.mdpi.com/article/10.3390/cryst14040357/s1, Figure S1. Chemical composition of LC the corresponding weight ratios. Figure S2. Effect of different UV glues on LCE shape mor Figure S3. De-assembly of an LCE film into pixels. Figure S4. Thickness of UV glue layer. Fig Stress–strain curve for LCE without (a) and with (b) UV glue. Figure S6. Polarized optical scopic images of uniform (a) and splay (b) aligned liquid crystal elastomers. Figure S7. Shap phing of LCE sample fabricated 3 months prior. Figure S8. Polarized optical images of LCEs at different temperatures. Figure S9. Bending angle of LCE film measured with respect to heating–cooling cycles. Figure S10. Normalized curvature with respect to thickness of two layers. Movie S1. Effect of UV glue on shape morphing of LCE. Movie S2. Effect of pixel geometry on shape morphing of LCE. Movie S3. Reversibility experiment for LCEs.

Author Contributions: Y.G. and T.Z. conceived of the research. X.Z. performed most of the experiments. X.Z., L.L., J.S., R.D., X.X., B.X. and G.W. analyzed the data. X.Z., T.Z. and Y.G. wrote the manuscript with input from all the other authors. All authors have read and agreed to the published version of the manuscript.

Funding: This work was supported by the National Natural Science Foundation of China through project 52375560.

Data Availability Statement: The original contributions presented in the study are included in the article and Supplementary Materials, further inquiries can be directed to the corresponding authors.

Conflicts of Interest: The authors declare no conflicts of interest.

rences

Tajbakhsh, A.R.; Terentjev, E.M. Spontaneous thermal expansion of nematic elastomers. *Eur. Phys. J. E* **2001**, *6*, 181–188. [CrossRef]

Küpfer, J.; Finkelmann, H. Nematic liquid single crystal elastomers. *Makromol. Chem. Rapid Commun.* **1991**, *12*, 717–726. [CrossRef]

Hebner, T.S.; Korner, K.; Bowman, C.N.; Bhattacharya, K.; White, T.J. Leaping liquid crystal elastomers. *Sci. Adv.* **2023**, *9*, eade1320. [CrossRef] [PubMed]

Li, Y.; Liu, Y.; Luo, D. Polarization Dependent Light-Driven Liquid Crystal Elastomer Actuators Based on Photothermal Effect. *Adv. Opt. Mater.* **2021**, *9*, 2001861. [CrossRef]

Zhang, J.; Guo, Y.; Hu, W.; Soon, R.H.; Davidson, Z.S.; Sitti, M. Liquid Crystal Elastomer-Based Magnetic Composite Films for Reconfigurable Shape-Morphing Soft Miniature Machines. *Adv. Mater.* **2021**, *33*, 2006191. [CrossRef] [PubMed]

Shin, J.; Kang, M.; Tsai, T.; Leal, C.; Braun, P.V.; Cahill, D.G. Thermally Functional Liquid Crystal Networks by Magnetic Field Driven Molecular Orientation. *ACS Macro Lett.* **2016**, *5*, 955–960. [CrossRef] [PubMed]

Yao, Y.; Waters, J.T.; Shneidman, A.V.; Cui, J.; Wang, X.; Mandsberg, N.K.; Li, S.; Balazs, A.C.; Aizenberg, J. Multiresponsive polymeric microstructures with encoded predetermined and self-regulated deformability. *Proc. Natl. Acad. Sci. USA* **2018**, *115*, 12950–12955. [CrossRef] [PubMed]

Schuhladen, S.; Preller, F.; Rix, R.; Petsch, S.; Zentel, R.; Zappe, H. Iris-Like Tunable Aperture Employing Liquid-Crystal Elastomers. *Adv. Mater.* **2014**, *26*, 7247–7251. [CrossRef] [PubMed]

McConney, M.E.; Martinez, A.; Tondiglia, V.P.; Lee, K.M.; Langley, D.; Smalyukh, I.I.; White, T.J. Topography from Topology: Photoinduced Surface Features Generated in Liquid Crystal Polymer Networks. *Adv. Mater.* **2013**, *25*, 5880–5885. [CrossRef] [PubMed]

Guo, Y.; Shahsavan, H.; Sitti, M. 3D Microstructures of Liquid Crystal Networks with Programmed Voxelated Director Fields. *Adv. Mater.* **2020**, *32*, 2002753. [CrossRef] [PubMed]

Xia, Y.; Cedillo-Servin, G.; Kamien, R.D.; Yang, S. Guided Folding of Nematic Liquid Crystal Elastomer Sheets into 3D via Patterned 1D Microchannels. *Adv. Mater.* **2016**, *28*, 9637–9643. [CrossRef] [PubMed]

Li, M.-H.; Keller, P. Artificial muscles based on liquid crystal elastomers. *Philos. Trans. R. Soc. Math. Phys. Eng. Sci.* **2006**, *364*, 2763–2777. [CrossRef] [PubMed]

Thomsen, D.L.; Keller, P.; Naciri, J.; Pink, R.; Jeon, H.; Shenoy, D.; Ratna, B.R. Liquid Crystal Elastomers with Mechanical Properties of a Muscle. *Macromolecules* **2001**, *34*, 5868–5875. [CrossRef]

Yu, Z.; Wang, Y.; Zheng, J.; Sun, S.; Fu, Y.; Chen, D.; Cai, W.; Wang, D.; Zhou, H.; Li, D. Fast-Response Bioinspired Near-Infrared Light-Driven Soft Robot Based on Two-Stage Deformation. *ACS Appl. Mater. Interfaces* **2022**, *14*, 16649–16657. [CrossRef] [PubMed]

Huang, Z.; Tsui, G.C.-P.; Deng, Y.; Tang, C.-Y.; Yang, M.; Zhang, M.; Wong, W.-Y. Bioinspired near-infrared light-induced ultrafast soft actuators with tunable deformation and motion based on conjugated polymers/liquid crystal elastomers. *J. Mater. Chem. C* **2022**, *10*, 12731–12740. [CrossRef]

Palagi, S.; Mark, A.G.; Reigh, S.Y.; Melde, K.; Qiu, T.; Zeng, H.; Parmeggiani, C.; Martella, D.; Sanchez-Castillo, A.; Kapernaum, N.; et al. Structured light enables biomimetic swimming and versatile locomotion of photoresponsive soft microrobots. *Nat. Mater.* **2016**, *15*, 647–653. [CrossRef] [PubMed]

Wani, O.M.; Zeng, H.; Priimagi, A. A light-driven artificial flytrap. *Nat. Commun.* **2017**, *8*, 15546. [CrossRef] [PubMed]

Kotikian, A.; Truby, R.L.; Boley, J.W.; White, T.J.; Lewis, J.A. 3D Printing of Liquid Crystal Elastomeric Actuators with Spatially Programed Nematic Order. *Adv. Mater.* **2018**, *30*, 1706164. [CrossRef] [PubMed]

19. Peng, X.; Wu, S.; Sun, X.; Yue, L.; Montgomery, S.M.; Demoly, F.; Zhou, K.; Zhao, R.R.; Qi, H.J. 4D Printing of Freesta Liquid Crystal Elastomers via Hybrid Additive Manufacturing. *Adv. Mater.* **2022**, *34*, 2204890. [CrossRef] [PubMed]
20. de Haan, L.T.; Sánchez-Somolinos, C.; Bastiaansen, C.M.W.; Schenning, A.P.H.J.; Broer, D.J. Engineering of Complex Ord the Macroscopic Deformation of Liquid Crystal Polymer Networks. *Angew. Chem. Int. Ed.* **2012**, *51*, 12469–12472. [Cro [PubMed]
21. Jiang, M.; Guo, Y.; Yu, H.; Zhou, Z.; Turiv, T.; Lavrentovich, O.D.; Wei, Q. Low *f* -Number Diffraction-Limited Panchara Berry Microlenses Enabled by Plasmonic Photopatterning of Liquid Crystal Polymers. *Adv. Mater.* **2019**, *31*, 1808028. [Cro [PubMed]
22. Babakhanova, G.; Turiv, T.; Guo, Y.; Hendrikx, M.; Wei, Q.-H.; Schenning, A.P.H.J.; Broer, D.J.; Lavrentovich, O.D. Liquid c elastomer coatings with programmed response of surface profile. *Nat. Commun.* **2018**, *9*, 456. [CrossRef] [PubMed]
23. Tabrizi, M.; Ware, T.H.; Shankar, M.R. Voxelated Molecular Patterning in Three-Dimensional Freeforms. *ACS Appl. Interfaces* **2019**, *11*, 28236–28245. [CrossRef]
24. Zeng, H.; Martella, D.; Wasylczyk, P.; Cerretti, G.; Lavocat, J.G.; Ho, C.; Parmeggiani, C.; Wiersma, D.S. High-Resoluti Direct Laser Writing for Liquid-Crystalline Elastomer Microstructures. *Adv. Mater.* **2014**, *26*, 2319–2322. [CrossRef] [PubM
25. Guo, Y.; Zhang, J.; Hu, W.; Khan, M.T.A.; Sitti, M. Shape-programmable liquid crystal elastomer structures with arbitrary dimensional director fields and geometries. *Nat. Commun.* **2021**, *12*, 5936. [CrossRef] [PubMed]
26. Wu, Y.; Zhang, S.; Yang, Y.; Li, Z.; Wei, Y.; Ji, Y. Locally controllable magnetic soft actuators with reprogrammable contra derived motions. *Sci. Adv.* **2022**, *8*, eabo6021. [CrossRef]
27. Lahikainen, M.; Zeng, H.; Priimagi, A. Reconfigurable photoactuator through synergistic use of photochemical and pho mal effects. *Nat. Commun.* **2018**, *9*, 4148. [CrossRef] [PubMed]
28. Jiang, Z.; Xiao, Y.; Tong, X.; Zhao, Y. Selective Decrosslinking in Liquid Crystal Polymer Actuators for Optical Reconfigu of Origami and Light-Fueled Locomotion. *Angew. Chem.* **2019**, *131*, 5386–5391. [CrossRef]
29. Lavrentovich, O.D. Prepatterned liquid crystal elastomers as a step toward artificial morphogenesis. *Proc. Natl. Acad. Sci* **2018**, *115*, 7171–7173. [CrossRef] [PubMed]
30. Zeng, H.; Wani, O.M.; Wasylczyk, P.; Kaczmarek, R.; Priimagi, A. Self-Regulating Iris Based on Light-Actuated Liquid C Elastomer. *Adv. Mater.* **2017**, *29*, 1701814. [CrossRef]
31. Pei, Z.; Yang, Y.; Chen, Q.; Terentjev, E.M.; Wei, Y.; Ji, Y. Mouldable liquid-crystalline elastomer actuators with exchang covalent bonds. *Nat. Mater.* **2014**, *13*, 36–41. [CrossRef] [PubMed]
32. Ware, T.H.; Perry, Z.P.; Middleton, C.M.; Iacono, S.T.; White, T.J. Programmable Liquid Crystal Elastomers Prepared by Ene Photopolymerization. *ACS Macro Lett.* **2015**, *4*, 942–946. [CrossRef] [PubMed]
33. Wu, J.; Yao, S.; Zhang, H.; Man, W.; Bai, Z.; Zhang, F.; Wang, X.; Fang, D.; Zhang, Y. Liquid Crystal Elastomer Metama with Giant Biaxial Thermal Shrinkage for Enhancing Skin Regeneration. *Adv. Mater.* **2021**, *33*, 2106175. [CrossRef] [PubM

crystals

MDPI

w

nm Plasmons: Properties, Applications, and Tuning with Help Liquid Crystals

r Yu. Reshetnyak [1,2], Igor P. Pinkevych [1,*], Michael E. McConney [3], Timothy J. Bunning [3] and Dean R. Evans [3]

1 Physics Faculty, Taras Shevchenko National University of Kyiv, 01601 Kyiv, Ukraine; victor.reshetnyak@gmail.com
2 School of Physics and Astronomy, University of Leeds, Leeds LS2 9JT, UK
3 Air Force Research Laboratory, Materials and Manufacturing Directorate, Wright-Patterson Air Force Base, Dayton, OH 45433, USA
* Correspondence: ipinkevych@gmail.com

Abstract: This article provides a brief overview of the research on localized optical states called Tamm plasmons (TPs) and their potential applications, which have been extensively studied in recent decades. These states arise under the influence of incident light at the interface between a metal film and a medium with the properties of a Bragg mirror, or between two media with the properties of a Bragg mirror. The localization of the states in the interfacial region is a consequence of the negative dielectric constant of the metal and the presence of a photonic band gap of the Bragg reflector. Optically, TPs appear as resonant reflection dips or peaks in the transmission and absorption spectra in the region corresponding to the photonic band gap. The relative simplicity of creating a Tamm structure and the significant sensitivity of TPs to its parameters make them attractive for applications. The formation of broadband and tunable TP modes in hybrid structures containing, in particular, rugate filters and porous distributed Bragg reflectors are considered. Considerable attention is paid to TP designs that include liquid crystals, which allow for the remote tuning of the TP spectrum without the mechanical restructuring of the system. The application of TPs in sensors, thermal emitters, absorbers, laser generation, and the experimental capabilities of TP-liquid crystal devices are also discussed.

Keywords: Tamm plasmon; Tamm plasmon polariton; optical Tamm state; liquid crystal; reflectance; sensors; absorbers; thermal emitters

nic Editor: Ingo Dierking
d: 2 November 2024
: 15 December 2024
ed: 20 December 2024
ed: 27 January 2025

n: Reshetnyak, V.Y.; ych, I.P.; McConney, M.E.; g, T.J.; Evans, D.R. Tamm ns: Properties, Applications, ning with Help of Liquid s. *Crystals* **2025**, *15*, 138. /doi.org/10.3390/ 020138

1. Introduction

Surface electromagnetic waves propagating along a flat interface between two dissimilar media have been studied for quite a long time, and the results of many of these studies are reflected in a large amount of literature (see, for example, [1,2]). The conditions that ensure the surface nature of such waves are very different. For example, the so-called Dyakonov or Dyakonov–Voigt surface waves propagate along a plane interface between isotropic and uniaxial dielectric media when both media are non-dissipative [3–5]. The propagation of waves along the interface without the diversion of their energy into the volume of the media occurs due to the difference in the symmetry of these media, and only in a small range of directions parallel to the plane of the interface.

At the interface between metal and dielectric media, well-known surface plasmon (SP) waves are formed [6,7]. The attenuation of the SP waves with distance from the interface into the bulk of the metal is a consequence of the negative dielectric permittivity of the metal, while the attenuation into the bulk of the dielectric has the same nature as at total

internal reflection. It should be noted that due to the momentum conservation la
direct optical excitation of SPs in planar structures is impossible without the use a pri
diffraction grating, which is not always convenient in applications (Figure 1).

Surface electromagnetic waves can also propagate between linear and non
media [8], in homogeneous layers of chiral or optically active materials [9]. The
reports of surface light waves along the interface of optical mirrors with different
vector space topologies (topological photonic states) [10,11].

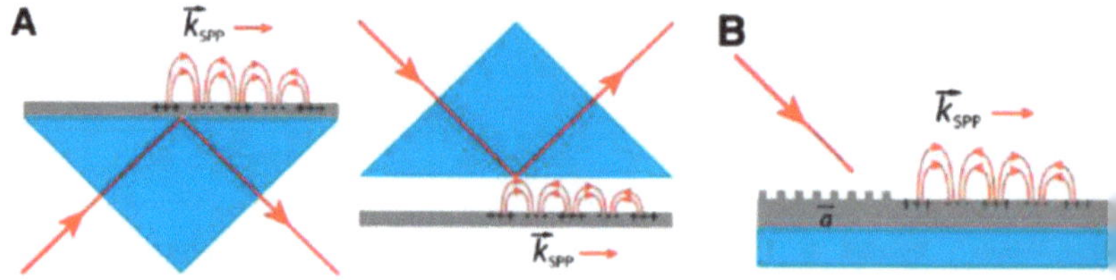

Figure 1. Schemes illustrating methods of optical excitation of SPs [12]. (**A**) Prism coupling sch
Kretschmann configuration (left) and in Otto configuration (right). (**B**) Grating coupling sche

In recent decades, much attention has been focused on the study of localized o
states that arise at the interface between a thin metal film and a dielectric medium wi
properties of a Bragg mirror (Figure 2a), or at the interface between two dielectric r
with the Bragg mirror properties (Figure 2b). A distributed Bragg reflector (DBR) is u
chosen as a medium with the properties of a Bragg mirror.

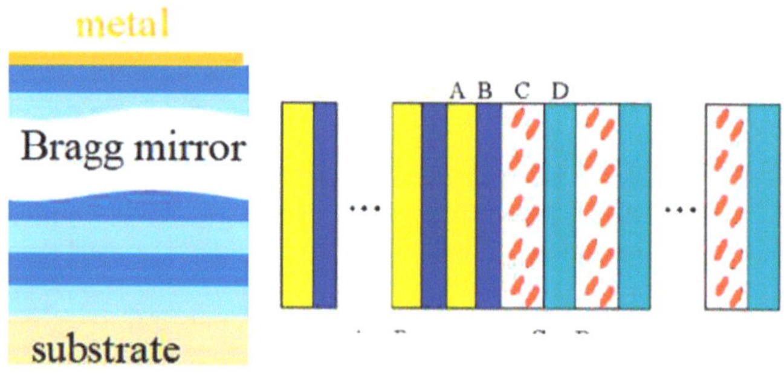

(**a**) (**b**)

Figure 2. Schematic of system for excitation of localized optical states: (**a**) system metal-DBR a
system with two DBRs [13].

As in the case of conventional SPs, the damping of such localized optical
with distance from the interface in a metal bulk is a consequence of the metal neg
permittivity, and the damping in the DBR occurs due to the existence of the DBR ph
band gap. Such localized states are called optical Tamm states (OTSs) [14–17] or T
plasmons (TPs), since they can be associated with the simultaneous excitation of plas
in the metal film [18–22]. These states were named by analogy with the electronic
predicted by Tamm [23], which can arise in the electronic band gap on the crystal su
The Tamm plasmon (TP) is often also referred to as the Tamm plasmon polariton (T
emphasize the strong relationship between the incident light and the plasmon oscil
In contrast to conventional SPs, TPs can be excited by an electromagnetic wave at any
of incidence without an additional prism or grating, and can be formed in both t
and TM polarizations. TPs manifest themselves optically in the form of resonance d

the reflection or peaks in the transmission and absorption spectra in the spectral region corresponding to the DBR band gap.

For TP applications, it is extremely important to be able to control the spectral characteristics of the TP, such as the wavelength, magnitude, and width of the dip/peak that arises because of TP in the reflection/absorption (transmission) spectra of the TP structure. These characteristics depend on the values of the TP-structure parameters, namely the permittivity and thickness of the metal film and DBR layers, as well as the corresponding spacers, if present. Changing these values, which requires the mechanical restructuring of the TP system, is not very convenient for applications; therefore, a number of TP structures have been proposed that incorporate liquid crystals (LCs), as nematic or cholesteric LCs. An important advantage of such TP structures is the possibility of the distant control of the LC refractive index using external electric or magnetic fields, which makes it possible to remotely tune the spectrum associated with the TP without the mechanical restructuring of the TP system.

Next, a brief overview of the main results of theoretical and experimental studies of TP structures and their resonance properties, as well as their potential applications, are presented. In the following, for simplicity, we predominantly use the term TP, except in some cases where the more general term OTS seems more appropriate.

2. Basic Schemes Used to Excite Tamm Plasmons and Study Their Properties

2.1. The Formation of a Confined Electromagnetic Mode Localized at the Interface Between a Metal and a Distributed Bragg Reflector

The theory of electromagnetic modes localized at the interface between metal and DBR was first formulated in [18] using a procedure similar to that previously described in [24]. A DBR has alternate planar dielectric layers of thicknesses a and b, and refractive indices n_A and n_B, respectively, such that $n_A a = n_B b = \pi c/2\omega_0$, where ω_0 is the Bragg frequency. A layer of type A in the DBR is adjacent to the metal. It was shown that localized electromagnetic modes could exist with a frequency below the plasma frequency of the metal and close to the DBR Bragg frequency if

$$r_M r_{DBR} = 1 \tag{1}$$

In Equation (1), r_M is the amplitude reflection coefficient for the wave incident on the metal from the medium with refractive index n_A, and r_{DBR} is the amplitude reflection coefficient of the wave incident from the medium with refractive index n_A on the DBR starting with a layer of the same refractive index (n_A).

The reflection coefficient r_M is given by the usual Fresnel formula $r_M = (n_A - n_M)/(n_A + n_M)$, where n_M is the refractive index of the metal. Using the Drude model for metal, one can obtain

$$r_M = \exp\left[i\left(\pi + \frac{2n_A\omega}{\sqrt{\varepsilon_b}\omega_p}\right)\right] \tag{2}$$

where ε_b is the background dielectric constant and ω_p is the plasma frequency.

The reflectance r_{DBR} can be obtained using the transfer matrix method [25]. If we assume that the DBR has a large number of layers, then a wave that has a frequency sufficiently close to the Bragg frequency ω_0 has a reflection coefficient of

$$r_{DBR} = \pm\exp[i\beta(\omega - \omega_0)/\omega_0] \tag{3}$$

In Equation (3), the negative (positive) sign corresponds to the case $n_A > n_B$ (n_A and

$$\beta = \frac{\pi n_A}{|n_A - n_B|}$$

The analysis of Equations (1)–(4) shows that the solution corresponding to a loca photonic state can be obtained near the center of the first DBR band gap, where the refl is large, and therefore, the radiative decay of the state will be small if $n_A > n_B$. In this Equation (1) reduces to

$$\pi + \beta\frac{\omega - \omega_0}{\omega_0} + \pi + \frac{2n_A\omega}{\sqrt{\varepsilon_b}\omega_p} = 2\pi l$$

where l = 0 or an integer. For a solution close to the Bragg frequency, ω_0, l = 0 and Equ (5) gives the following angular frequency of TP:

$$\omega \approx \frac{\omega_0}{(1 + 2n_A\omega_0/\sqrt{\varepsilon_b}\beta\omega_p)}$$

In particular, in the case of an interface between gold (for which $\hbar\omega_p = 8.9$ eV) quarter-wave GaAs/AlAs Bragg reflector (with a Bragg frequency given by $\hbar\omega_0 =$ Equation (6) predicts a TP frequency of $\omega_{TPP} \approx (0.95\,/\hbar)$ eV.

Figure 3 shows the calculated reflection (Figure 3a), transmission, and absor (Figure 3b) spectra of a 14-period DBR on a semi-infinite gold layer for different ang incidence [18]. For frequencies that are not eigenfrequencies of the structure, the refle is determined by the reflection coefficient of the metal, which is close to unity. A eigenfrequencies, the absorption increases, and each TP appears as a narrow dip i reflection spectrum and a peak in the transmission and absorption spectra. As noted a a TP can be formed with either TE or TM polarization and can be optically excited wi the use of prisms or gratings.

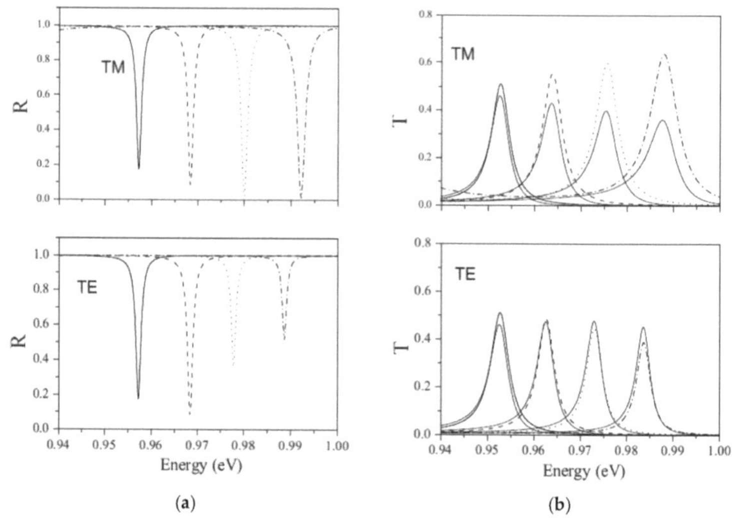

Figure 3. (**a**) Reflection spectra of the AlAs/GaAs DBR on a semi-infinite gold layer for TE- an polarized light for various angles of incidence: normal incidence (solid line), 30° (dashed line (dotted line), and 60° (dash-dotted line). (**b**) Transmission spectra of the free-standing AlAs/ DBR covered by a 30 nm-thick gold film using TE and TM polarized light at various ang incidence: normal incidence (solid line), 30° (dashed line), 45° (dotted line), and 60° (dash-d line). Thin solid lines show absorption spectra for each case [18].

TPs can be observed in the transmission spectra, despite the structure consisting of two objects (metal film and DBR) that are almost opaque to the radiation at the frequency of the TP. The reason for the appearance of the peaks in the transmission is similar to those for a Fabry–Perot cavity [25]. When the frequency of the incident radiation is equal to the frequency of the eigenmode, energy accumulates inside the structure and the magnitude of the field inside the structure increases compared to the incident wave. Thus, the flux of radiation incident on the DBR or metal film at the interface is increased by a factor corresponding to the specific eigenmode, and in turn the transmission becomes substantial.

In [26], the hard-mirror model for a one-dimensional cavity formed between a DBR and a metal mirror is applied to study TPs. It is also shown that the effective index model can describe TP modes that are laterally confined by the presence of a metallic disk deposited on the DBR.

2.2. Obtaining a Broadband TP

TPs in a one-dimensional multilayer structure composed of a finite DBR covered by a thin metallic layer have been theoretically and experimentally investigated in paper [22]. The DBR unit cell comprises two types of alternating dielectric layers with high (H) and low (L) refractive indices, n_H, n_L, and thicknesses, d_H, d_L, respectively. The one-unit cell on the top of the DBR can be truncated continuously to provide a smooth transition from the HLH to the LHL configuration. The DBR is placed in an air environment with refractive index $n_{air} = 1$. Porous quartz (SiO_2) with refractive indices of $n_L = 1.22$ and $n_H = 1.45$ as the DBR material is considered. The alternating layer thicknesses are $d_L = 150$ nm and $d_H = 127$ nm. The total number of layers is $N = 101$ (Figure 4).

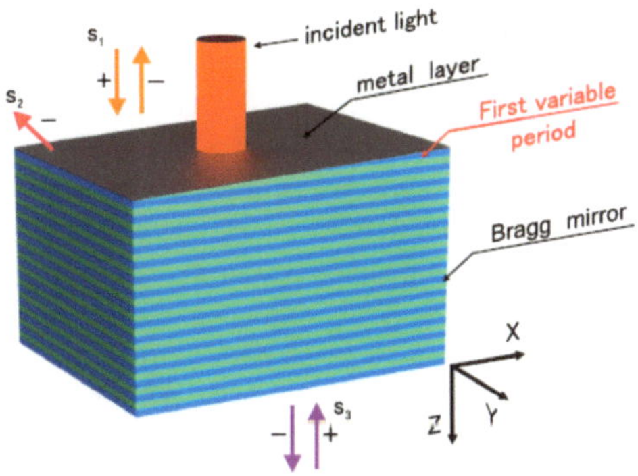

Figure 4. Schematic of the structure consisting of a Bragg mirror and a metal layer [22].

The absorptance, reflectance, and transmittance of the structure are calculated using the temporal coupled mode theory [27,28]. The resulting reflectance spectra for the HLH and LHL configurations for different metal layer materials are shown in Figure 5. According to the numerical calculation of the reflectance spectra, the use of aluminum and silver layers deposited onto the LHL configuration of the structure (Figure 5a) leads to an increase in the TP quality factor, which is observed as a resonance curve narrowing. For a chromium layer deposited onto the DBR with the LHL configuration, the TP makes it possible to completely suppress reflection and maximize absorption.

The transmittance spectra of the LHL and HLH structures measured at different chromium layer thicknesses show that transmission bands are not formed in the band gap for both the LHL and HLH configurations. It is also concluded that 95% of the incident light at the TP wavelength is only absorbed in the LHL structure. The transition from one configuration to the other allows for changing the reflection regime and ensuring the broad-band absorption. It is also shown that the variation of the metal film ma permits control of the width of the TP resonance line. In particular, the use of a meta of chromium allows for a maximum width of the spectral line.

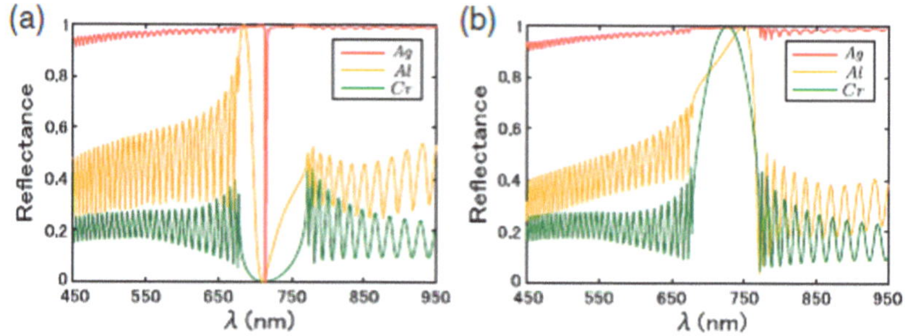

Figure 5. Reflectance spectra of (a) LHL and (b) HLH structures with silver ($d_{Ag} = 50$ nm), alun ($d_{Al} = 5$ nm), and chromium ($d_{Cr} = 6$ nm) layers [22].

2.3. *Tunability and Optimization of Coupling Efficiency in TAMM Plasmon Modes*

In paper [29], the structure for obtaining TP is considered as a dissymmetric reso enclosed by a metal mirror on one side and a DBR on the other. Based on the general t of open resonators [30], resonant transmittance and reflectance take the following fo

$$T = \frac{4\Gamma^1_{rad}\Gamma^2_{rad}}{\left(\Gamma^1_{rad} + \Gamma^2_{rad} + \Gamma_{dis}\right)^2},\ R = \frac{\left(\Gamma^2_{rad} - \Gamma^1_{rad} + \Gamma_{dis}\right)^2}{\left(\Gamma^1_{rad} + \Gamma^2_{rad} + \Gamma_{dis}\right)^2}$$

where Γ_{dis} is the damping factor due to intrinsic (ohmic) loss, and Γ^1_{rad} and Γ^2_{rad} a radiative loss factors for the two barriers. The transmittance T may reach unity on non-absorbing, symmetric structures ($\Gamma_{dis} = 0$, $\Gamma^1_{rad} = \Gamma^2_{rad}$). This situation occ dielectric cavities surrounded by two identical DBRs, since the response of the c improves with the reflectivity of the two mirrors. The quality factor increases wi number of layers, yielding a field enhancement inside the cavity and a more com transmission at resonance. The introduction of a metal element [30] changes the situ since a 100% transmission is no longer possible due to light absorption by the lossy stru The reflection may still reach 0%, under the critical coupling condition $\Gamma_{dis} = \Gamma^1_{rad}$ - When the transmission is prohibited ($\Gamma^2_{rad} = 0$) due to total internal reflection [31], c coupling coincides with complete light absorption. Changing structural parameter incident conditions changes the balance between radiative and non-radiative loss cha leading to the different behavior of the optical response. The conditions leading disappearance of the reflection coefficient of the Tamm structure due to the formatio TP have been studied.

The Tamm structure is presented in Figure 6. An experimentally relevant configu in the near-infrared region (gold with DBR formed with alternating layers of AlA GaAs) is considered. The complex dielectric function for the gold layer is taken fron To obtain the resonance condition associated with TP, it is necessary to find a pole complex reflectance associated with the multilayer structure. Fresnel equations were at successive interfaces of the multilayer structure [33].

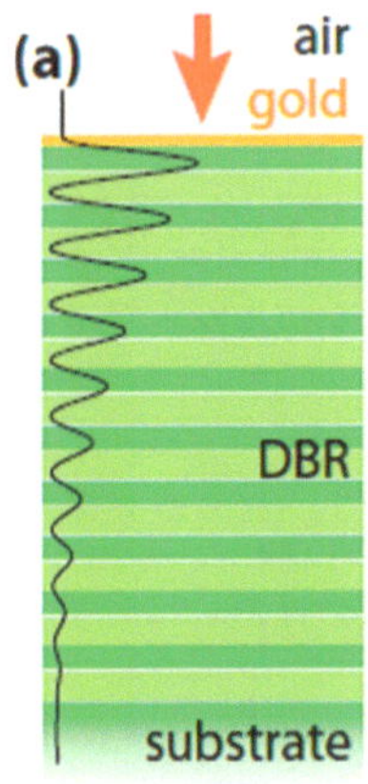

Figure 6. Schematic view of the Tamm structure: a GaAs/AlAs DBR with a gold film. The letter (**a**) indicates the pattern of light intensity distribution in the structure. Reprinted with permission from [29] © Optical Society of America.

Figure 7 shows the different situations considered. The left panels (a, b, and c) consider incidence from the air–Au side, presenting the far-field optical response (reflectance R, absorbance A, and transmittance T) at normal incidence for three different Tamm structures. Three situations are considered: (a) an unoptimized structure, where light only partially couples to the Tamm mode (R $\neq$ 0), (b) a critical coupling situation with R = 0, and (c) perfect absorption (A = 1). The same layout is adopted for the right panels (d, e, and f), where structures were selected to illustrate the reverse case of light incident from the substrate–DBR side. This allows a direct comparison between the two possible experimental configurations.

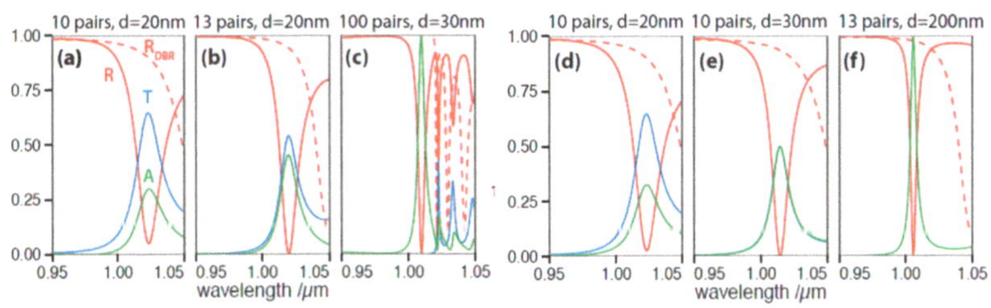

Figure 7. Far-field response for different TP structures. Left panels consider light incident from the substrate-DBR side, and the right panels consider the reverse case of incidence from the air–gold side. (**a**,**d**) Unoptimized structure, (**b**,**e**) critical coupling, and (**c**,**f**) perfect absorption. Reprinted with permission from [29] © Optical Society of America.

Some features can be noted from the results obtained. First, a broad range of parameters (number of layers, N; metal thickness, d; and spacer thickness, s) can be tuned to reach critical coupling conditions (Figure 7b,c,e,f). Second, the reflectivity of both mirrors (Au layer and DBR) affects the spectral position of the TP, as well as the linewidth of the mode. As the DBR (or Au) thickness increases, the TP mode blue-shifts and gets narrower as a consequence of reduced radiative damping. Third, the influence of the direction of incidence on the optical response. Cases (a) and (d) are for the exact same structure, only the direction of incidence differs. The far-field properties are qualitatively similar, with a negligible shift in the TP resonance for these cases.

The asymmetry in the response is better revealed in the optimized conditions (Figure 7b allowing for critical coupling and perfect absorption. When light is incident from the ai side, an optimum and finite metal thickness is required to allow the coupling of inc light with the TP mode. The DBR, however, may extend to infinity, as it prevents rad leakage through transmission. On the other hand, with incidence from the DBR a finite number of layers enables optimum coupling to the TP, while the metal m optically opaque. Optimization for various configurations at total absorption has also considered.

In [34], a TP structure is studied using a standard quarter-wavelength DBR comp of eight pairs of TiO_2 and SiO_2 with a Bragg wavelength of 500 nm. The TiO_2 and layers were deposited onto a glass substrate. The thicknesses of TiO_2 and SiO_2 are 50. 86.7 nm, respectively. A 30 nm-thick silver film is coated on top of the DBR.

Expansion of the top H layer in the simulation was shown to shift the TP wavele from 584 to 504 nm. Measured results agree well with these simulation results. The of the metal film material on the TP was studied for TP structures using gold, silver aluminum, designed for resonance wavelengths of 500 and 700 nm. It is shown that m with a higher ratio of the imaginary part to the real part of the target wavelength g narrower TRR resonance.

A number of articles propose the use of a magnetic field in Tamm structures example, it was shown in [35] that the spectral position of the TP can be controlle changing the magnetization of the magneto-optical layer placed between the metal and the DBR. In [36], a magnetic film added to the Tamm structure leads to the formati a hybrid state that includes a surface plasmon, a magnetic mode, and an OTS. This h state appears as a resonant peak in the transmission spectrum of the system.

2.4. *Influence of Rugate Filters on the Spectral Manifestation of Tamm Plasmon*

In [37], a rugate filter (RF) is used to study TP, which replaces the multilayer B reflector. RFs are thin dielectric films with a smooth periodic refractive index pr giving rise to spectral band gaps similar to Bragg mirrors that have a rectangular ste refractive index profile [38]. The RF has a smooth refractive index profile, which imp many characteristics of optical devices compared to dielectric multilayer Bragg mirro particular, RFs provide a photonic band gap without significant ripples in the refle spectrum outside the band gap and without its higher harmonics; this enables the possi to overlay multiple harmonic waves (giving rise to multiple spectral notches) [38–41

In Figure 8, a structure composed of an RF with a periodic dielectric function a the z-axis and a metal layer adjacent to the RF is shown. A light beam, polarized alon x-axis, is normally incident on the metal layer along the z-axis and propagates throug metal and adjacent dielectric RF.

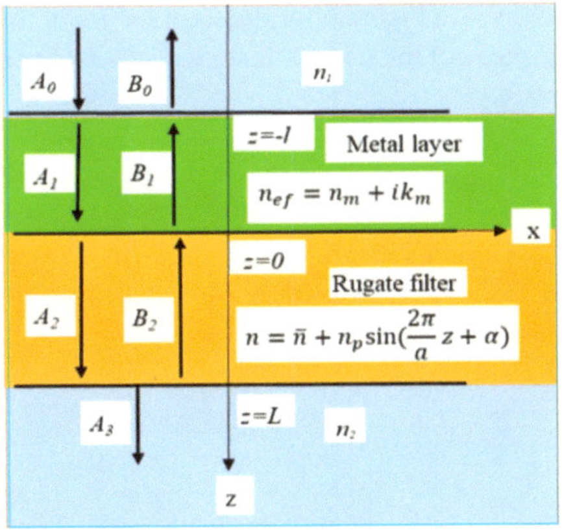

Figure 8. Scheme of the "metal layer–RF" structure with directions of the light beams propagating in the system and the refractive indices of the constituent layers [37].

In the area above the metal layer, $z \leq -l$, the electromagnetic field of the incident and reflected beams is described by the electric and magnetic vectors:

$$E_0(z) = A_0 \exp(ikn_1 z) + B_0 \exp(-ikn_1 z),$$
$$H_0(z) = \frac{n_1}{\mu_0 c}[A_0 \exp(ikn_1 z) - B_0 \exp(-ikn_1 z)], \tag{8}$$

where $k = \omega/c$, n_1 is a refractive index of the medium in front of the metal layer, and A_0, B_0 are the amplitudes of the incident and reflected beams, respectively. The electric and magnetic vectors of the wave field in the metal layer can be written analogously with the complex refractive index of the metal: $n_{ef} = n_m + ik_m$.

In the RF area, the dielectric tensor is periodic along the z-axis and can be expanded in a Fourier series. Then, solving the wave equation in the RF area, one can use the coupled wave method [42,43] and present the electric and magnetic vectors of the electromagnetic field in the form of a superposition of the forward and backward waves with amplitudes that are the slowly varying functions of z satisfying the Kogelnik equations [41]. Solving these equations, one can obtain the corresponding solutions when a wavelength of the incident beam is close to the Bragg wavelength of RF [44].

In the area below the RF, there is only an outgoing wave described by the electric and magnetic vectors as follows:

$$E_3(z) = A_3 \exp(ikn_2 z), \quad H_3(z) = \frac{n_2}{\mu_0 c} A_3 \exp(ikn_2 z) \tag{9}$$

where n_2 is a refractive index of the medium in the area $z \geq L$.

By solving the boundary conditions for the electric and magnetic vectors at $z = -l$, $z = 0$, and $z = L$, one can obtain expressions for the reflectance, $R = |B_0/A_0|^2$, and transmittance, $T = (n_2/n_1)|A_3/A_0|^2$, of the system.

For numerical calculations, the structure "Ag layer–RF" is used with the RF refractive index in the following form:

$$n(z) = \overline{n} + n_p \sin\left(\frac{2\pi}{a} z + \alpha\right) \tag{10}$$

For the complex refractive index of Ag, frequency dispersion is taken according to [45].

The reflectance and transmittance spectra of the system composed of the RF film with a refractive index profile described by Equation (10) and the Ag layer placed at the top of the RF are shown in Figure 9 for different values of the phase, α. Narrow dips reflectance and peaks of the transmittance have been obtained by exciting the TP at t layer–RF interface.

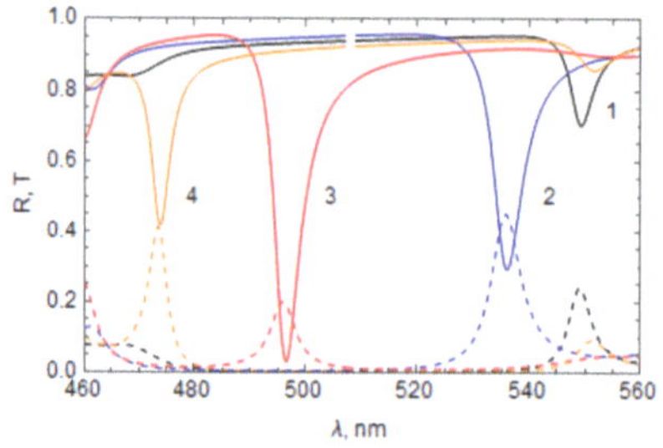

Figure 9. Reflectance and transmittance spectra of the "Ag layer–RF" system for different phase values, α: $\alpha = -\pi/2$(1, black), 0(2, blue), $\pi/2$(3, red), and π(4, orange). Reflectance s are shown with solid lines and transmittance spectra with dashed lines. The Ag film thick 45 nm [37].

It is shown that parameters of the harmonic profile of the RF refractive index icantly affect the TP wavelength and magnitude of the TP dips/peaks. The influe the metal layer thickness and the external medium refractive index on the positio magnitude of the TP dips/peaks was also studied.

Recently, black phosphorus (BP) has attracted attention as a material that c obtained in the form of atomically thin layers with unique physical properties. Th dimensional BP layers have an electrostatically controlled density of charge carriers a a result of the corrugated structure of each monolayer, they have highly anisotropic in- electronic and optical properties [46–49]. The 2d BP charge carriers have high mc and a short (sub-picosecond) carrier lifetime, which make 2d BP a potential materi devices operating in the terahertz (THz) frequency regime [50–54]. THz frequenci widely used in information and communication technologies, biomedical imaging, o switches, detectors, filters, and various sensors [55–57]. On this basis, 2d BP is consid promising candidate for optical applications in the infrared and terahertz regions spectrum and, in particular, for studying surface plasmon polaritons [58–60]. Rece multi-layer photonic structure with 2d BP was used to excite TPs in the THz region [

2.5. Excitation and Tuning of Optical Tamm States in a Hybrid Structure with Liquid Crys

In [62], an RF (described in Section 2.4) used to excite TPs at the interface wi 2d BP film is theoretically studied. The 2d BP film is separated from the RF by a crystal (LC) layer. As it is known [63], LCs easily change their state under the influe external fields, which makes it possible to use them in photonic structures to change properties. Therefore, the LC layer provides an additional degree of freedom to infl TPs. Figure 10 shows an influence of the LC refractive index on the reflection coeff when the RF band gap is located in the wavelength region centered near 300 μm (Figur and in the wavelength region centered near 30 μm (Figure 10b). The dependence of t wavelength on the LC refractive index makes it possible to control the spectral po of the reflection dip associated with the TP using an external electric field to reorie LC director. It is also shown that the TP wavelength is a periodic function of the LC thickness and refractive index, and two or more TPs can be obtained simultaneously RF band gap region.

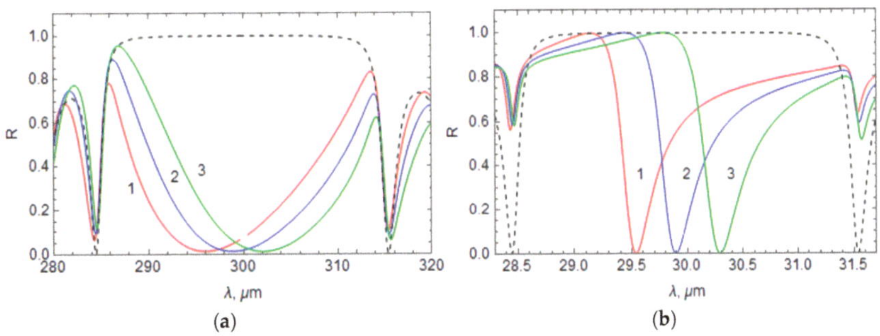

Figure 10. Influence of the LC refractive index n_{LC} on the reflection coefficient: $n_{LC} = 1.5$ (1), 1.6 (2), and 1.7 (3). (**a**) 1 THz region, and (**b**) 10 THz region [62].

Layers of nematic LCs are used to create holographic polymer-liquid crystal gratings (HPLCGs). HPLCG has a form of periodically alternating polymer-rich and LC-rich layers and can be used as a DBR. The recording of such holographic gratings by the interference of two intersecting coherent laser beams in a light-sensitive mixture of monomer and LC is carried out by various methods [64–69].

In papers [70,71], the excitation of OTSs in a hybrid structure containing a metal film and a HPLCG is studied. Au and Ag were taken as the metals. The HPLCG contained layers of LC 5CB or E7. It is shown that in such a hybrid structure, a narrow dip in the reflection spectrum appears in the region of the grating band gap, which is associated with the OTS formation. The spectral position of the reflection dip depends on the parameters of the polymer and LC. The type of metal and its thickness affect the value of the OTS reflection dip, but they do not affect its spectral position.

In Figure 11, the reflectance spectra of the hybrid structure with the Au film and HPLCG containing LC 5CB (Figure 11a) and LC E7 (Figure 11b) are shown in the region of the HPLCG band gap at different values of the LC director angle $\overline{\theta}$. It is seen that the reflectance dip corresponding to the OTS in the Au-HPLCG structure appears in the band gap for both LCs (5CB and 7E) used in the HPLCG. As the angle $\overline{\theta}$ increases, the OTS dip spectral position shifts to the short wavelength side along with a shift in the HPLCG band gap. This shift is due to a decrease in the HPLCG average refractive index and, consequently, a decrease in the HPLCG Bragg wavelength associated with the HPLCG band gap.

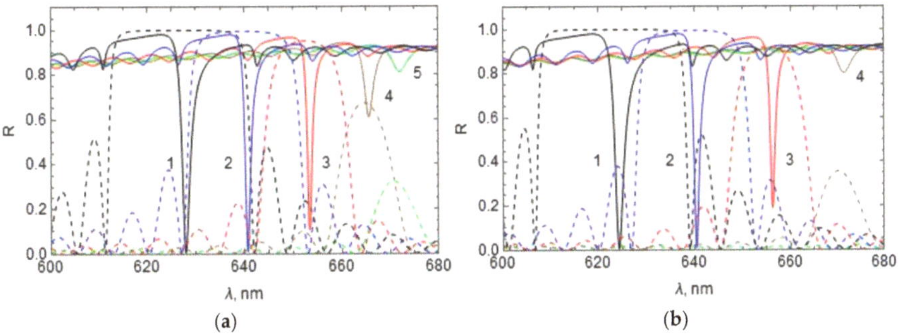

Figure 11. Reflectance spectra of the hybrid structures. (**a**) Au-HPLCG (5CB) and (**b**) Au-HPLCG (7E) for different values of the LC director angle: (1) $\overline{\theta} = \pi/2$, (2) $\pi/3$, (3) $\pi/4$, (4) $\pi/6$, and (5) $\pi/9$. The dashed line shows the position of the HPLCG band gap corresponding to the angle $\overline{\theta}$ [70].

The same dependence on the LC director angle also occurs for the Ag-HPLC stru
The obtained dependence of the OTS wavelength on the LC director orientation ma
possible to control the spectral position of the reflection dip using an external elect
magnetic field. In particular, it becomes possible to switch from OTSs in the band gap
four-layer HPLCG to OTSs in the band gap of the two-layer HPLCG, and vice versa

The controllable switching behavior of OTSs based on nematic LCs is consi
in [13]. The heterostructure is composed of two one-dimensional photonic crystals wi
a metallic film. The left photonic crystal is composed of isotropic dielectric layers A
and the right photonic crystal is composed of nematic LCs and isotropic dielectric la

Figure 12 shows the transmission spectra of the left photonic crystal alone (
1), the right photonic crystal alone (curve 2), and the composed heterostructure (cu
obtained using the Berreman transfer matrix method [72]. We can see that each pho
crystal exhibits a band gap, the two band gaps overlap in the range from 1.896 to 2.0
and there exists a transmission peak at 2.004 eV within the overlapping band gaps.
peak is due to the occurrence of OTSs at the interface of two photonic crystals.

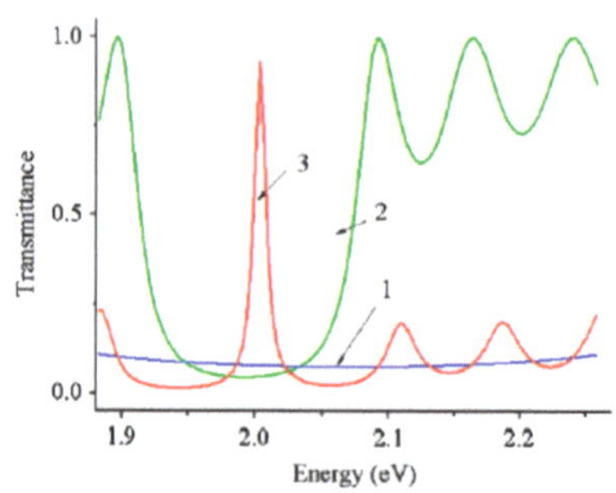

Figure 12. The transmission spectra of the left photonic crystal alone, the right photonic crystal
and the heterostructure are denoted by 1, 2, and 3, respectively [13].

By changing the external electric and magnetic fields or temperature, the orien
of LC molecules can be modified to offer an effective way to control the OTS. Theor
analysis and numerical modeling show that, by choosing the appropriate paramet
is possible not only to change the frequency position of the OTS, but also to realiz
disappearance of the OTS.

In cholesteric liquid crystals (CLC), circularly polarized light incident along the
helical axis undergoes a Bragg reflection if the circular polarization of light coincides
the CLC helix, while light with the opposite circular polarization does not experie
Bragg reflection [62]. This makes it possible to use CLC as Bragg mirrors in pho
structures for exciting and studying OTS.

In [73], a structure comprising a left-handed cholesteric (LCLC), a right-ha
cholesteric (RCLC), and a metal film are proposed to excite OTSs (Figure 13). Th
layers' thicknesses are L = 2 μm, the helical pitch is p = 0.4 μm, and the ordinary
extraordinary refractive indices are n_o = 1.4 and n_e = 1.6, respectively. The metal laye
made of silver, for which the real part of the permittivity is negative at the waveleng
the CLC band gap. The silver film thickness was d_m = 50 nm.

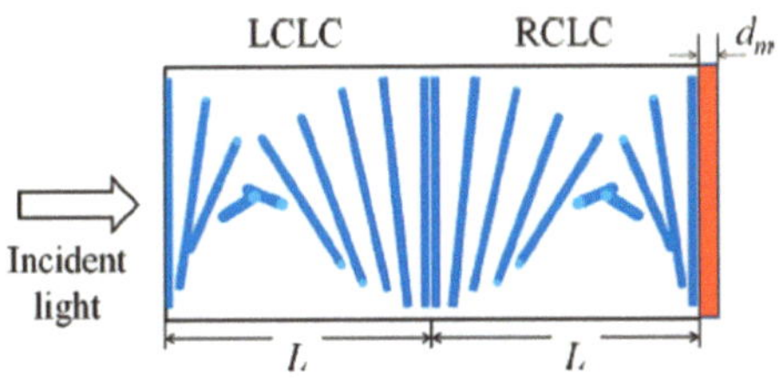

Figure 13. Schematic of the structure consisting of two oppositely handed CLCs and a metal layer [73].

Figure 14 shows the calculated transmission spectrum for circularly polarized light. At wavelengths of λ = 557, 591, and 632 nm, transmission peaks corresponding to the localized optical states are observed. At $L = 2$ μm, these wavelengths lie in the CLC band gap. As it is seen from Figure 14, different polarizations of the waves passing through the CLC lead to different transmittances. The transmission spectrum of the system can be controlled by varying the angle between the directors at the boundary interface of the oppositely handed cholesterics. As the CLC layer thickness is increased, the number of peaks grows. Transmission peaks of circularly polarized light were obtained, which corresponds to the appearance of several OTSs; this is similar to what is demonstrated in structures with a multilayer Bragg mirror [74,75].

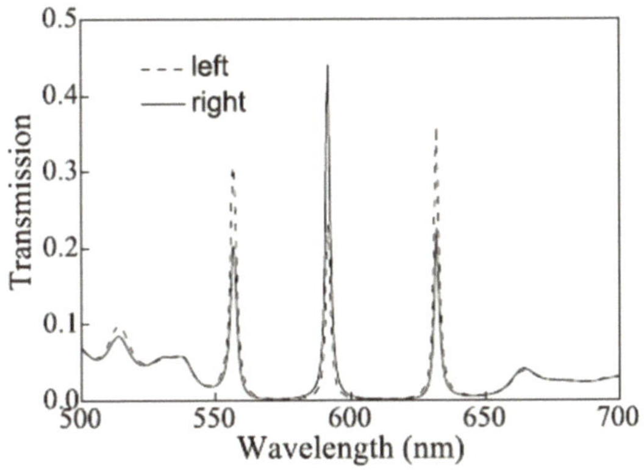

Figure 14. Transmission spectrum of the structure for circular polarizations [73].

In [76], the occurrence of an OTS in a system consisting of a metal film and two CLC layers with opposite directions of helical twist in the field of an incident linearly polarized light wave is studied. In contrast to [73], both CLC layers are thick enough to ensure the total reflection of waves with circular polarization coinciding with the CLC helix. Near the CLC, there is also a dielectric matching layer. It is shown that in such a system, the excitation of an OTS by linearly polarized light is possible. The influence of the CLC pitch, refractive indices, thicknesses of the dielectric index-matching layer, and metal film on the OTS manifestation in the reflection spectrum of the system is studied.

A series of theoretical works has recently proposed a new localized state of light, the chiral OTS [77–82]. The chiral OTS exists at the interface between a polarization-retaining anisotropic mirror (PPAM) and a CLC layer. To excite chiral OTSs, the metallic film cannot be used as the conventional TP structures, because the isotropic metal changes the polarization of light and the Bragg reflection of the CLC layer does not exist for some polarizations. PPAM preserves not only the chirality sign, but also the ellipticity magnitude upon reflection [83]. The simplest example of such a structure consists of alternating similar uniaxial dielectric layers with orthogonal directions of the optical axes. A schematic of the structure for the excitation of a chiral CLC is shown in Figure 15. Chiral OTS ca be excited in a structure with a quarter-wave plate between the metallic film and layer [78].

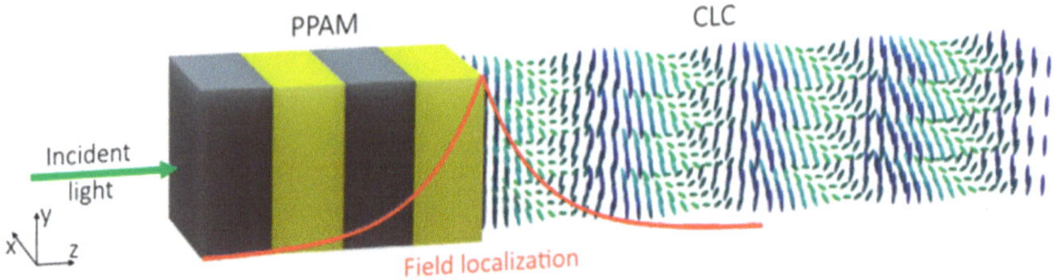

Figure 15. Schematic of a chiral OTS structure composed of a CLC layer and a polarization-ret anisotropic mirror (PPAM) [82].

The chiral OTS is manifested in the form of the resonances of polarized reflectio transmission. The replacement of a solid photonic crystal substrate with the fluid in chiral OTSs makes the system tunable due to the ease of controlling the CLC lay external stimuli. This allows the proposed structure to be used not only for optical but also for sensor devices.

2.6. Excitation of Optical Tamm States in a Hybrid Structure with Porous DBR

Conventional DBR mirrors are constructed from dielectric or semiconductor mat However, semiconductor DBRs have a low refractive index contrast and typically a mismatch between layers [84]. Dielectric DBRs characteristically provide a high refr index contrast between the layers, but there may be complications integrating them semiconductor technologies due to epitaxial difficulties [85]. At the same time, nanop GaN provides perfect lattice matching with refractive index contrasts as large as diel DBRs [85,86], while the porosity of nanoporous GaN can be controlled through d conditions and electrochemical etching [87]. Therefore, nanoporous GaN has been pro as a DBR material for TP structures, and the first observation of a TP mode occu at the interface between nanoporous GaN DBR and a thin Ag layer has already demonstrated [88]. A narrow dip in the DBR reflectance spectrum in the experimen Ag deposition indicates the presence of a TP mode at ~454 nm. The wavelengths of t mode are highly dependent on the thickness of the Ag layer and the properties of the such as the porosity, number of pairs, and the environment inside the pores. The simu results show a shift in the position of the TP mode by ~4–7 nm when the refractive of the nanoporous layer changes from 1.0 to 1.3, indicating the possibility of using structures for optical sensing applications.

In [89], porous silicon is used as a material for the fabrication of a photonic cryst Tamm structure. This nanomaterial is well suited for the fabrication of photonic cry because multilayers can be easily created by periodically varying the current during formation [90]. In such a structure, consisting of a Si photonic crystal coated wit a dip in the measured reflectance in the visible part of the spectrum is observed, v is associated with the excitation of TPs. To simulate the propagation of electromag waves in porous Si layers, the Bruggeman effective medium model [91] was used, the wavelength of visible light is at least an order of magnitude larger than the size pores [92]. The high sensitivity of the TP spectral parameters to changes in the refr index of the photonic crystal, determined by the porosity of Si, has been shown.

3. Applications of Tamm Plasmons

As follows from the results of the studies discussed in Section 2, the spectral characteristics of TPs are highly sensitive to most parameters of the hybrid structures in which they arise. This makes structures with TPs very promising for the design of sensors, lasers based on TPs, optical switches and filters, as well as selective thermal and light emitters.

3.1. *Tamm Plasmon Resonance-Based Sensors*

A scheme providing a potential way to realize refractive index sensing with a large measuring range and high sensitivity is proposed in [93]. The structure is formed by a Si/SiO_2 alternate-layered DBR coated with Ag film. In the reflection spectrum of the structure, a dip related to the formation of TP appears. It is shown that the wavelength and reflectivity of this dip are sensitive to changes in the refractive index of the surrounding medium.

An optical sensor based on TPs at the interface between a multilayer of porous SiO_2 and TiO_2 and a gold film is proposed in [94]. The transmission spectra of the structure reveal a narrow Tamm mode in the band gap of the DBR, the spectral position of which was controlled by an exposure to various solvents, thereby demonstrating the sensitivity of the device to changes in the refractive index. A TP based temperature sensor, which measures the temperature up to 185 °C, is demonstrated in [95]. The geometry for supporting TP modes is comprised of a one-dimensional DBR and a thin Ag layer. The DBR geometry includes 10 bilayers of Ta_2O_5 and SiO_2, which have been deposited on a glass substrate. In order to ascertain the temperature sensitivity of the configuration supporting a TP mode, the fabricated DBR was housed in an oven which provides homogeneous and isotropic heating to the "DBR + 30 nm Ag" film structure. The oven temperature could be adjusted from room temperature to 190 °C with a resolution of ± 0.1 °C. Depending on the temperature, the change in the reflectance spectrum of the Ag-DBR structure is shown in Figure 16 for normal light incidence. It is observed that the reflectivity minimum depicting TP-mode excitation decreases with an increasing oven temperature, resulting in a small (≤ 0.2 nm) shift in the resonant wavelength.

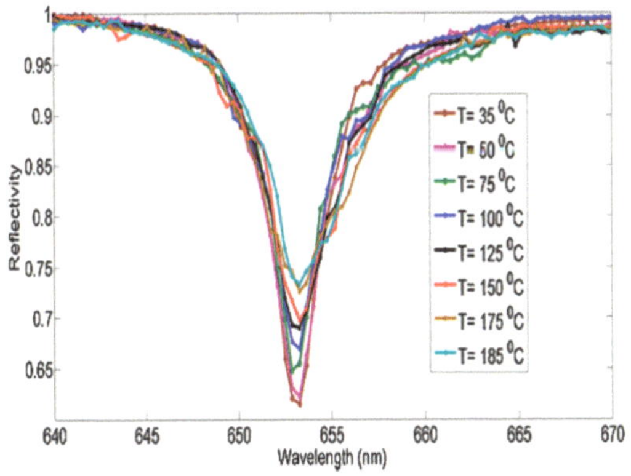

Figure 16. Measured reflection spectrum of "DBR + 30 nm Ag film" at different temperatures [95].

The reflection dip in the DBR band gap due to TP-mode excitation undergoes a change depending on the temperature. The measured temperature sensitivity of the structure is $7.8 \times 10^{-4}/°$ and covers the temperature range from 35 to 185 °C. The mechanisms of the influence of temperature on the TP spectral parameters are discussed. It is believed that the efficiency and sensitivity of the structure can be improved by using a metal, such as aluminum, which has a stronger temperature dependence of the refractive index.

In [96], sensors based on the temperature displacement of the TP resonance i near-infrared region are presented. The high sensitivity of 1.10 nm/K is due to the refra index change conditioned by the pyroelectric effect in the lithium niobate material, v is used as a one-dimensional photonic crystal. A polarization analysis of TP at th Ta_2O_5/SiO_2 bilayer system interface has been investigated for temperature sensing ir TP modes with different polarizations (TE and TM) are found to have different sensi characteristics that can be used depending on the desired application. In [98], the p multilayered photonic–plasmonic structure is developed, which supports OTSs o interface between metal and dielectric materials. The coupling of incident light t OTS results in light absorption within one of several narrow frequency bands, w is accompanied by a singular behavior of the phase of the electromagnetic field. demonstrated theoretically and experimentally how such phase changes can improv accuracy of temperature measurements by more than an order of magnitude.

A sensor based on a coupled hybrid mode TP is proposed for the detection of dif blood group components in [99]. The appearance of OTS in a three-dimensional pho crystal of an inverse opal located on a flat metal substrate was demonstrated in [100] it was shown that this structure can be an attractive sensing tool.

3.2. *Lasing and Amplification of Light in Hybrid Tamm Structures*

Recently, lasing in TP-based structures have been demonstrated [101–104]. In lasers, the role of the resonant cavity of conventional lasers is performed by the TP itself. This mode provides a stronger field confinement at the interface of the meta DBR, allowing the active layer to be thinner and achieve a lower operation thres In [101], a TP-laser was realized using an InGaInAs quantum well, which was embe in a TP structure consisting of a Bragg reflector formed by a stack of 40 AlAs/$Ga_{0.95}$ As pairs with a 45 nm silver layer on the top. In the optical range, a superlinear inc in emission was observed under the influence of pumping, indicating a laser effect. emission was experimentally demonstrated in [102] for Tamm structures with TP m confined under metallic microdisks. The polarized and spatially localized emissio Tamm laser from an anisotropic three-dimensional confinement of TP modes at the inte between DBR and a silver film is obtained in [103]. The spatial confinement is achiev patterning microrectangles in the top metallic layer. In result, the fundamental con Tamm mode splits into two modes, which are orthogonally polarized along the two of the structure. This weak splitting allows for the existence of a highly linearly pola laser emission at ~850 nm. The ZnO-based TP ultraviolet laser is realized in [104 lasing at a wavelength of 373 nm is observed under optical pumping. Relevant l characteristics such as the threshold energy, linewidth, and angular dispersion curv established and verified.

In addition to quantum walls, quantum dots are also introduced into TP tures [105]. In this case, the emission of single photons from the quantum dots is co to the confined TP mode, such that the spontaneous emission in the system is enha The possibility of controlling laser generation in a hybrid structure consisting of a Ag a layer of a nematic LC doped with a light-absorbing dye, and a DBR with a rectan refractive index profile is modeled in [106]. In [107], the efficiency of a third harm generation was studied in the metal/dielectric/photonic crystal multilayer structure shown that the third harmonic signal can be amplified by a factor of 16 by introduc monolayer of WS_2 into a dielectric spacer, which makes it possible to improve the nonl optical response in multilayer photonic microstructures.

3.3. Selective Thermal Emitters by Confined Tamm Plasmons

In a metal-DBR structure, TPs can be thermally excited and then produce strong thermal radiation at the resonant wavelength. An ultra-sharp emission peak in the mid-IR region was obtained in [108] for a TP structure, which comprises a DBR, consisting of Si and SiO_2 layers, and an adjacent Al film. Figure 17a shows the measured reflectance in the case of TP with the wavelength 4 μm, which has a dip value of <0.2 and a bandwidth of 110 nm. The measured emissivity spectrum using a heating current from 0.6 to 0.85 A (temperature ranging from ~115 to ~270 °C) is shown in Figure 17b. As the heating current increased, the emissivity peak shifted to the red by ~60 nm.

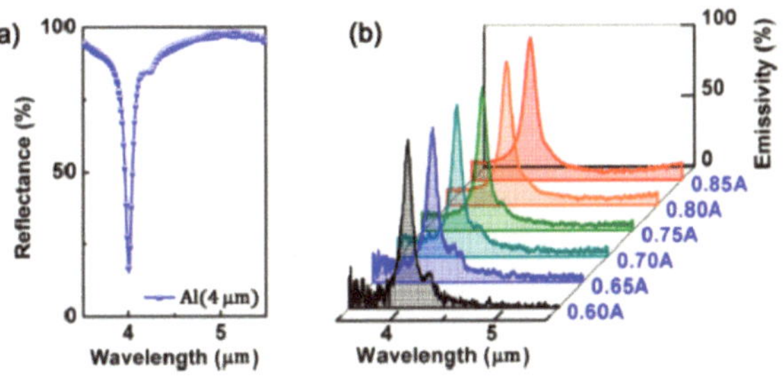

Figure 17. TP in the Al-DBR (Si/SiO_2) structure near a wavelength of 4 μm: (**a**) measured reflectance spectrum at room temperature, and (**b**) emissivity spectra where the heating current varies from 0.6 to 0.85 A [108].

High-intensity and narrowband thermal emission is demonstrated in [109] for a TP structure consisting of a DBR having a thick layer adjacent to the metallic film. A wide range of emission wavelengths (from 4.4 to 5.7 μm) is experimentally realized by adjusting the last DBR-layer thickness. The tunability of the thermal emission by only changing the DBR band gap was experimentally achieved in [110]. Various designs and modifications of TP selective thermal emitters have been proposed and discussed in [111–114].

3.4. Perfect Absorber Supported by Optical Tamm States

A significant number of works consider structures with TPs as the basis for obtaining perfect absorbers of electromagnetic radiation at selected wavelengths. Ref. [115] theoretically and experimentally investigated the perfect optical absorptance (in the visible region) of a photonic heterostructure constructed from a TiO_2/SiO_2 photonic crystal and a thick Ag film. Based on the measured experimental results, the three experimental structures achieved over 90% absorption at wavelengths of 489 nm, 604 nm, and 675 nm. Single and multiple bands near-perfect absorption were also realized in a similar TP structure, but with a thin Cu film [116].

An absorber based on a two-dimensional plasmonic metal–dielectric–metal waveguide is proposed in [117]. An air core is sandwiched by upper and lower semi-infinite metallic claddings, and a thin metallic layer followed by a dielectric photonic crystal is inserted in it (Figure 18). A photonic crystal consists of N periodically changing dielectric layers A and B. When a TM-polarized light wave is incident on the waveguide, as shown in Figure 18, a dip appears in the reflection spectrum in the form of zero reflection at a wavelength of 1550 nm. This dip arises due to the excitation of the OTS at the boundary between the thin metallic layer and the photonic crystal, and is spectrally located in the band gap of the photonic crystal. Since the OTS occurs in the band gap, the reflection dip can be tuned by varying the band gap.

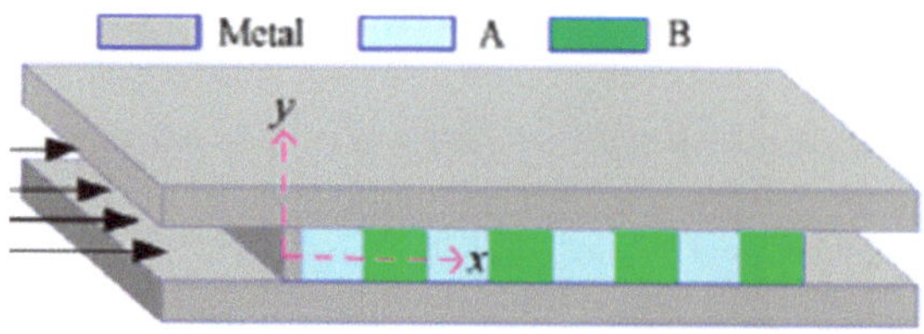

Figure 18. Scheme of the proposed near-perfect absorber: in the air core of a two-dimen metal-dielectric-metal waveguide there is a photonic crystal with N periodically varying die layers A (TiO_2) and B ($PSiO_2$) adjacent to a thin metal layer [117].

An effective method for achieving perfect absorption can be obtained by us system "graphene–one-dimensional DBR" [118]. The DBR is composed of two kir media layers, poly 4-methyl pentene-1 and SiO_2. Due to the metallic properties of grap the OTS appears in this structure in the THz range. When a layer of graphene is plac the top of the structure, the absorption peak appears at 0.956 THz and almost reaches with the full width at half-maximum about 0.025 THz. In the structure, which cons an ultrathin metal film coated on DBR and a graphene sheet in a silica spacer, stro confinement increases the light absorption in a monoatomic graphene layer by more 30 times [119].

3.5. Applications of Tamm Plasmon-Liquid Crystal Devices

The theoretical works discussed in Section 2 show the possibility of obtainin in hybrid systems containing LCs, and also establish a dependence of the TP sp parameters on the LC parameters. The obtained dependencies create a theoretical bas the use of LCs as an active element in TP devices.

In [120], the design of an LC-active TP device with a controlled LC phase and thickness was demonstrated to study the influence of LC parameters on the optical c teristics of the TP. The tunable TP device contains a thin nematic LC layer between a film and a photonic crystal (Figure 19).

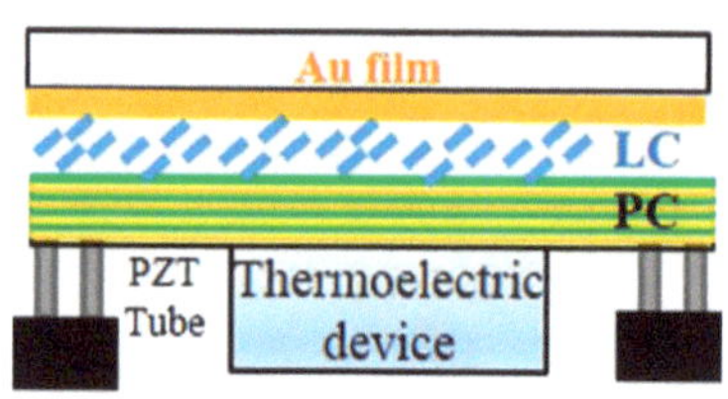

Figure 19. Scheme of an LC-active TP device with both a controllable LC phase and layer thic by using a thermoelectric device and piezoelectric tube (PZT), respectively [120].

Figure 20 illustrates the measured reflectance spectrum of the TP device unde different phases of LC 5CB, nematic and isotropic ($n_e \sim 1.68$, $n_o \sim 1.53$ at T ~ 30 °C λ = 633 nm, $n_{is} \sim 1.58$, T ~ 35 °C [121]).

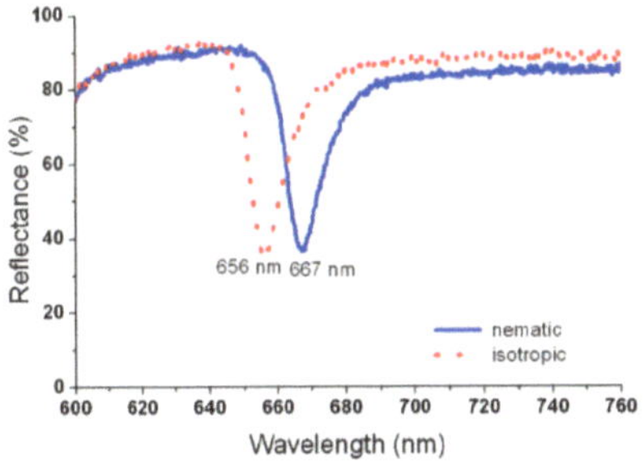

Figure 20. Reflectance spectra of TP device with an LC layer switched between the nematic (T ~ 30 °C) and isotropic (T ~ 35 °C) phases [120].

The spectral position of the reflection dip caused by the TP (λ_{TP}) can be also controlled by changing the LC layer thickness manually or electrically by piezoelectric (PZT) tubes. In [122], three different thicknesses of the LC layer were studied, corresponding to three λ_{TP}, if the LC is in the isotropic phase. It was found that the change of the LC layer thickness by about 140 nm produces a resonance shift of up to 91 nm, depending on the initial LC layer thickness and LC refractive index. The effect of the nematic LC layer thickness on the reflectance of the TP structure, if λ_{TP} is in the THz region, was studied in [62] (see Figure 21).

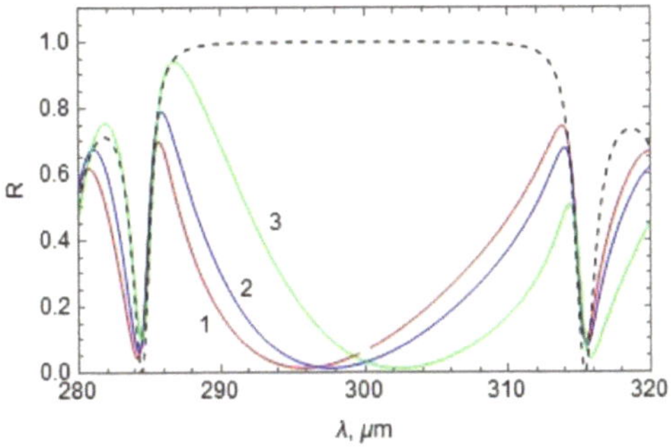

Figure 21. Influence of the nematic LC thickness d on the reflection coefficient of the TP structure in THz region: d(μm) = 1 (1), 5 (2), 15 (3) [62].

As mentioned above, TP-based devices can gain controllability from LC materials. However, TP devices with LC can be used for studying the anisotropy of solid films, such as alignment films in LC devices (LCDs). Anisotropy of a thin film can be created by a wide variety of methods. For conventional LCDs, the LC molecules are anchored on an anisotropic polyimide (PI) alignment film with a specific orientation and pretilt angle. The rubbing process on the PI film is adopted in the LCDs industry and it causes the physic-chemical anisotropy of the PI film with birefringence Δn in the azimuthal direction [123,124]. The inspection of the rubbing-induced anisotropy axis is crucial, which influences the performance of the LCDs. Inspecting the rubbing-induced anisotropy axis of PI film with a resolution better than 0.1 degree in the azimuthal angle is demanded for many LCDs end products. In [120,125], the lab-scale apparatus is demonstrated for measuring reflectance spectra of the TP device that allows one to determine the anisotropy axis of thin PI film (Figure 22).

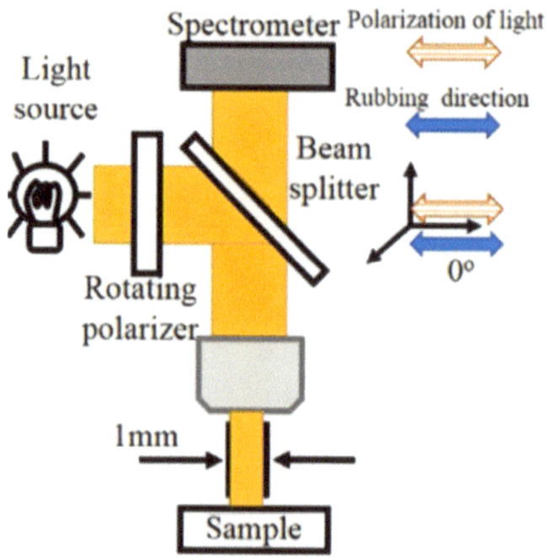

Figure 22. Apparatus for measuring reflectance spectra of the TP devices [125].

The reflectance spectra of the TP device at different azimuthal angles are meas by a spectrometer via the optical microscope. A white light source illuminates th sample from the metal side. An aluminum mirror is used for the reflectance reference polarization direction of the incident light through the optical microscope is controll rotating the polarizer for measuring the TP wavelength λ_{TP} at different azimuthal ang

A thin silverfilm~40 nm is deposited on the rubbed PI film. The polarization dire of the incident light through the optical microscope is rotated every 18 degrees to o the reflection spectra of TP devices at different φ. The TP can be generated by TE o waves, so it is easy to create the TP at various φ by rotating the incident polarizer experimental results of reflectance spectra at different incident polarization directic are shown in Figure 23.

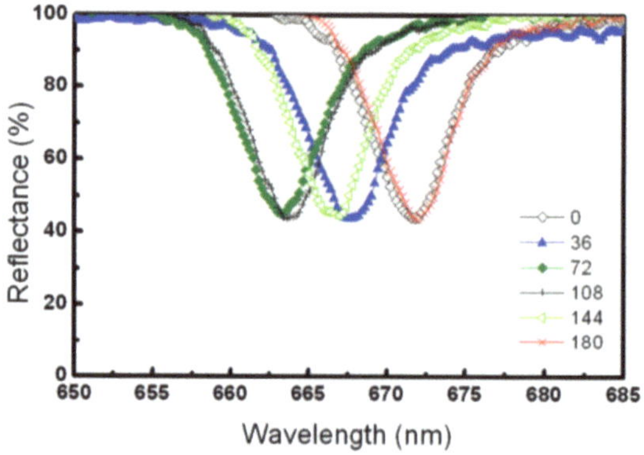

Figure 23. The experimental results of reflectance spectra of the TPP device for different inc polarization directions φ [120].

The initial azimuthal angle φ = 0 degree can be arbitrarily chosen, and it is assi as the rubbing direction in Figure 23. The experimental results of resonance λ_{TP} at diff φ are summarized in Figure 24, and it shows that the λ_{TP} changes with the inci polarization direction. The created TP resonance λ_{TP} is corresponding to the refra index of PI film at the specific φ. It has been shown that the rubbing process on PI film create an anisotropy axis and produce the birefringence Δn; therefore, the λ_{TP} is deper on φ. For PI materials, the anisotropy axis is parallel to the rubbing direction, and it out to be the alignment direction of the LC molecules on the PI film, called the easy axis. The anisotropy axis is easily identified as $\varphi \sim 0$ or $\varphi \sim 180$ degrees, as shown in Figure 24.

Using the present scheme, the resolution in azimuthal angle $\delta\varphi$ for determining the anisotropy axis can be estimated as 0.1. The resolution $\delta\varphi$ can be further improved by fine-rotating the polarizer several times around the anisotropy axis for improving the signal-to-noise ratio, or using a spectrometer with a higher accuracy than 0.01 nm.

Ref. [125] used a TP-resonant device shown in Figure 22 to study reflectance spectra of a sample comprising an LC layer located between a copper film and a TiO_2/SiO_2 DBR (see Figure 25).

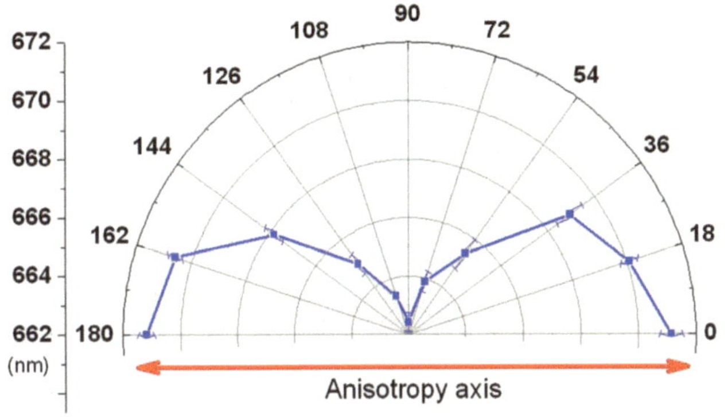

Figure 24. The experimental results of resonance λ_{TP} for different incident polarization directions φ [120].

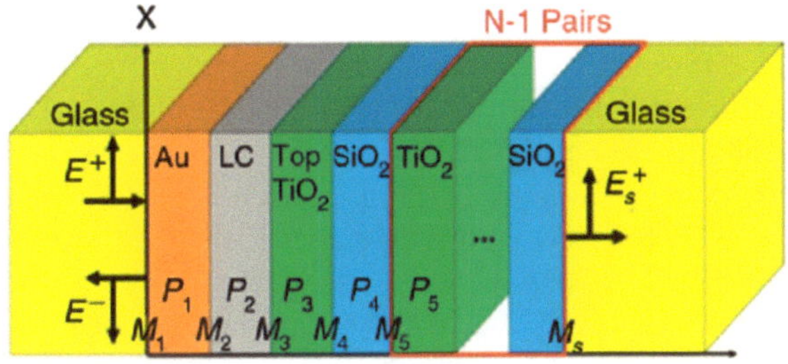

Figure 25. A sample comprising an LC layer sandwiched between a Au-covered glass and a TiO_2/SiO_2 DBR.

The DBR comprises eight paired layers of TiO_2 and SiO_2 deposited on a B270 glass substrate. The thickness of the TiO_2 top layer is about 205 nm. This layer undergoes buffing with a nylon cloth to achieve a homogeneous alignment of LC molecules. A thin gold film is applied to a depth of approximately 30 nm on a glass substrate using a thermal evaporator. The Au-coated substrate and the DBR are then assembled with a small gap between them to form a tunable TP device. In each run of the experiment, the width of the gap is varied, with the aim of optimizing the measured resonance wavelengths. The gap is then filled with LC molecules (5CB, $n_e \sim 1.68$, $n_o \sim 1.53$ at T ~ 30 °C, $n_{iso} \sim 1.58$, T_c ~35 °C).

The resonance wavelength of the TP device is tuned by changing the phase of the 5CB LC between the nematic phase and the isotropic phase. It is realized by changing the temperature of 5CB LC via a thermoelectric device attached to the bottom of the TP device. The voltage control of the thermoelectric device makes it possible to switch the LC layer in the TP device between the nematic phase (approximately 30 °C) and the isotropic phase (about 36 °C)

Figure 26 presents polarization-dependent measurements of the TP device, whe LC layer is operated in nematic phase and the cell gap is set at approximately 180 nn resonance wavelengths are 651.4 and 667.6 nm when the LC director angle values θ a and 0°, respectively. Therefore, the TP resonance can be tuned from 668 to 651 nm s by rotating the LC molecules from 0° to 90°.

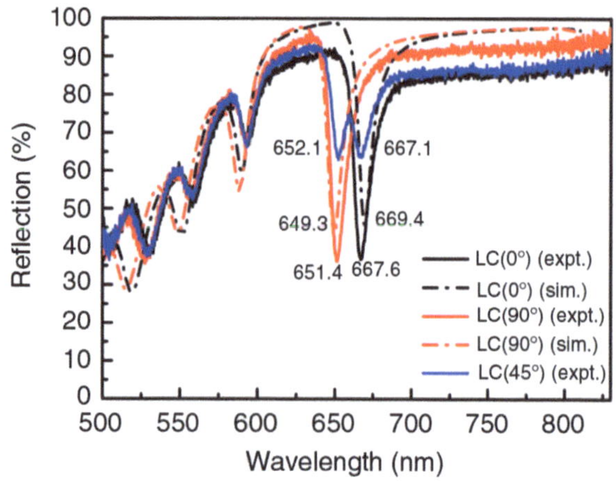

Figure 26. Reflectance spectra of the TP devices measured at three angles θ, where the LC lay the nematic phase.

The TP resonance wavelength can be tuned by changing the phase of the LC changing the LC gap. Figure 27 shows two resonance wavelengths related to two dif LC gaps.

Recently, it was experimentally shown that TPs in the near-IR range can be exci the interface between a DBR and a nanopatterned metal film acting as a non-diffra optical metasurface [126]. The pattern of a square array of 500 nm Ag disks with the p of 600 nm is fabricated by a focused ion beam (see Figure 28).

The metasurface, due to its discrete structure, provided closer external access spatially limited weak fields of the Tamm plasmon. Placing a dielectric such as I (birefringence = 0.23) in direct contact with the outside of the metasurface resulte red shift of the TP wavelength by as much as 35 nm, while there was no spectral upon LC deposition onto a continuous metallic film. In addition, the TP wavelength c tuned within a 10 nm range by changing the LC refractive index above the metasu with an externally applied electric field. One can improve the efficiency of the tuni employing LCs with a higher optical birefringence (some of the currently availabl exhibit birefringence as high as 0.4–0.8 [127,128]). The demonstrated ability to contr spectral location of TPs opens a feasible path to exploiting this resonant optical sta optical switching, the enhancement of optical non-linearity, lasing, and light emissic

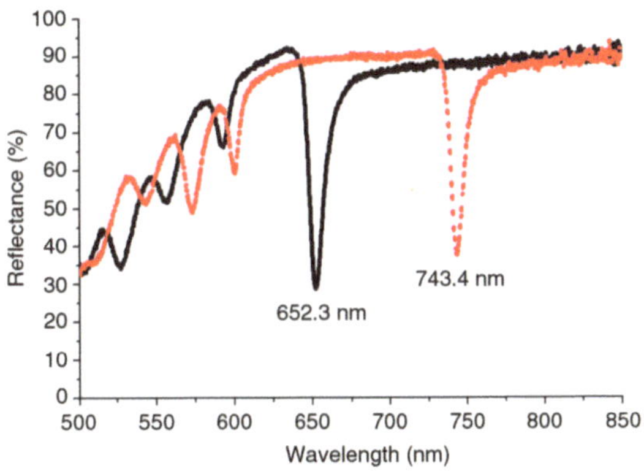

Figure 27. Reflectance spectra of the TP devices at two different LC gaps, wherein the LC layer is in the isotropic phase.

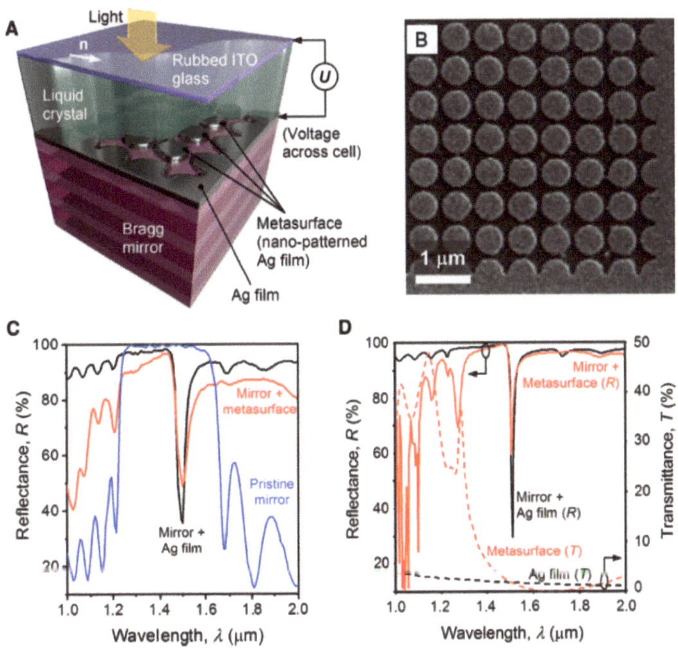

Figure 28. Design of a TP at the interface with a metasurface [123]. (**A**) Schematic of the structure used in experiments. The white arrow indicates the direction of rubbing, which controlled the LC alignment in the cell. (**B**) Scanning electron micrograph of a fragment of the metasurface fabricated on top of a DBR. (**C**) Experimental reflectivity spectra of the DBR acquired while it was in a pristine state (blue), after it was interfaced with a 37 nm-thick continuous Ag film (black), and after the Ag film was nanopatterned to become a metasurface (red). (**D**) Calculated spectra of the continuous Ag film and metasurface. Solid curves show the reflectivity of the film (black) and metasurface (red) placed atop a Bragg mirror, while dashed curves show the transmission of the film (black) and metasurface (red) residing on a niobium pentoxide substrate.

Author Contributions: Conceptualization, V.Y.R.; formal analysis, M.E.M.; investigation, I.P.P. and V.Y.R.; writing—original draft preparation, I.P.P.; writing—review and editing, D.R.E. and T.J.B.; and supervision, D.R.E. All authors have read and agreed to the published version of the manuscript.

Funding: This research received no external funding.

Data Availability Statement: Data are contained within the article.

Acknowledgments: The authors are grateful to Mariacristina Rumi (AFRL) for helpful discus
V.Y.R. and I.P.P. thank Azimuth Corporation (USA) for the support received.

Conflicts of Interest: The authors declare no conflicts of interest.

References

1. Boardman, A.D. *Electromagnetic Surface Modes*; Wiley: Chicester, UK, 1982.
2. Polo, J.A., Jr.; Mackay, T.G.; Lakhtakia, A. *Electromagnetic Surface Waves: A Modern Perspective*; Elsevier: Waltham, MA, USA
3. D'yakonov, M.I. New type of electromagnetic wave propagating at an interface. *Sov. Phys. JETP* **1988**, *67*, 714–716.
4. Takayama, O.; Crasovan, L.C.; Johansen, S.K.; Mihalache, D.; Artigas, D.; Torner, L. Dyakonov surface waves: A re *Electromagnetics* **2008**, *28*, 126–145. [CrossRef]
5. Mackay, T.G.; Zhou, C.; Lakhtakia, A. Dyakonov–Voigt surface waves. *Proc. R. Soc. A* **2019**, *475*, 20190317. [CrossRef] [Pub
6. Maier, S.A. *Plasmonics: Fundamentals and Applications*; Springer: New York, NY, USA, 2007.
7. Pohl, D.W. *Near-Field Optics and Surface Plasmon Polaritons*; Kawata, S., Ed.; Springer: Berlin, Germany, 2001.
8. Ponath, H.E.; Stegeman, G.I. *Nonlinear Surface Electromagnetic Phenomena*; North Holland Publisher: Amsterdam, The Nether 1991.
9. Engheta, N.; Pelet, P. Surface waves in chiral layers. *Opt. Lett.* **1991**, *16*, 723–725. [CrossRef]
10. Lu, L.; Joannopoulos, J.D.; Solja, M. Topological photonics. *Nat. Photonics* **2014**, *8*, 821–829. [CrossRef]
11. Deng, J.; Dong, H.; Zhang, C.; Wu, Y.; Yuan, J.; Zhu, X.; Jin, F.; Li, H.; Wang, Z.; Cai, H.; et al. Observing the quantum topolc light. *Science* **2022**, *378*, 6623. [CrossRef]
12. Ye, F.; Merlo, J.M.; Burns, M.J.; Naughton, M.J. Optical and electrical mappings of surface plasmon cavity modes. *Nanopho* **2014**, *3*, 33–49. [CrossRef]
13. Luo, J.; Xu, P.; Gao, L. Controllable switching behavior of optical Tamm state based on nematic liquid crystal. *Solid State Co* **2011**, *151*, 993–995. [CrossRef]
14. Kavokin, A.V.; Shelykh, I.A.; Malpuech, G. Lossless interface modes at the boundary between two periodic dielectric struc *Phys. Rev. B* **2005**, *72*, 233102. [CrossRef]
15. Vinogradov, A.P.; Dorofeenko, A.V.; Erokhin, S.G.; Inoue, M.; Lisyansky, A.A.; Merzlikin, A.M.; Granovsky, A.B. Surface peculiarities in one-dimensional photonic crystal interfaces. *Phys. Rev. B* **2006**, *74*, 045128. [CrossRef]
16. Goto, T.; Dorofeenko, A.V.; Merzlikin, A.M.; Baryshev, A.V.; Vinogradov, A.P.; Inoue, M.; Lisyanskii, A.A.; Granovsky, A.B. O Tamm states in one-dimensional magnetophotonic structures. *Phys. Rev. Lett.* **2008**, *101*, 113902. [CrossRef] [PubMed]
17. Vetrov, S.Y.; Bikbaev, R.G.; Timofeev, I.V. Optical Tamm states at the interface between a photonic crystal and a nanocom with resonance dispersion. *J. Exp. Theor. Phys.* **2013**, *117*, 988–998. [CrossRef]
18. Kaliteevski, M.; Iorsh, I.; Brand, S.; Abram, R.A.; Chamberlain, J.M.; Kavokin, A.V.; Shelykh, I.A. Tamm plasmon-polar Possible electromagnetic states at the interface of a metal and a dielectric Bragg mirror. *Phys. Rev. B* **2007**, *76*, 165415. [Cros
19. Shelykh, I.A.; Kaliteevski, M.A.; Kavokin, A.V.; Brand, S.; Abram, R.A.; Chamberlain, J.M.; Malpuech, G. Interface photonic at the boundary between a metal and a dielectric Bragg mirror. *Phys. Stat. Solidi A* **2007**, *204*, 522–525. [CrossRef]
20. Sasin, M.E.; Seisyan, R.P.; Kaliteevski, M.A.; Brand, S.; Abram, R.A.; Chamberlain, J.M.; Egorov, A.Y.; Vasil'ev, A.P.; Mikhrir Kavokin, A.V. Tamm plasmon polaritons: Slow and spatially compact light. *Appl. Phys. Lett.* **2008**, *92*, 251112. [CrossRef]
21. Afinogenov, B.I.; Bessonov, V.O.; Nikulin, A.A.; Fedyanin, A.A. Observation of hybrid state of Tamm and surface pla polaritons in one-dimensional photonic crystals. *Appl. Phys. Lett.* **2013**, *103*, 061112. [CrossRef]
22. Vyunishev, A.M.; Bikbaev, R.G.; Svyakhovskiy, S.E.; Timofeev, I.V.; Pankin, P.S.; Evlashin, S.A.; Vetrov, S.A.; Myslivets, Arkhipkin, V.G. Broadband Tamm plasmon polariton. *J. Opt. Soc. Am. B* **2019**, *36*, 2299. [CrossRef]
23. Tamm, I. O vozmozhnoi sviazi elektronov na poverkhnostiakh kristalla. *Zh. Eksp. Teor. Fiz.* **1933**, *3*, 34. (In Russian)
24. Kavokin, A.V.; Kaliteevski, M.A. Excitionic light reflection and absorption in semiconductor microcavities at oblique inci *Solid State Commun.* **1995**, *95*, 859–862. [CrossRef]
25. Yeh, P. *Optical Waves in Layered Media*; Wiley: New York, NY, USA, 1988.
26. Adams, M.; Cemlyn, B.; Henning, I.; Parker, M.; Hrbord, E.; Oulton, R. Model for confined Tamm plasmon devices. *J. Op Am. B* **2019**, *36*, 125–130. [CrossRef]
27. Joannopoulos, J.D.; Johnson, S.G.; Winn, J.N.; Meade, R.D. *Photonic Crystals: Molding the Flow of Light*, 2nd ed.; Princeton Univ Press: Princeton, NJ, USA, 2008.
28. Haus, H.A. *Waves and Fields in Optoelectronics*; Prentice-Hall: Englewood Cliffs, NJ, USA, 1984.
29. Auguié, B.; Bruchhausen, A.; Fainstein, A. Critical coupling to Tamm plasmons. *J. Opt.* **2015**, *17*, 035003. [CrossRef]
30. Bliokh, K.; Bliokh, Y.; Freilikher, V.; Savel'ev, S.; Nori, F. Colloquium: Unusual resonators: Plasmonics, metamaterials, and ra media. *Rev. Mod. Phys.* **2008**, *80*, 1201–1213. [CrossRef]

Herminghaus, S.; Klopfleisch, M.; Schmidt, H.J. Attenuated total reflectance as a quantum interference phenomenon. *Opt. Lett.* **1994**, *19*, 293–295. [CrossRef] [PubMed]

Ru, E.C.L.; Etchegoin, P.G. *Principles of Surface Enhanced Raman Spectroscopy and Related Plasmonic Effects*; Elsevier: Amsterdam, The Netherlands, 2009.

Amirand, A.; Vukusic, P. Elucidating the stop bands of structurally colored systems through recursion. *Am. J. Phys.* **2013**, *81*, 253–257. [CrossRef]

Chang, C.-Y.; Chen, Y.-H.; Tsai, Y.-L.; Kuo, H.-C.; Chen, K.-P. Tunability and optimization of coupling efficiency in Tamm plasmon modes. *IEEE J. Sel. Top. Quantum Electron.* **2015**, *21*, 4600206. [CrossRef]

Mathmann, B.; Haidar, Q.; Talbi, A.; Tiercelin, N.; Mir, A.; Boudouti, E.H.E.; Djafari-Rouhani, B.; Lévêque, G.; Akjouj, A.; Dusch, Y. Optical Tamm states in 2d nanostructured magnetophotonic structures. *Plasmonics* **2024**. [CrossRef]

Fang, Y.-T.; Ni, Y.-X.; He, H.-Q.; Hu, J.-X. Effect of hybrid state of surface plasmon–polaritons, magnetic defect mode and optical Tamm state on nonreciprocal propagation. *Opt. Commun.* **2014**, *320*, 99–104. [CrossRef]

Reshetnyak, V.Y.; Pinkevych, I.P.; Bunning, T.J.; Evans, D.R. Influence of rugate filters on the spectral manifestation of Tamm plasmon-polaritons. *Materials* **2021**, *14*, 1282. [CrossRef]

Bovard, B.G. Rugate filter theory: An overview. *Appl. Opt.* **1993**, *32*, 5427–5442. [CrossRef]

Southwell, W.H.; Hall, R.L. Rugate filter sidelobe suppression using quintic and rugated quintic matching layers. *Appl. Opt.* **1989**, *28*, 2949–2951. [CrossRef]

Lorenzo, E.; Oton, C.J.; Capuj, N.E.; Ghulinyan, M.; Navarro-Urrios, D.; Gaburro, Z.; Pavesi, L. Porous silicon-based rugate filters. *Appl. Opt.* **2005**, *44*, 5415–5421. [CrossRef] [PubMed]

Jupé, M.; Lappschies, M.; Jensen, L.; Starke, K.; Ristau, D. Laser-induced damage in gradual index layers and rugate filters. *Proc. SPIE* **2006**, *6403*, 640311. [CrossRef]

Kogelnik, H. Coupled wave theory for thick hologram gratings. *Bell Syst. Tech. J.* **1969**, *48*, 2909–2947. [CrossRef]

Yariv, A.; Yeh, P. *Optical Waves in Crystals: Propagation and Control of Laser Radiation*; John Wiley & Sons: Hoboken, NJ, USA, 2003; pp. 177–201.

Karpov, S.Y.; Stolyarov, S.N. Propagation and transformation of electromagnetic waves in one-dimensional periodic structures. *Phys. Uspekhi* **1993**, *36*, 1–22. [CrossRef]

Rakić, A.D.; Djurišic, A.B.; Elazar, J.M.; Majewski, M.L. Optical properties of metallic films for vertical-cavity optoelectronic devices. *Appl. Opt.* **1998**, *37*, 5271–5283. [CrossRef]

Castellanos-Gomez, A.; Vicarelli, L.; Prada, E.; Island, J.O.; Narasimha-Acharya, K.L.; Blanter, S.I.; Groenendijk, D.J.; Buscema, M.; Steele, G.A.; Alvarez, J.V.; et al. Isolation and characterization of few-layer black phosphorus. *2D Materials* **2014**, *1*, 025001. [CrossRef]

Xia, F.; Wang, H.; Jia, Y. Rediscovering black phosphorus as an anisotropic layered material for optoelectronics and electronics. *Nat. Commun.* **2014**, *5*, 4458. [CrossRef]

Wang, X.; Jones, A.M.; Seyler, K.L.; Tran, V.; Jia, Y.; Zhao, H.; Wang, H.; Yang, L.; Xu, X.; Xia, F. Highly anisotropic and robust excitons in monolayer black phosphorus. *Nat. Nanotechnol.* **2015**, *10*, 517–521. [CrossRef]

Lan, S.; Rodrigues, S.; Kang, L.; Cai, W. Visualizing optical phase anisotropy in black phosphorus. *ACS Photonics* **2016**, *3*, 1176–1181. [CrossRef]

Liu, H.; Neal, A.T.; Zhu, Z.; Luo, Z.; Xu, X.; Tománek, D.; Ye, P.D. Phosphorene: An unexplored 2d semiconductor with a high hole mobility. *ACS Nano* **2014**, *8*, 4033–4041. [CrossRef]

Mittendorff, M.; Suess, R.J.L.; Murphy, T.E. Optical gating of black phosphorus for terahertz detection. *Nano Lett.* **2017**, *17*, 5811–5816. [CrossRef] [PubMed]

Doha, M.H.; Batista, J.I.S.; Rawwagah, A.F.; Thompson, J.P.; Fereidouni, A.; Watanabe, K.; Taniguchi, T.; El-Shenawee, M.; Churchill, H. Integration of multi-layer black phosphorus into photo-conductive antennas for THz emission. *J. Appl. Phys.* **2020**, *128*, 063104. [CrossRef]

Viti, L.; Hu, J.; Coquillat, D.; Knap, W.; Tredicucci, A.; Politano, A.; Vitiello, M.S. Black phosphorus terahertz photodetectors. *Adv. Mater.* **2015**, *27*, 5567–5572. [CrossRef]

Guo, W.; Dong, Z.; Xu, Y.; Liu, C.; Wei, D.; Zhang, L.; Shi, X.; Guo, C.; Xu, H.; Chen, G.; et al. Sensitive terahertz detection and imaging driven by the photo thermoelectric effect in ultrashort-channel black phosphorus devices. *Adv. Sci.* **2020**, *7*, 1902699. [CrossRef]

Federici, J.; Moeller, L. Review of terahertz and subterahertz wireless communications. *J. Appl. Phys.* **2010**, *107*, 111101. [CrossRef]

Jepsen, P.U.; Cooke, D.G.; Koch, M. Terahertz spectroscopy and imaging–modern techniques and applications. *Laser Photonics Rev.* **2011**, *5*, 124–166. [CrossRef]

Tonouchi, M. Cutting-edge terahertz technology. *Nat. Photonics* **2007**, *1*, 97–105. [CrossRef]

Low, T.; Roldan, R.; Wang, H.; Xia, F.; Avouris, P.; Moreno, L.M.; Guinea, F. Plasmons and screening in monolayer and multilayer black phosphorus. *Phys. Rev. Lett.* **2014**, *113*, 106802. [CrossRef]

59. Liu, Z.; Aydin, K. Localized surface plasmons in nanostructured monolayer black phosphorus. *Nano Lett.* **2016**, *16*, 3457 [CrossRef]
60. Jin, F.; Roldán, R.; Katsnelson, M.I.; Yuan, S. Plasmonics in strained monolayer black phosphorus. *Phys. Rev. B* **2015**, *92*, [CrossRef]
61. Pan, J.; Zhu, W.; Zheng, H.; Yu, J.; Chen, Y.; Guan, H.; Lu, H.; Zhong, Y.; Luo, Y.; Chen, Z. Exploiting black phosp based-Tamm plasmons in the terahertz region. *Opt. Express* **2020**, *28*, 13443. [CrossRef] [PubMed]
62. Reshetnyak, V.Y.; Pinkevych, I.P.; Bunning, T.J.; McConney, M.E.; Slagle, J.E.; Evans, D.R. Tunable spectral manifestation of plasmon-polaritons in a hybrid structure with 2d black phosphorus in the terahertz range. *Liq. Cryst.* **2023**, *50*, 36–44. [Cro
63. De Gennes, P.G.; Prost, J. *The Physics of Liquid Crystals*; Clarendon Press: Oxford, UK, 1993.
64. Caputo, R.; De Sio, L.; Veltri, A.; Umeton, C.; Sukhov, A.V. Development of a new kind of switchable holographic grating m liquid-crystal films separated by slices of polymeric material. *Opt. Lett.* **2004**, *29*, 1261–1263. [CrossRef] [PubMed]
65. Holmstrom, S.A.; Natarajan, L.V.; Tondiglia, V.P.; Sutherland, R.L.; Bunning, T.J. Mechanical tuning of holographic po dispersed liquid crystal reflection gratings. *Appl. Phys. Lett.* **2004**, *85*, 1949–1951. [CrossRef]
66. Jazbinšek, M.; Drevensek-Olenik, I.; Zgonik, M.; Fontecchio, A.K.; Crawford, G.P. Characterization of holographic pc dispersed liquid crystal transmission gratings. *J. Appl. Phys.* **2001**, *90*, 3831. [CrossRef]
67. Lucchetta, D.E.; Criante, L.; Simoni, F. Optical characterization of polymer dispersed liquid crystals for holographic recor *Appl. Phys.* **2003**, *93*, 9669–9674. [CrossRef]
68. Liu, K.; Xu, H.; Hu, H.; Gan, Q.; Cartwright, A.N. One-step fabrication of graded rainbow-colored holographic photopc reflection grating. *Adv. Mater.* **2012**, *24*, 1604–1609. [CrossRef]
69. De Sio, L.; Lloyd, P.F.; Tabiryan, N.V.; Bunning, T.J. Hidden gratings in holographic liquid crystal polymer-dispersed liquid films. *ACS Appl. Mater. Interfaces* **2018**, *10*, 13107–13112. [CrossRef]
70. Reshetnyak, V.Y.; Pinkevych, I.P.; Godman, N.P.; Bunning, T.J.; Evans, D.R. Optical Tamm states in a hybrid structure holographic polymer-liquid crystal grating. *Liq. Cryst.* **2023**, *50*, 45–53. [CrossRef]
71. Reshetnyak, V.Y.; Pinkevych, I.P.; Bunning, T.J.; Evans, D.R. Excitation and tuning of optical Tamm states in a hybrid str with a metal film adjacent to a four-layer polymer-liquid crystal stack. *Photonics* **2024**, *11*, 278. [CrossRef]
72. Berreman, D.W. Optics in stratified and anisotropic media: 4 × 4-matrix formulation. *J. Opt. Soc. Am.* **1972**, *62*, 502–510. [Cro
73. Pyatnov, M.V.; Vetrov, S.Y.; Timofeev, I.V. Localised optical states in a structure formed by two oppositely handed chol liquid crystal layers and a metal. *Liq. Cryst.* **2017**, *44*, 674–678. [CrossRef]
74. Zhou, H.; Yang, G.; Wang, K.; Long, H.; Lu, P. Multiple optical Tamm states at a metal-dielectric mirror interface. *Opt. Lett* *35*, 4112. [CrossRef] [PubMed]
75. Fei, Y.; Liu, Y.; Dong, D.; Gao, K.; Ren, S.; Fan, Y. Multiple adjustable optical Tamm states in one-dimensional ph quasicrystals with predesigned band gaps. *Opt. Exp.* **2018**, *26*, 34872–34879. [CrossRef]
76. Reshetnyak, V.Y.; Pinkevych, I.P.; Bunning, T.J.; McConney, M.E.; Evans, D.R. Spectral manifestation of optical Tamm stat metal-cholesteric liquid crystals stack. *Phys. Rev. E* **2023**, *107*, 014702. [CrossRef]
77. Vetrov, S.Y.; Pyatnov, M.V.; Timofeev, I.V. Surface modes in photonic cholesteric liquid crystal—Phase plate-metal structur *Lett.* **2014**, *39*, 2743–2746. [CrossRef]
78. Vetrov, S.Y.; Pyatnov, M.V.; Timofeev, I.V. Spectral and polarization properties of a 'cholesteric liquid crystal-phase film structure. *J. Opt.* **2016**, *18*, 015103. [CrossRef]
79. Pyatnov, M.V.; Vetrov, S.Y.; Timofeev, I.V. Localized optical modes in a defect-containing liquid-crystal structure adjacent metal. *J. Opt. Soc. Am. B* **2017**, *34*, 2011–2017. [CrossRef]
80. Timofeev, I.V.; Pankin, P.S.; Vetrov, S.Y.; Arkhipkin, V.G.; Lee, W.; Zyryanov, V.Y. Chiral optical Tamm states: Temporal cc mode theory. *Crystals* **2017**, *7*, 113. [CrossRef]
81. Pyatnov, M.V.; Timofeev, I.V.; Vetrov, S.Y.; Rudakova, N.V. Coupled chiral optical Tamm states in cholesteric liquid cı *Photonics* **2018**, *5*, 30. [CrossRef]
82. Rudakova, N.V.; Timofeev, I.V.; Bikbaev, R.G.; Pyatnov, M.V.; Vetrov, S.Y.; Lee, W. Chiral optical Tamm states at the int between an all-dielectric polarization-preserving anisotropic mirror and a cholesteric liquid crystal. *Crystals* **2019**, [CrossRef]
83. Plum, E.; Zheludev, N.I. Chiral mirrors. *Appl. Phys. Lett.* **2015**, *106*, 221901. [CrossRef]
84. Ng, H.M.; Moustakas, T.D.; Chu, S.N.G. High reflectivity and broad bandwidth AlN/GaN distributed Bragg reflectors gro molecular-beam epitaxy. *Appl. Phys. Lett.* **2000**, *76*, 2818–2820. [CrossRef]
85. Tawara, T.; Gotoh, H.; Akasaka, T.; Kobayashi, N.; Saitoh, T. Low-threshold lasing of InGaN vertical-cavity surface-emitting with dielectric distributed Bragg reflectors. *Appl. Phys. Lett.* **2003**, *83*, 830–832. [CrossRef]
86. Mishkat-Ul-Masabih, S.; Luk, T.S.; Rishinaramangalam, A.; Monavarian, M.; Nami, M.; Feezell, D. Nanoporous distributed reflectors on free-standing nonpolar m-plane GaN. *Appl. Phys. Lett.* **2018**, *112*, 041109. [CrossRef]

Zhang, C.; Park, S.H.; Chen, D.; Lin, D.-W.; Xiong, W.; Kuo, H.-C.; Lin, C.-F.; Cao, H.; Han, J. Mesoporous GaN for photonic engineering—Highly reflective GaN mirrors as an example. *ACS Photonics* **2015**, *2*, 980–986. [CrossRef]
Lheureux, G.; Monavarian, M.; Anderson, R.; Decrescent, R.A.; Bellessa, J.; Symonds, C.; Schuller, J.A.; Speck, J.S.; Nakamura, S.; DenBaars, S.P. Tamm plasmons in metal/nanoporous GaN distributed Bragg reflector cavities for active and passive optoelectronics. *Opt. Express* **2020**, *28*, 17934. [CrossRef]
Juneau-Fecteau, A.; Fréchette, L.G. Tamm plasmon-polaritons in a metal coated porous silicon photonic crystal. *Opt. Mater. Express* **2018**, *8*, 2774. [CrossRef]
Vincent, G. Optical properties of porous silicon superlattices. *Appl. Phys. Lett.* **1994**, *64*, 2367–2369. [CrossRef]
Volk, J.; Grand, T.L.; Bársony, I.; Gombkötő, J.; Ramsden, J.J. Porous silicon multilayer stack for sensitive refractive index determination of pure solvents. *J. Phys. D Appl. Phys.* **2005**, *38*, 1313. [CrossRef]
Salonen, J.; Björkqvist, M.; Laine, E.; Niinistö, L. Effects of fabrication parameters on porous p+-type silicon morphology. *Phys. Status Solidi (A)* **2000**, *182*, 249–254. [CrossRef]
Zhang, W.L.; Wang, F.; Rao, Y.J.; Jiang, Y. Novel sensing concept based on optical Tamm plasmon. *Opt. Express* **2014**, *22*, 14524–14528. [CrossRef] [PubMed]
Auguié, B.; Fuertes, M.C.; Angelomé, P.C.; Abdala, N.L.; Soler Illia, G.J.A.A.; Fainstein, A. Tamm plasmon resonance in mesoporous multilayers: Toward a sensing application. *ACS Photonics* **2014**, *1*, 775–780. [CrossRef]
Kumar, S.; Maji, P.S.; Das, R. Tamm-plasmon resonance based temperature sensor in a Ta_2O_5 /SiO_2 based distributed Bragg reflector. *Sens. Actuators A* **2017**, *260*, 10–15. [CrossRef]
Ahmed, A.M.; Mehaney, A. Novel design of wide temperature ranges sensor based on Tamm state in a pyroelectric photonic crystal with high sensitivity. *Phys. E* **2021**, *125*, 114387. [CrossRef]
Maji, P.S.; Maity, A.R. Polarization modal analysis of Tamm plasmon at the metal-DBR interface for temperature sensing. *Opt. Quantum Electron.* **2022**, *54*, 349. [CrossRef]
Tsurimaki, Y.; Tong, J.; Boriskin, V.N.; Semenov, A.; Ayzatskiy, M.; Machekhin, Y.; Chen, G.; Boriskina, S. Topological engineering of interfacial optical Tamm states for highly sensitive near-singular-phase optical detection. *ACS Photonics* **2018**, *5*, 929–938. [CrossRef]
Maji, P.S.; Shukla, M.K.; Das, R. Blood component detection based on miniaturized self-referenced hybrid Tamm-plasmon-polariton sensor. *Sens. Actuators B* **2018**, *255*, 729–734. [CrossRef]
Mudi, R.; Carpentiero, A.; Bollani, M.; Barozzi, M.; Debnath, K.; Chiappini, A.; Bhaktha, B.N.S. Inverse Opal Optical Tamm State for Sensing Applications. *arXiv* **2024**, arXiv:2405.10701. [CrossRef]
Symonds, C.; Lemaıtre, A.; Senellart, P.; Jomaa, M.H.; Aberra Guebrou, S.; Homeyer, E.; Brucoli, G.; Bellessa, J. Lasing in a hybrid GaAs/silver Tamm structure. *Appl. Phys. Lett.* **2012**, *100*, 121122. [CrossRef]
Symonds, C.; Lheureux, G.; Hugonin, J.P.; Greffet, J.J.; Laverdant, J.; Brucoli, G.; Lemaitre, A.; Senellart, P.; Bellessa, J. Confined Tamm plasmon lasers. *Nano Lett.* **2013**, *13*, 3179–3184. [CrossRef] [PubMed]
Lheureux, G.; Azzini, S.; Symonds, C.; Senellart, P.; Lemaître, A.; Sauvan, C.; Hugonin, J.-P.; Greffet, J.-J.; Bellessa, J. Polarization-controlled confined Tamm plasmon lasers. *ACS Photonics* **2015**, *2*, 842–848. [CrossRef]
Xu, W.-H.; Chou, Y.-H.; Yang, Z.-Y.; Liu, Y.-Y.; Yu, M.-W.; Huang, C.-H.; Chang, C.-T.; Huang, C.-Y.; Lu, T.-C.; Lin, T.-R. Tamm Plasmon-polariton ultraviolet lasers. *Adv. Photonics Res.* **2022**, *3*, 2100120. [CrossRef]
Braun, T.; Baumann, V.; Iff, O.; Hofling, S.; Schneider, C.; Kamp, M. Enhanced single photon emission from positioned InP/GaInP quantum dots coupled to a confined Tamm-plasmon mode. *Appl. Phys. Lett.* **2015**, *106*, 041113. [CrossRef]
Yakovkin, I.I.; Ledney, M.F.; Reshetnyak, V.Y.; Pakamoryte, I.; Hands, P.J.W. Modeling of laser generation in a Fabry–Pérot-Tamm structure with a nematic liquid crystal layer. *J. Appl. Phys.* **2024**, *135*, 213102. [CrossRef]
Zhao, J.; Lu, H.; Zheng, J.; Li, D.; Zhang, Y.; Gan, X.; Zhao, J. Strong enhancement of third harmonic generation from a Tamm plasmon multilayer structure with WS_2. *Opt. Lett.* **2024**, *49*, 3130–3133. [CrossRef]
Yang, Z.-Y.; Ishii, S.; Yokoyama, T.; Dao, T.D.; Sun, M.-G.; Pankin, P.S.; Timofeev, I.V.; Nagao, T.; Chen, K.-P. Narrowband wavelength selective thermal emitters by confined Tamm plasmon polaritons. *ACS Photonics* **2017**, *4*, 2212–2219. [CrossRef]
Wang, Z.; Clark, J.K.; Ho, Y.-L.; Vilquin, B.; Daiguji, H.; Delaunay, J.-J. Narrowband thermal emission from Tamm plasmons of a modified distributed Bragg reflector. *Appl. Phys. Lett.* **2018**, *113*, 161104. [CrossRef]
Yang, Z.-Y.; Ishii, S.; Yokoyama, T.; Dao, T.D.; Sun, M.-G.; Nagao, T.; Chen, K.-P. Tamm plasmon selective thermal emitters. *Opt. Lett.* **2016**, *41*, 4453–4456. [CrossRef]
Chen, C.-Y.; Tsai, M.-W.; Jiang, Y.-W.; Ye, Y.-H.; Chang, Y.-T.; Lee, S.-C. Coupling of surface plasmons between two silver films in a plasmonic thermal emitter. *Appl. Phys. Lett.* **2007**, *91*, 243111. [CrossRef]
Botros, J.; Ali, M.O.; Tait, R.N.; Amaya, R.; Gupta, S. Direct thermal emission testing of aperiodic dielectric stack for narrowband thermal emission at mid-IR. *J. Appl. Phys.* **2020**, *127*, 114502. [CrossRef]
Hsiao, H.-H.; Xu, B.-T. Selective multi-wavelength infrared emission by stacked gap-plasmon thermal emitters. *Nanotechnology* **2021**, *32*, 165201. [CrossRef] [PubMed]

114. He, M.; Nolen, J.R.; Nordlander, J.; Cleri, A.; McIlwaine, N.S.; Tang, Y.; Lu, G.; Folland, T.G.; Landman, B.A.; Maria, J.-P Deterministic inverse design of Tamm plasmon thermal emitters with multi-resonant control. *Nat. Mater.* **2021**, *20*, 1663 [CrossRef] [PubMed]
115. Lu, G.; Zhang, K.; Zhao, Y.; Zhang, L.; Shang, Z.; Zhou, H.; Diao, C.; Zhou, X. Perfect optical absorbers by all-dielectric ph crystal/metal heterostructures due to optical Tamm state. *Nanomaterials* **2021**, *11*, 3447. [CrossRef] [PubMed]
116. Liu, B.; Lu, G.; Cui, L.; Li, J.; Sun, F.; Liu, F.; Li, Y.; Yang, T.; Du, G. Experimental investigation of multiple near-perfect absor in sandwich structures containing thin metallic films. *Opt. Express* **2017**, *25*, 13271–13277. [CrossRef]
117. Gong, Y.; Liu, X.; Lu, H.; Wang, L.; Wang, G. Perfect absorber supported by optical Tamm states in plasmonic waveguid *Express* **2011**, *19*, 18394. [CrossRef]
118. Wang, X.; Jiang, X.; You, Q.; Guo, J.; Dai, X.; Xiang, Y. Tunable and multichannel terahertz perfect absorber due to Tamm s plasmons with graphene. *Photonics Res.* **2017**, *5*, 536–542. [CrossRef]
119. Lu, H.; Gan, X.; Jia, B.; Mao, D.; Zhao, J. Tunable high-efficiency light absorption of monolayer graphene via Tamm pla polaritons. *Opt. Lett.* **2016**, *41*, 4743–4746. [CrossRef]
120. Jeng, S.-C. Applications of Tamm plasmon-liquid crystal devices. *Liq. Cryst.* **2020**, *47*, 1223–1231. [CrossRef]
121. Li, J.; Wen, C.-H.; Gauza, S.; Lu, R.B.; Wu, S.-T. Refractive indices of liquid crystals for display applications. *J. Disp. Technol* *1*, 51–61. [CrossRef]
122. Cheng, H.-C. Tunable Tamm Plasmon Devices. Master's Thesis, National Chiao Tung University, Tainan, Taiwan, 2016.
123. Takatoh, K.; Hasegawa, M.; Koden, M.; Itoh, N.; Hasegawa, M. *Alignment Technologies and Applications of Liquid Crystal D* Taylor & Francis: Oxford, UK, 2005.
124. Chigrinov, V.G.; Kozenkov, V.M.; Kwok, H.S. *Photoalignment of Liquid Crystalline Materials: Physics and Applications*; John W Sons Ltd.: West Sussex, UK, 2008.
125. Cheng, H.-C.; Kuo, C.-Y.; Hung, Y.-J.; Chen, K.-P.; Jeng, S.-C. Liquid-Crystal Active Tamm-Plasmon Devices. *Phys. Rev. Appl* *9*, 064034. [CrossRef]
126. Buchnev, O.; Belosludtsev, A.; Reshetnyak, V.; Evans, D.R.; Fedotov, V.A. Observing and controlling a Tamm plasmon interface with a metasurface. *Nanophotonics* **2020**, *9*, 897–903. [CrossRef]
127. Dabrowski, R.; Dziaduszek, J.; Ziolek, A.; Szczuciński, L.; Stolarz, Z.; Sasnouski, G.; Bezborodov, V.; Lapanik, W.; Gauza, S S.T. Low viscosity, high birefringence liquid crystalline compounds and mixtures. *Opto-Electron. Rev.* **2007**, *15*, 47–51. [Cro
128. Arakawa, Y.; Kang, S.; Tsuji, H.; Watanabe, J.; Konishi, G.I. The design of liquid crystalline bisolane-based materials with extr high birefringence. *RSC Adv.* **2016**, *6*, 92845–92851. [CrossRef]

crystals

e

nsity Functional Theory Calculations for Interpretation of ra-Red Spectra of Liquid Crystalline Chiral Compound

sandra Deptuch [1,*], Natalia Górska [2], Michaela Murzyniec [2], Monika Srebro-Hooper [2], James Hooper [2], lalena Dziurka [2] and Magdalena Urbańska [3]

1 Institute of Nuclear Physics, Polish Academy of Sciences, Radzikowskiego 152, PL-31342 Kraków, Poland
2 Faculty of Chemistry, Jagiellonian University, Gronostajowa 2, PL-30387 Kraków, Poland; natalia.gorska@uj.edu.pl (N.G.); monika.srebro@uj.edu.pl (M.S.-H.); james.hooper@uj.edu.pl (J.H.)
3 Institute of Chemistry, Military University of Technology, Kaliskiego 2, PL-00908 Warsaw, Poland; magdalena.urbanska@wat.edu.pl
* Correspondence: aleksandra.deptuch@ifj.edu.pl

Abstract: The experimental IR spectra of (*S*)-4′-(1-methylheptyloxycarbonyl) biphenyl-4-yl 4-[2-(2,2,3,3,4,4,4-heptafluorobutoxy) ethyl-1-oxy]-2-fluorobenzoate in the crystal phase are analyzed with the help of dispersion-corrected density functional theory (DFT+D3) calculations for isolated molecular monomer and dimer models, and a periodic model computed at the extended density functional tight-binding (xTB) level of theory. It is found that the frequency scaling coefficients obtained with the results of the molecular calculations are good matches for the crystal phase, being close to 1. The molecular and periodic models both confirm that varied intra- and intermolecular interactions are crucial in order to reproduce the broadened shape of the experimental band related to C=O stretching; the key factors are the conjugation of the ester groups with the aromatic rings and the varied intermolecular chemical environments, wherein the C=O group that bridges the biphenyl and F-substituted phenyl groups seems particularly sensitive. The C=O stretching vibrations are investigated as a function of temperature, covering the range of the crystal, smectic C_A*, smectic C* and isotropic liquid phases. The structure changes are followed based on the X-ray diffraction patterns collected in the same temperatures as the IR spectra. The experimental and computational results taken together indicate that the amount of weak C=O...H-C hydrogen bonds between the molecules in the smectic layers decreases with increasing temperature.

Keywords: smectic liquid crystal; FT-IR spectroscopy; DFT calculations; scaling coefficient; Fourier transform

n: Deptuch, A.; Górska, N.; iec, M.; Srebro-Hooper, M.; , J.; Dziurka, M.; Urbańska, M. Functional Theory tions for Interpretation of ed Spectra of Liquid ine Chiral Compound. 2024, 14, 645. https:// /10.3390/cryst14070645

nic Editor: Ingo Dierking

d: 24 June 2024
: 10 July 2024
ed: 11 July 2024
ed: 13 July 2024

1. Introduction

Theoretical infra-red (IR) spectra simulated for isolated molecules by density functional theory (DFT) calculations are usually in agreement with the experimental spectra to the extent that enables the assignment of the observed absorption bands to intramolecular vibrations [1–6]. This is especially beneficial for liquid crystals, whose molecules consist of numerous atoms and consequently usually require a long computing time. On the other hand, the introduction of more molecules into calculations, by using dimers, trimers or periodic boundary conditions, provides the opportunity to include the effect of intermolecular interactions [5–7]. This paper presents the detailed analysis of the IR spectra of smectogenic (*S*)-4′-(1-methylheptyloxycarbonyl) biphenyl-4-yl 4-[2-(2,2,3,3,4,4,4-heptafluorobutoxy)ethyl-1-oxy]-2-fluorobenzoate [8–10], denoted as 3F2HPhF6 (Figure 1). On heating in the crystal phase, 3F2HPhF6 undergoes phase transitions in the sequence of Cr (325.7 K) SmC_A* (364.5 K) SmC* (372.3 K) Iso [10]. The Cr and Iso notations correspond to the crystal and isotropic liquid phases, while SmC_A* and SmC* denote the chiral tilted smectic phases: antiferroelectric smectic C_A* and ferroelectric smectic C*. The tilt angle in the SmC_A* phase of 3F2HPhF6 and similar compounds with a partially fluorinated terminal chain takes values close to 45°; fore, they are investigated as components of the orthoconic liquid crystalline mi to be applied in displays [8,11–14]. The experimental IR spectra of this compoun been previously reported only for the solution in CCl_4 and in the crystal phase in temperature, and the full band assignment had not been performed [9,10]. Herei band assignment is based on the DFT+D3/BLYP-def2SVP calculations, perform an isolated molecule in two different conformations, and, to include the intermol interactions, for head-to-head and head-to-tail dimers (Figure 1). The DFT mol models are complemented with explicitly periodic models of an idealized tilted sn phase at the extended density functional tight-binding (xTB) level of theory, prin to obtain insight into how the intermolecular interactions of the crystal environ may broaden the C=O band shape. The temperature dependence of selected absor bands is discussed. The interpretation of the IR spectra is supported by the X-ray d tion (XRD) measurements, which enable the observation of the structural changes temperature [15–18].

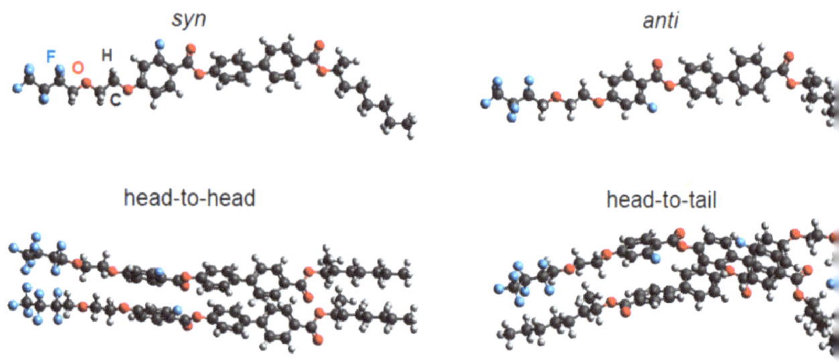

Figure 1. 3F2HPhF6 molecule in the *syn* and *anti* conformations, and two types of dimers con of molecules in the *anti* conformations, optimized with DFT+D3/BLYP-def2SVP. The symb atoms, given for the *syn* conformation, apply also for other models.

2. Materials and Methods

The mesogenic compound (*S*)-4′-(1-methylheptyloxycarbonyl) biphenyl-4-y (2,2,3,3,4,4,4-heptafluorobutoxy) ethyl-1-oxy]-2-fluorobenzoate was synthesized acc to the general method presented in [10,19].

The experimental IR spectra for the 3F2HPhF6/KBr tablet were measured upo ing at 273–393 K in the wavenumber range of 480–4000 cm^{-1} with a resolution of 2 using the Bruker VERTEX 70v vacuum spectrometer (Bruker, Billerica, MA, USA) equ with an Advanced Research System DE-202A cryostat and ARS-2HW water-cooled h compressor (ARS, Macungie, PA, USA).

The simulated IR spectra were calculated with the DFT method in Gaussian0 using the def2SVP basis set [21] and the BLYP exchange-correlation functional [with Grimme's semi-empirical dispersion corrections and Becke–Johnson dampi 25]. The calculations were performed for an isolated molecule (a hockey-stick-sh model, introduced in [7]) in two conformations differing in the position of the F (Figure 1). In the *syn* conformation, the F atom and a neighboring C=O grou on the same side of the benzene ring, while in the *anti* conformation they are o opposite sides. Additionally, the IR spectra were calculated for two dimers cons of molecules in the *anti* conformation, arranged in the head-to-head and head-t manner. The molecular models were visualized in Avogadro [26]. The IR data analyzed in OriginPro.

The periodic lattice was built by initially placing the molecules (in the more stab conformation) on a hexagonal lattice with 6 Å spacings between neighboring mol (the rigid core was initially aligned to be perpendicular to this hexagonal lattice), and then rotating the rigid core by 45° towards the plane of the lattice; the resulting models loosely represent idealized layers of a tilted smectic phase. Finally, a supercell, with cell vectors **a** and **b**, was constructed from the primitive hexagonal lattice vectors, **u** and **v**, via the following cell transformation: $\mathbf{a} = 1\mathbf{u} - 2\mathbf{v}$ and $\mathbf{b} = 3\mathbf{u} + 1\mathbf{v}$; this generated a simulation cell with seven molecules and ensured that no molecules were in direct contact with their own mirror images. It also created a model with intermolecular nearest-neighbor interactions that consisted exclusively of head-to-head relative orientations; such a setup most easily allowed us to create a close-packed distribution of 3F2HPhF6 molecules. The geometry optimizations and molecular dynamics simulations were run with the DFTB module of the Amsterdam Modelling Suite (AMS) [27], version 2023.1, and all simulations sampled only the gamma k-point and used the GFN1-xTB method [28]. After being equilibrated with NVE (constant particle number, volume and energy) and NVT (constant particle number, volume and temperature) simulations with the Berendsen thermostat, the final molecular dynamics simulations were performed at 250 K with the default Nose-Hoover thermostat that is implemented in the AMS program (the oscillatory period of the thermostat was set to 400 fs). To assess the dynamic contributions to the C=O vibrational band shape, the discrete Fourier transform was applied to the time evolution of each C=O distance in the molecular dynamics simulations, similar in spirit to what has been described before in the literature [29,30]. The production run lasted 6 ps, resulting in a frequency resolution of ~8 cm^{-1} after the discrete Fourier transforms were applied.

The XRD patterns of the flat sample of 3F2HPhF6 were collected upon heating at 273–393 K in the angular 2θ range of 2–30° using the Bragg–Brentano geometry. The measurements were performed with CuKα radiation (λ = 1.5406 Å) with an X'Pert PRO (PANalytical, Malvern, UK) diffractometer with a TTK-450 (AntonPaar, Graz, Austria) temperature stage. The XRD results were analyzed in WinPLOTR 7.20 Oct2019 [31] and OriginPro 2020b.

3. Results and Discussion

3.1. Band Assignment

The simulated IR spectra, calculated with DFT+D3/BLYP-def2SVP, and the experimental IR spectrum of the 3F2HPhF6 compound in the crystal phase, measured at 253 K, are compared in Figure 2. The particular parts of the 3F2HPhF6 molecule are denoted by letters (a–j), as shown in Figure 3, and the detailed band assignments based on the simulated spectra are gathered in Tables 1–3. The notations of the vibrational motions in the tables are as follows: β—in-plane deformations of the aromatic ring, γ—out-of-plane deformations of the aromatic ring, δ—scissoring, ν—stretching, ρ—rocking, τ—twisting, ω—wagging. The absorption bands in the wavenumber range of 500–1000 cm^{-1} are assigned to the in-plane and out-of-plane deformations of the aromatic rings, scissoring vibrations in the terminal chains, wagging of the CF_2 and CH_2 groups, and twisting of the CH_2 groups. Most of the bands in the 1000–1400 cm^{-1} range originate from the in-plane deformations of the aromatic rings, wagging and twisting of the CH_2 groups and stretching of the C–C and C–O bonds in the terminal chains and in the spacer between the benzene ring and biphenyl. There are also bands assigned to the CH_2 rocking and C-F stretching. The 1400–1650 cm^{-1} range contains bands attributed to the in-plane deformations of the benzene rings and CH_2 wagging and scissoring. Three absorption bands between 1650 and 1750 cm^{-1} are related to the stretching of two double C=O bonds. The bands with the wavenumbers of 2800–3000 cm^{-1} originate from the C-H stretching vibrations in the terminal chains.

There are four differences between the DFT+D3/BLYP-def2SVP simulated spectra for the isolated molecule in the *syn* and *anti* conformations, which are the most significant:

(1) The proximity of the F atom and C=O group in the *syn* conformation leads to the strengthening and shifting of the band related to the in-plane deformations of the

aromatic rings and C–C and C–O stretching in the non-chiral chain to lower wavenun
1008.8 cm^{-1} (*syn*) and 1013.5 cm^{-1} (*anti*).

(2) The location of the strongest absorption band is different for each conform
For *syn*, it is the band at 1197.3 cm^{-1}, related to the in-plane deformations of the aro
rings and C–C and C–O stretching in the spacer between the benzene and biphenyl
For *anti*, the strongest band is at 1247.6 cm^{-1} and has the same origin as for *syn*, in add
of the C–C and C–O stretching of the bonds between the biphenyl part and the chiral c
as well as the CH_2 twisting in the chiral chain.

(3) The simulated spectrum of *syn* contains a strong band at 1365.6 cm^{-1} ar
from the in-plane deformation of the fluorinated benzene ring and CH_2 wagging i
non-chiral chain, which is absent for *anti*.

(4) The bands related to the C=O stretching are strongly split for *syn* and lo
at 1719.1 cm^{-1} and 1756.5 cm^{-1} for C=O groups in the molecular core and close t
chiral center, respectively. Meanwhile, for *anti*, these bands are close to each oth
1718.5 cm^{-1} and 1719.9 cm^{-1}, and each of them is related to the stretching of both
groups, anti-phase at a lower wavenumber and in-phase at a higher wavenumber.

To make an assignment for the experimental absorption bands purely from the re
of the isolated molecule, it is necessary to consider the results for both the *syn* anc
conformations. This is especially visible for the C=O stretching bands: the experim
spectra contain three such bands, while the calculations for the isolated molecule pre
only two bands in this region; this matter was discussed also in [9]. Meanwhile
simulated spectra for the dimers enable the full assignment of the experimental IR spec
Namely, there are four C=O vibrations predicted for each dimer. For the head-to-
model, the positions of the bands are close to the ones obtained for the isolated molec
the *syn* conformation. For the head-to-tail model, the splitting of the C=O stretching b
is also visible but their wavenumbers are shifted towards lower values than those of
models (Figure 2). Noteworthy, the splitting of the C=O stretching bands in the di
is not caused by the proximity of the C=O group in the molecular core and the F
substituted in the benzene ring, as it is for the *syn* model. In both dimers, the F atoms
one molecule are not in close contact with the C=O group from the neighboring mol
The closest F–O contact, where O belongs to the C=O group, is 2.7 Å in the *syn* mode
4.2 Å in the *anti* model. In the head-to-head dimer, the closest F–O contact is also 4
between atoms within the same molecule, and the closest F–O contact between atoms
different molecules is slightly larger, at 4.5 Å. For the head-to-tail dimer, the closest
contact is between atoms from different molecules and equals 3.8 Å. In this last cas
F atom is located in the terminal chain, not in the aromatic core. This indicates tha
splitting of the C=O stretching bands does not have to be caused only by the proxim
the C=O group and F atom, as the interactions with other neighboring atoms may h
similar effect.

Figure 4 shows the plots of the experimental vs. calculated wavenumbers wit
linear fits performed with an intercept fixed to zero to determine the scaling factor, eq
the slope of the fitted line [32], that can thus be used to assess the quality of a reprodu
of the measured frequencies by the computations. The scaling factor 0.994(2) is clo
1 for results for the isolated molecule in both conformations, taken as one dataset
the dimers, the scaling factor is 0.998(3) and 0.996(3) for the head-to-head and head-t
model, respectively. For all the linear fits, the coefficient of determination R^2, de
according to [33], is also close to 1, which indicates good agreement with the assu
linear dependence. The linear fits performed in the ranges of <1000 cm^{-1}, 1000–2000 c
and >2000 cm^{-1} show that the calculated peak positions are mainly underestimated b
2000 cm^{-1} and overestimated above 2000 cm^{-1} (Table 4).

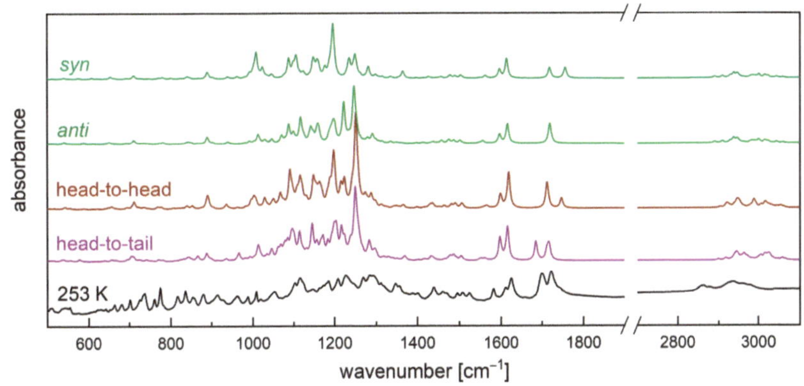

Figure 2. IR spectrum of 3F2HPhF6 measured in the crystal phase in 253 K, compared with simulated IR spectra for the isolated molecule in *syn* and *anti* conformations, and for two dimers.

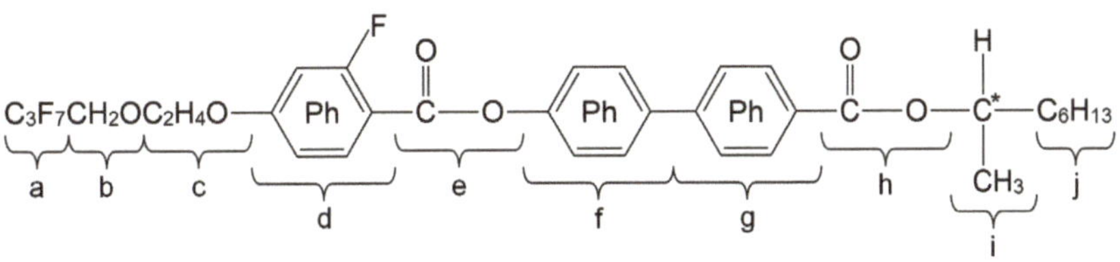

Figure 3. Notation of parts of the 3F2HPhF6 molecule used in assignment of IR absorption bands.

Table 1. Band assignments of the IR spectrum measured for 3F2HPhF6 in the crystal phase at 253 K (exp), based on DFT+D3/BLYP-def2SVP calculations for the isolated molecule in *anti* and *syn* conformations (calc). The wavenumbers are given in cm^{-1}. See the text for the notations of the vibrational motions.

exp	calc, *syn*	calc, *anti*	Description
511	497.3	496.2	δCCC(j)
542	539.7	540.8	δCF_2(a), δCOC(b,c), β_{asym}Ph(d)
554	552.5	554.7	γPh(f,g)
622	607.9		β_{asym}Ph(d,f,g)
630		610.4	β_{asym}Ph(d,f,g)
643	627.5	628.2	$\gamma Ph(d)^{syn}$, $\beta_{asym}Ph(d)^{anti}$, β_{asym}Ph(f,g)
653	640.7	641.1	δCF_2(a), δCOC(b,c), β_{asym}Ph(f,g)
664		651.1	β_{asym}Ph(d,f,g)
681	654.2		β_{asym}Ph(d,f,g)
702	664.1	662.2	γPh(d)
725	693.2	693.3	γPh(f,g)
736	711.0	711.1	ωCF_2(a), δCCO(a,b), β_{asym}(d)
760	770.5	772.3	γPh(f,g)
774	780.5	781.3	ωCF_2(a), νCC(a,b), δCCO(a,b)
816	800.5		γPh(d,f,g)
835	839.5		τCH_2(c), γPh(d)
854	841.5	841.6	γPh(f,g), τC^*HCH_3(i), τCH_2(j)
880	846.1	849.5	$\gamma Ph(f,g)^{syn}$, $\gamma Ph(d)^{anti}$
914	890.3	890.1	ωCF_2(a), ν_{sym}CCO(a,b), ωCH_2(b)
926	903.5	902.9	ν_{asym}OC*C(h,i,j)
962	939.5	939.6	τCH_2(b)
987	992.0	991.6	β_{asym}(f,g)

Table 1. *Cont.*

exp	calc, *syn*	calc, *anti*	Description
1008	1008.8	1013.5	ν_{sym}COC(b,c), ν_{sym}CCO(c), β_{asym}Ph(d,g), β_{sym}Ph(f)^{syn}, β_{asym}Ph(f
1020	1024.3	1028.8	ν_{sym}COC(b,c), ν_{asym}CCO(c), β_{asym}Ph(d,f,g)
1052	1046.8	1046.8	ωC*HCH_3(i), ωCH_2(j)
1103	1087.8	1087.8	β_{asym}Ph(g), ν_{asym}COC*(h,i), ωCH_2(j)
1116	1106.5	1117.1	β_{asym}Ph(d)
1146	1123.4	1123.5	$\nu_{asym}CF_2$(a), ρCH_2(c)
1164		1141.7	β_{sym}Ph(f)
1173	1147.4	1147.7	δCCC(a,b), ν_{asym}COC(b,c), β_{sym}Ph(f)
1185	1157.2	1157.1	β_{sym}Ph(g), ρCH_2(j)
1208	1176.4		ν_{asym}COC(c,d), β_{asym}Ph(d)
1228	1196.2	1196.1	ν_{asym}CCC(a,b), τCH_2(b,c), β_{asym}Ph(d,f)^{syn}
	1197.3	1200.2	β_{asym}Ph(d,f), ν_{asym}CCO(d,e), ν_{asym}COC(e,f)
1269		1222.8	β_{asym}Ph(d,f), ν_{asym}CCO(d,e), ν_{asym}COC(e,f)
1285		1247.6	β_{asym}(d,f,g), ν_{asym}CCO(d,e), ν_{asym}CCO(g,h), τCH_2(j)
1292	1249.9		β_{asym}(g), ν_{asym}CCO(g,h), τCH_2(j)
1311	1281.5	1291.3	ωCH_2(b,c), β_{asym}Ph(d,f,g)
1347	1334.9	1335.2	ωCH_2(j)
1356		1349.6	ωCH_2(b,c), β_{asym}Ph(d)
1389	1365.6		ωCH_2(b,c), β_{asym}Ph(d)
1400	1398.0	1398.3	β_{asym}Ph(f,g)
1439	1427.5		δCH_2(b), ωCH_2(c), β_{asym}Ph(d)
1461		1437.7	δCH_2(b,c), β_{asym}Ph(d)
1468	1456.9	1458.3	δCH_2(b,c), β_{asym}Ph(d)
1495	1478.9	1475.8	β_{asym}Ph(f,g)
1507	1490.7	1487.6	δCH_2(c), β_{asym}Ph(d)
1523	1504.4	1502.5	β_{asym}Ph(f,g)
1581	1562.2	1555.9	β_{asym}Ph(d)
1611	1596.4	1596.6	β_{sym}Ph(f,g)
1625	1613.7	1615.6	β_{sym}Ph(d,f)
1700	1719.1	1718.5	νC=O(h)^{syn}, anti-phase νC=O(e,h)^{anti}
1722		1719.9	in-phase νC=O(e,h)
1737	1756.5		νC=O(e)
2861	2937.6	2938.0	$\nu_{sym}CH_2$(c)
	2938.0	2938.1	$\nu_{sym}CH_2$(j)
2878	2945.2	2945.9	$\nu_{sym}CH_3$(j)
	2948.1	2948.2	νC*H(i), $\nu_{sym}CH_2$(j), $\nu_{asym}CH_2$(j)
2936	3001.7	3001.9	$\nu_{sym}CH_2$(i), νC*H(i), $\nu_{asym}CH_2$(j)
2960	3016.8	3016.8	$\nu_{asym}CH_2$(j)
2976	3024.7	3024.7	$\nu_{asym}CH_3$(j)

Table 2. Band assignments of the IR spectrum measured for 3F2HPhF6 in the crystal phase at (exp), based on DFT+D3/BLYP-def2SVP calculations for the head-to-head dimer of molecules *anti* conformations (calc). The wavenumbers are given in cm^{-1}. See the text for the notations vibrational motions.

exp	calc	Description
511	505.8	γPh(f,g)
542	541.8	β_{asym}Ph(d)
554	554.7	β_{asym}Ph(d)
622	588.0	β_{asym}Ph(d)
630	609.5	γPh(d,f)
643	621.0	γPh(d), β_{asym}Ph(f,g)
653	640.0	β_{asym}Ph(f,g)
664	649.5	γPh(d,f), β_{asym}Ph(g)

Table 2. *Cont.*

exp	calc	Description
681	657.0	β_{asym}Ph(d,f,g)
	660.4	γPh(d)
702	668.5	γPh(d)
725	693.6	γPh(f,g)
	695.4	γPh(f,g)
736	709.8	ωCF_2(a), δCCO(a,b)
	711.4	ωCF_2(a), δCCO(a,b)
760	767.9	γPh(f,g)
774	776.5	ωCF_2(a), νCC(a,b), δCCO(a,b)
	777.0	ωCF_2(a), νCC(a,b), δCCO(a,b)
816	819.2	τCH_2(c), γPh(d)
835	839.2	τCH_2(c), γPh(d)
854	852.9	γPh(f,g)
880	854.2	γPh(f,g)
914	887.6	ωCF_2(a), ν_{sym}CCO(a,b), ωCH_2(b)
	891.0	ωCF_2(a), ν_{sym}CCO(a,b), ωCH_2(b)
926	892.5	ν_{asym}OC*C(h,i,j)
	903.5	ν_{asym}OC*C(h,i,j)
962	935.8	τCH_2(b)
987	995.5	β_{asym}(f,g)
1008	1002.5	ν_{sym}COC(b,c), ν_{sym}CCO(c), β_{asym}Ph(d,f)
	1007.7	ν_{sym}COC(b,c), ν_{sym}CCO(c), β_{asym}Ph(d)
1020	1029.1	ν_{sym}COC(b,c), ν_{asym}CCO(c), β_{asym}Ph(d,f,g)
1052	1050.6	ωC^*HCH_3(i), ωCH_2(j)
1103	1090.3	β_{asym}Ph(g), ν_{asym}COC*(h,i)
1116	1115.7	β_{asym}Ph(d)
1146	1120.7	$\nu_{asym}CF_2$(a), ν_{asym}COC(b,c), ρCH_2(c)
1164	1147.8	δCCC(a,b), ν_{asym}COC(b,c), β_{sym}Ph(f)
1173	1154.4	ν_{asym}COC(c,d), β_{asym}Ph(d), β_{sym}Ph(f)
1185	1163.7	β_{sym}Ph(g)
	1168.4	ν_{asym}CCO(c), ν_{asym}COC(c,d), β_{asym}Ph(d), β_{sym}Ph(f,g)
1208	1188.0	$\nu_{asym}CF_2$(a), τCH_2(c)
1228	1197.6	ν_{asym}CCC(a,b), τCH_2(b,c)
	1198.7	τCH_2(c), β_{asym}Ph(d,f), ν_{asym}COC(e,f)
1269	1215.7	β_{asym}Ph(d,f), ν_{asym}CCO(d,e)
	1224.1	β_{asym}Ph(d,f), ν_{asym}CCO(d,e)
1285	1241.8	$\nu_{asym}CF_3$(a), $\nu_{sym}CF_2$(a), νCC(a), τCH_2(b,c)
	1243.7	β_{asym}Ph(d,f,g), ν_{asym}CCO(g,h), τC^*HCH_3(i)
1292	1251.5	τCH_2(b,c), β_{asym}(d,f,g), ν_{asym}CCO(d,e), ν_{asym}CCO(g,h), τC^*HCH_3(i)
1311	1288.2	ωCH_2(b,c), β_{asym}Ph(d)
1347	1344.2	β_{asym}Ph(f,g), ωCH_2(i), ωC^*HCH_3(j)
1356	1350.7	ωCH_2(b,c), β_{asym}Ph(d)
1389	1365.6	ωCH_2(b,c), β_{asym}Ph(d)
1400	1398.1	β_{asym}Ph(f,g)
1439	1430.6	δCH_2(b), ωCH_2(c), β_{asym}Ph(d)
1461	1437.0	δCH_2(b,c), β_{asym}Ph(d)
1468	1464.4	δCH_2(b,c), β_{asym}Ph(d)
1495	1481.3	β_{asym}Ph(f,g)
1507	1491.1	δCH_2(c), β_{asym}Ph(d)
1523	1505.5	β_{asym}Ph(f,g)
	1506.8	β_{asym}Ph(f,g)
1581	1565.7	β_{asym}Ph(d)
1611	1598.6	β_{sym}Ph(f,g)
1625	1618.6	β_{asym}Ph(d)
1700	1712.0	νC=O(h)
1722	1713.5	anti-phase νC=O(e,h)
	1714.6	in-phase νC=O(e,h)
1737	1747.3	νC=O(e)

Table 2. *Cont.*

exp	calc	Description
2861	2919.9	$\nu_{sym}CH_2$(b,c)
	2924.1	$\nu_{sym}CH_2$(b,c)
	2946.7	$\nu_{sym}CH_2$(j)
2878	2950.6	$\nu_{asym}CH_2$(j), $\nu_{sym}CH_3$(j)
	2952.3	$\nu_{asym}CH_2$(b,c)
2936	2988.7	$\nu_{asym}CH_2$(j)
2960	3006.7	$\nu_{asym}CH_2$(b,c)
2976	3017.1	$\nu_{asym}CH_2$(j)

Table 3. Band assignments of the IR spectrum measured for 3F2HPhF6 in the crystal phase at (exp), based on DFT+D3/BLYP-def2SVP calculations for the head-to-tail dimer of molecules *anti* conformations (calc). The wavenumbers are given in cm^{-1}. See the text for the notations vibrational motions.

exp	calc	Description
511	505.6	γPh(d,f,g)
542	535.6	δCF_2(a), δCOC(b,c), β_{asym}Ph(d)
	539.4	β_{asym}Ph(d), γPh(f,g)
554	556.0	β_{asym}Ph(d), γPh(f,g)
622	577.6	δCOC(c,d), β_{asym}Ph(d)
630	610.6	γPh(d,f)
643	620.1	γPh(d), β_{asym}Ph(f,g)
653	638.2	γPh(d), β_{asym}Ph(f,g)
664	656.8	γPh(d), β_{asym}Ph(f,g)
681	663.4	γPh(d)
702	668.9	γPh(d)
725	693.9	γPh(f,g)
	695.0	γPh(f,g)
736	704.5	ωCF_2(a), δCCO(a,b)
	709.8	ωCF_2(a), δCCO(a,b)
760	770.0	γPh(f,g)
	770.6	γPh(f,g)
774	779.8	ωCF_2(a), δCCO(a,b)
816	823.9	γPh(f,g), ν_{sym}OC*C(h,i,j), τCH_2(j)
835	840.5	γPh(d,f,g)
854	845.9	γPh(f,g)
880	866.1	ωCF_2(a), ν_{sym}CCO(a,b), τCH_2(b,c)
914	888.2	ωCF_2(a), ν_{sym}CCO(a,b), ωCH_2(b)
926	894.9	ν_{asym}OC*C(h,i,j)
	899.9	ν_{asym}OC*C(h,i,j)
962	936.1	$\nu_{asym}CF_2$(a), ρCH_2(b)
987	992.3	β_{asym}(f,g)
1008	1012.8	ν_{sym}CCO(c), β_{asym}Ph(d)
1020	1014.4	ν_{sym}COC(b,c), β_{asym}Ph(d,f,g)
1052	1045.2	ωC*HCH$_3$(i), ωCH_2(j)
1103	1092.8	β_{asym}Ph(d)
	1096.4	β_{asym}Ph(d)
1116	1114.1	β_{asym}Ph(d)
1146	1122.1	$\nu_{asym}CF_2$(a), ρCH_2(c)
1164	1145.0	$\nu_{asym}CF_3$(a), $\nu_{sym}CF_2$(a), ρCH_2(b,c), β_{asym}Ph(d), β_{sym}Ph(f)
1173	1156.7	β_{sym}Ph(f), ρCH_2(j)
1185	1166.0	β_{sym}Ph(g)
	1171.4	$\nu_{asym}CF_3$(a), $\nu_{sym}CF_2$(a), ν_{asym}CCC(a,b), β_{asym}Ph(d)
1208	1183.2	$\nu_{asym}CF_3$(a), δCCC(a), τCH_2(b,c)
1228	1192.8	$\nu_{asym}CF_3$(a), $\nu_{sym}CF_2$(a), ν_{asym}CCC(a,b), τCH_2(b,c)
	1197.7	$\nu_{asym}CF_2$(a), τCH_2(b)

Table 3. *Cont.*

exp	calc	Description
1269	1216.5	τCH_2(c), β_{asym}Ph(d), ν_{asym}CCO(d,e), ν_{asym}COC(e,f)
	1224.0	β_{asym}Ph(d,f), ν_{asym}CCO(d,e)
1285	1246.6	β_{asym}(d,f), ν_{asym}CCO(d,e), ν_{asym}CCO(g,h), τC^*HCH_3(i), τCH_2(j)
1292	1250.0	β_{asym}(d,f), ν_{asym}CCO(d,e), ν_{asym}CCO(g,h), τC^*HCH_3(i), τCH_2(j)
1311	1283.3	τCH_2(b,c,j), β_{asym}Ph(d,f,g)
1347	1339.0	ν_{asym}CCO(g,h), ωCH_2(i), ωC^*HCH_3(j)
1356	1351.0	ωCH_2(c), τCH_2(c), β_{asym}Ph(d)
1389	1369.8	β_{asym}Ph(d)
1400	1404.4	β_{asym}Ph(f,g)
1439	1432.2	β_{asym}Ph(d), δCH_2(i,j)
1461	1438.1	β_{asym}Ph(d), δCH_2(i,j)
1468	1464.4	δCH_2(b,c), β_{asym}Ph(d)
1495	1479.0	β_{asym}Ph(f,g)
1507	1486.3	δCH_2(c), β_{asym}Ph(d,f,g)
1523	1503.8	β_{asym}Ph(f,g)
1581	1560.8	β_{asym}Ph(d)
1611	1596.5	β_{asym}Ph(f,g)
	1597.1	β_{asym}Ph(f,g)
1625	1614.8	β_{sym}Ph(d,f,g)
	1615.5	β_{sym}Ph(d,f,g)
1700	1684.9	νC=O(h)
1722	1711.4	νC=O(e)
	1715.4	anti-phase νC=O(e,h)
1737	1718.2	in-phase νC=O(e,h)
2861	2944.8	$\nu_{asym}CH_2$(b,c)
	2945.3	$\nu_{sym}CH_2$(j)
	2956.3	$\nu_{asym}CH_2$(j), $\nu_{sym}CH_3$(j)
2878	2963.8	$\nu_{sym}CH_2$(b,c)
	2964.8	$\nu_{asym}CH_2$(b), $\nu_{asym}CH_2$(j), $\nu_{sym}CH_3$(j)
2936	3006.8	$\nu_{sym}CH_2$(b,c)
2960	3016.5	$\nu_{asym}CH_2$(j)
2976	3027.9	$\nu_{asym}CH_2$(j)

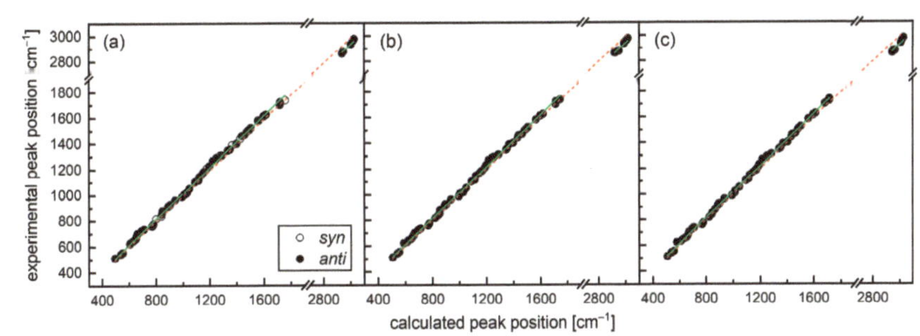

Figure 4. Experimental peak positions determined from the IR spectrum of 3F2HPhF6 in the crystal phase at 253 K vs. peak positions calculated for isolated molecules in two conformations: (**a**) head-to-head dimer (**b**) and head-to-tail dimer (**c**). Dashed and solid lines indicate linear fits in the whole spectral range and in the separate ranges, respectively.

As a next step with the molecular models, we built a periodic model of 3F2HPhF6 that esthetically represents one of its tilted smectic phases (see computational details and Figure 5a,b). Figure 5c,d show the overlaid discrete Fourier transforms of the evolutions of every respective C=O(h) (blue) and C=O(e) (red) bond distance in the system. Specifically, if the positions of the signal maxima are compared with the predicted C=O vibrational frequency from an isolated monomer (indicated by the dashed lines in Figure 5c,d periodic models further support the key observations from the molecular cluster m about how the influences of intermolecular interactions cause a general decrease i splitting of the observed frequencies. In addition, the xTB results indicate that the C groups are particularly sensitive to the nature of the splitting. Although we are cauti interpret the sources of the observed splitting without further exploring how the r setup and choice of methodology influence the splitting, we observed that the main ence between the "lowest" and "highest" frequency C=O(e) bonds were that the l frequency one (near 1700 cm^{-1}) engaged in mostly C=O. . .H-C contacts and some C= contacts, whereas the highest frequency one (near 1740 cm^{-1}) was different in two (I) it is bound to a F-substituted phenyl ring that rotated by 180° during pre-equilib (the C=O(e) group rotated with it in such a way as to preserve its local *anti* symmetry respect to the F substituent; this gave the molecule a distinct rotamer vs. all of the molecules in the cell), and (II) it creates an intermolecular lone pair. . .π contact [34] (~ with a nearby biphenyl group.

Table 4. The scaling coefficients between experimental and calculated IR absorption peak pos and the corresponding coefficients of determination R^2 of the linear fits (*in italics*).

Model	Full Range	$<$1000 cm^{-1}	1000–2000 cm^{-1}	$>$2000 cm
Isolated	0.994(2)	1.019(4)	1.011(2)	0.978(1)
molecule	*0.99961*	*0.99966*	*0.99988*	*0.99999*
Head-to-head	0.998(3)	1.020(4)	1.011(2)	0.980(2)
dimer	*0.99962*	*0.99964*	*0.99986*	*0.99998*
Head-to-tail	0.996(3)	1.018(5)	1.012(2)	0.975(2)
dimer	*0.99953*	*0.99955*	*0.99991*	*0.99998*

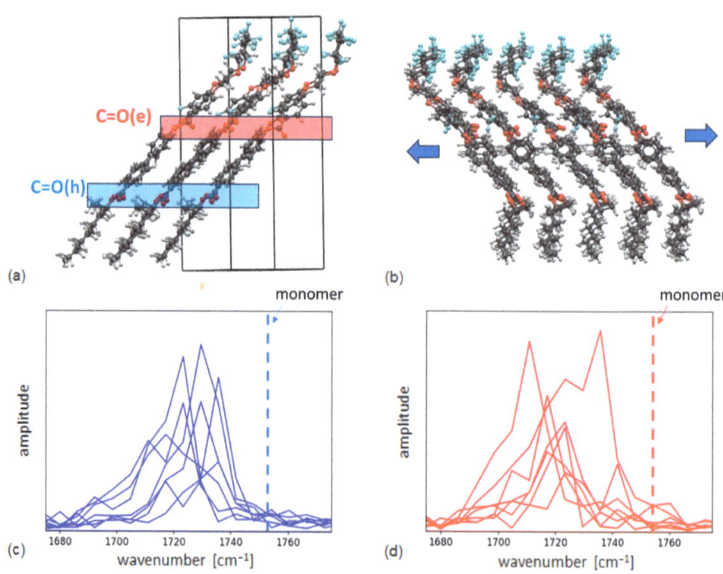

Figure 5. (**a**) A side view of the 7-molecule simulation cell that was used to model periodic laye tilted smectic phase for 3F2HPhF6; all of the molecules were initially assigned the *anti* configu and the initial distribution of the C=O(h) and C=O(e) carbonyl groups is indicated. (**b**) A dif perspective of the model that shows deviation after optimization from the underlying hocke shape. (**c**,**d**) The overlaid discrete Fourier transforms of every C=O(h) (**c**) and C=O(e) (**d**) t carbonyl group in the molecular dynamics simulations; the wavenumber of the C=O vibration normal mode analysis on an individual molecule is indicated by the dashed line.

3.2. Temperature Evolution of Vibrational Spectra and Structure

The XRD patterns and IR spectra were registered in the same temperatures to facilitate their comparison (Figure 6) and their selected fragments were analyzed closely (Figures 7–9). The XRD results are in agreement with the phase sequence obtained in [10] by differential scanning calorimetry (DSC). At 253–323 K, the diffraction patterns consisting of the sharp peaks are typical for the crystal phase (Figure 6a). Above the Cr → SmC_A* transition, at 333–373 K, there are only three sharp diffraction peaks—the main peak at 2θ = 3.2° and its second and third harmonics at 6.4° and 9.6°—related to the smectic layer order, while at higher angles, with a middle at 2θ ≈ 18°, there is a wide maximum arising from the short-range order in the smectic layers [15–19]. The SmC_A* and SmC* phases are characterized, respectively, by the anticlinic and synclinic order of the tilt angle in the neighbor smectic layers [15]. This structural change at the SmC_A* → SmC* transition does not lead to qualitative changes in the conventional XRD patterns (only the resonant XRD method gives different patterns for SmC*, SmC_A* and various sub-phases [35]). The sharp peaks from the smectic layers are absent at the patterns collected at 383 and 393 K, which signals the occurrence of the SmC* → Iso transition. The position of the first diffraction peak appearing in the XRD patterns at low angles is related by the Bragg equation $d = \lambda/2\sin\theta$ [15] to the layer spacing d in the crystal and smectic phases. The decrease in the tilt angle of the molecules in the SmC*, SmC_A* phases [8] leads to the increasing smectic layer spacing upon heating. Another important parameter is the integrated intensity of the low-angle peak. The abrupt changes in the layer spacing and integrated intensity indicate the phase transitions [17,18]. For 3F2HPhF6, the Cr → SmC_A* transition can be noticed in the $d(T)$ plot as a decrease in the layer spacing, while the SmC* → SmC_A* transition is not accompanied by any significant change in d. However, both transitions are clearly visible in the temperature dependence of the integrated intensity (Figure 8a).

The phase transition which influences the IR spectrum to the most extent is the melting of the crystal phase, which leads to a change in the shape and positions of some absorption bands (Figure 6b). Further transitions are not clearly visible in the IR spectra, which look practically the same for the SmC_A*, SmC* and Iso phases. As reported in other publications [1,2,5,6,9], the absorption bands in the 1700–1750 cm^{-1} range, assigned to the νC=O vibrations, show a noticeable sensitivity to changes in temperature, phase transitions, intermolecular interactions and molecular conformation; therefore, they are investigated as a function of temperature (Figures 7 and 8b,c). The νC=O bands are designated as I, II and III in order of their increasing wavenumber. According to the assignments presented in Tables 1–3 and the computational results from [9], band I is related to the stretching of the C–O bond close to the chiral center, while bands II and III are more likely related to the stretching of the C=O bonds located within the aromatic core. We consider this a sound guiding principle, but we further highlight that the results from the periodic models (in Figure 5) suggest that the shifts in the νC=O frequencies that are induced by non-homogeneous intermolecular interactions can overlap with the magnitudes of the shifts that are induced by intramolecular effects. The Cr → SmC_A* transition is connected with the change in the integrated intensity of all the νC=O absorption bands, while the significant change in the wavenumber occurs only for band I. There is no certain signature of the SmC_A* → SmC* transition, probably because the surroundings of the C=O groups do not differ in both smectic phases. In the isotropic liquid phase at 393 K, bands I and III decrease and band II increases in intensity; a decrease in the wavenumber of band II and an increase in the wavenumber of band III is also visible. At 383 K, the parameters of the νC=O bands are not very different to the values for the SmC* phase, despite the XRD results confirming that 3F2HPhF6 is in the isotropic liquid phase at this temperature. The explanation for this could be the presence of small domains with a preserved smectic order in the sample, which survived a few degrees above the SmC* → Iso transition temperature (resembling the so-called cybotactic clusters reported for the nematic phase [36]).

Since the computational results indicate that intermolecular interactions influence the wavenumbers of the νC=O bands, it is reasonable to compare their values with the average distance w between the long axes of molecules in the smectic layers and in the isot liquid. The w value is determined from the position of the wide maximum at 2θ in the XRD patterns [15,16]. For 3F2HPhF6, w equals 4.8–4.9 Å in the smectic phase 5.0–5.1 Å in the isotropic liquid phase, increasing with increasing temperature. I temperature range of 333–373 K in the smectic phases, there is a roughly linear incre the wavenumbers of the νC=O bands with an increasing intermolecular distance, wit slope equal to 11–15 cm^{-1}/Å (Figure 9). The larger the distances between the molecu the smectic layers, the weaker the intermolecular interaction. The computational re show that the proximity of the C=O group and the F atom shifts the νC=O band tow higher wavenumbers [9]. However, according to the literature [4,37–39], the preser the hydrogen bonds involving the O atom from the C=O group has an inverse effec shifts the νC=O band towards lower wavenumbers. This means that the intermole interactions have various effects on the νC=O bands, depending on the arrangement molecules. In our computational results, the theoretical positions of the νC=O bands lated for the *anti/syn* set of isolated molecules (1718.5–1756.5 cm^{-1}) are higher than obtained for the head-to-head (1712.0–1747.3 cm^{-1}) and head-to-tail (1684.9–1718.2 c dimers. The theoretical νC=O band with the lowest position of 1684.9 cm^{-1} is related stretching of the C=O(h) group making a short contact with the C-H atoms from the C group of the neighbor molecule; this is curiously similar to what we observed for the lo frequency C=O(e) contribution in Figure 5d, which also involved a contact with a n CH_2(b) group. The H(b)...O(h) and C(b)...O(h) distances in the molecular mode 2.2 Å and 3.2 Å, respectively, and the C-H(b)...O(h) angle equals 155.2°. These param are similar to the experimental values obtained for the C-H...O hydrogen bonds i crystal structures of other liquid crystals [40,41]. Taking this all into account, we con that the blueshift of the νC=O bands with the increasing intermolecular distance w increasing temperature is caused by the decreasing amount of hydrogen bonds bet molecules. In the isotropic liquid phase, the positions of the νC=O absorption bands various dependences on w: band III still blueshifts with increasing w, while bands II slightly redshift. The explanation of their relationship is not as straightforward the smectic phases because the orientational disorder in the isotropic liquid enables more configurations of neighboring molecules.

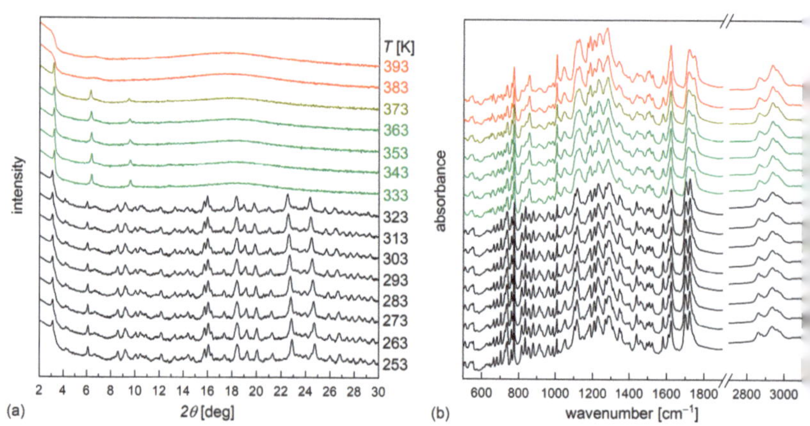

Figure 6. XRD patterns (**a**) and IR spectra (**b**) of 3F2HPhF6 collected upon heating. The sequence is as follows: 253–323 K—crystal (black), 333–363 K—SmC_A* (green), 373 K—SmC* (ye 383–393 K—isotropic liquid (red).

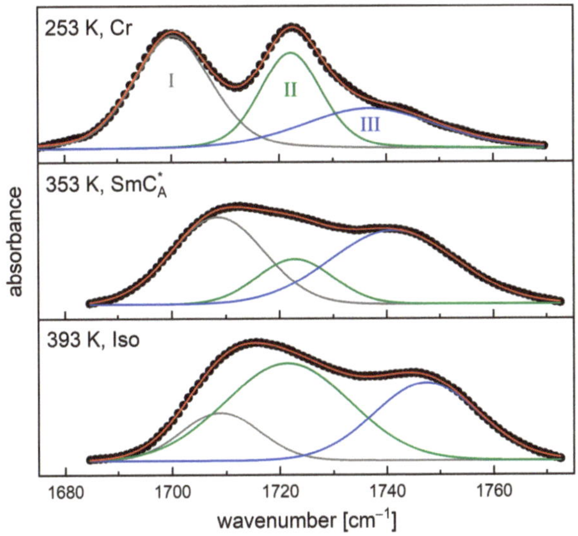

Figure 7. IR absorption bands related to the C=O stretching with the representative fitting results of the pseudo-Voigt (crystal phase) and Gaussian (smectic and isotropic liquid phases) peak functions after background subtraction.

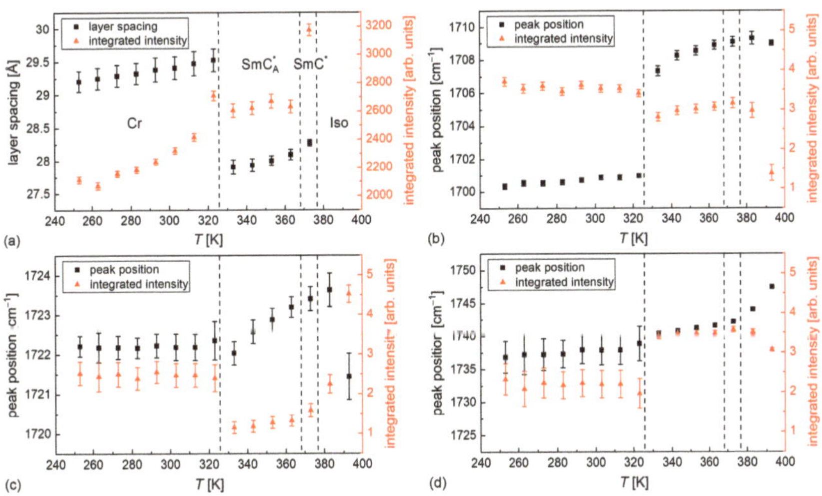

Figure 8. Layer spacing in the crystal and smectic phases of 3F2HPhF6, and the integrated intensity of the low-angle diffraction peak (**a**) combined with the positions and integrated intensities of the νC=O bands I (**b**), II (**c**), III (**d**). The vertical lines indicate the phase transition temperatures from [10], determined by DSC.

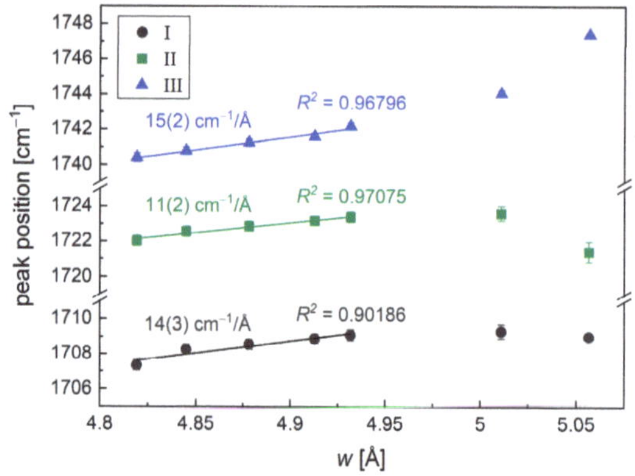

Figure 9. Relationship between the average distances between molecules within the smectic and in isotropic liquid determined by XRD, and the positions of the νC=O bands obtaine experimental IR spectra. For the results in the smectic phases, the linear fits are performed a slopes and coefficients of determination are given.

4. Conclusions

The intramolecular vibrations of the 3F2HPhF6 compound were investigat FT-IR spectroscopy and density functional theory calculations for the isolated mol dimers and periodic model. The detailed band assignments in the wavenumber of 500–3000 cm^{-1} show that the introduction of dimers into the DFT+D3 calcul gives a better agreement between the experimental and calculated spectra, i.e., the s factor is closer to 1. However, the calculations for isolated molecules are also suff as long as they include two possible positions of the fluorine substituent in the aro core and the neighboring C=O group (*syn* and *anti* conformations). The results f periodic model obtained with the xTB method provide additional support for thre observations: (1) how the inclusion of non-covalent interactions directly influenc predicted C=O stretching frequencies, (2) a major origin of the shifts in frequencies r with the influences of the shortest non-covalent contacts that the C=O groups parti in, and (3) it is important how the molecules ultimately position themselves vs. neighbors (concerning both their relative orientations and their adopted rotamers) is similar in spirit to the influence of the *anti* vs. *syn* monomer isomers, but rather p out the (dynamic) intermolecular aspect of this effect. The blueshift of the absor bands related to the C=O stretching upon heating, correlated with the increase average intermolecular distances obtained from the XRD patterns, indicates the decre amount of weak C-H. . .O=C hydrogen bonds in the smectic C_A* and smectic C* p The shift of the C=O stretching band towards lower wavenumbers upon hydrogen formation is confirmed by one of the dimer models, and the calculated parameters C-H. . .O=C bond are in agreement with the experimental results reported for other crystalline compounds.

Author Contributions: Conceptualization, A.D.; formal analysis, A.D.; investigation, A.D., M.M., M.S.-H., J.H. and M.D.; resources, M.U. and J.H.; writing—original draft preparation writing—review and editing, N.G., M.M., M.S.-H., J.H., M.D. and M.U. All authors have rea agreed to the published version of the manuscript.

Funding: The Bruker VERTEX 70v FT-IR spectrometer with Advanced Research System DE cryostat and ARS-2HW compressor were purchased thanks to the European Regional Develo Fund in the framework of the Innovative Economy Operational Program (contract no. POIG.02 12-023/08).

Data Availability Statement: The original contributions presented in the study are included article; further inquiries can be directed to the corresponding author.

Acknowledgments: We gratefully acknowledge Poland's high-performance Infrastructure PLGrid Academic Computer Centre Cyfronet AGH for providing computer facilities and support within computational grant no. PLG/2023/016486.

Conflicts of Interest: The authors declare no conflicts of interest.

ences

Drzewicz, A.; Juszyńska-Gałązka, E.; Zając, W.; Piwowarczyk, M.; Drzewiński, W. Non-isothermal and isothermal cold crystallization of glass-forming chiral smectic liquid crystal (S)-4′-(1-methyloctyloxycarbonyl) biphenyl-4-yl 4-[7-(2,2,3,3,4,4,4-heptafluorobutoxy) heptyl-1-oxy]-benzoate. *J. Mol. Liq.* **2020**, *319*, 114153. [CrossRef]

Singh, S.; Singh, H.; Srivastava, A.; Tandon, P.; Deb, R.; Debnath, S.; Rao, N.V.S.; Ayala, A.P. Study of phase transitions in a bent-core liquid crystal probed by infrared spectroscopy. *Vib. Spectrosc.* **2016**, *86*, 24–34. [CrossRef]

Osiecka, N.; Galewski, Z.; Juszyńska-Gałązka, E.; Massalska-Arodź, M. Studies of reorganization of the molecules during smectic A-smectic C phase transition using infrared spectroscopy and generalized two-dimensional correlation analysis. *J. Mol. Liq.* **2016**, *224*, 677–683. [CrossRef]

Drużbicki, K.; Mikuli, E.; Ossowska-Chruściel, M.D. Experimental (FT-IR, FT-RS) and theoretical (DFT) studies of vibrational dynamics and molecular structure of 4-n-pentylphenyl-4′-n-octyloxythiobenzoate (8OS5). *Vib. Spectrosc.* **2010**, *52*, 54–62. [CrossRef]

Drużbicki, K.; Kocot, A.; Mikuli, E.; Ossowska-Chruściel, M.D.; Chruściel, J. Temperature-Dependent Infrared Spectroscopy Studies of a Novel Antiferroelectric Liquid-Crystalline Thiobenzoate. *J. Phys. Chem. B* **2012**, *116*, 11332–11343. [CrossRef] [PubMed]

Juszyńska-Gałązka, E.; Zając, W. Mesomorphic behaviour and vibrational dynamics of nCFPB liquid crystalline homologues. *Phase Trans.* **2019**, *92*, 1077–1088. [CrossRef]

Kocot, A.; Loska, B.; Arakawa, Y.; Mehl, G.H.; Merkel, K. Study of the Experimental and Simulated Vibrational Spectra Together with Conformational Analysis of Thioester Cyanobiphenyl-Based Liquid Crystals. *Int. J. Mol. Sci.* **2022**, *23*, 8005. [CrossRef] [PubMed]

Deptuch, A.; Marzec, M.; Jaworska-Gołąb, T.; Dziurka, M.; Hooper, J.; Srebro-Hooper, M.; Fryń, P.; Fitas, J.; Urbańska, M.; Tykarska, M. Influence of carbon chain length on physical properties of 3FmHPhF homologues. *Liq. Cryst.* **2019**, *46*, 2201–2212. [CrossRef]

Drzewicz, A.; Bombalska, A.; Tykarska, M. Impact of molecular structure of smectogenic chiral esters ($3FmX_1X_2r$) on vibrational dynamics as seen by IR and Raman spectroscopy. *Liq. Cryst.* **2019**, *46*, 754–771. [CrossRef]

Żurowska, M.; Dąbrowski, R.; Dziaduszek, J.; Garbat, K.; Filipowicz, M.; Tykarska, M.; Rejmer, W.; Czupryński, K.; Spadło, A.; Bennis, N.; et al. Influence of alkoxy chain length and fluorosubstitution on mesogenic and spectral properties of high tilted antiferroelectric esters. *J. Mater. Chem.* **2011**, *21*, 2144–2153. [CrossRef]

Dąbrowski, R.; Gąsowska, J.; Otón, J.; Piecek, W.; Przedmojski, J.; Tykarska, M. High tilted antiferroelectric liquid crystalline materials. *Displays* **2004**, *25*, 9–19. [CrossRef]

Rastogi, A.; Agrahari, K.; Srivastava, A.; Manohar, R. Effect of graphene oxide dispersion in antiferroelectric liquid crystal mixture in the verge of SmC* to SmC_A* phase transition. *Chin. J. Phys.* **2020**, *67*, 91–106. [CrossRef]

Agrahari, K.; Nautiyal, V.K.; Vimal, T.; Pandey, S.; Kumar, S.; Manohar, R. Modification in different physical parameters of orthoconic antiferroelectric liquid crystal mixture via the dispersion of hexanethiol capped silver nanoparticles. *J. Mol. Liq.* **2021**, *332*, 115840. [CrossRef]

Tykarska, M.; Kurp, K.; Mironov, S.; Rychłowicz, N.; Karcz, J.; Dziaduszek, J.; Kula, P.; Stulov, S.; Bubnov, A. Tuning of self-organizing and electro-optical behaviour for orthoconic ferroelectric liquid crystal by non-chiral dopants. *J. Mol. Liq.* **2024**, *409*, 125426. [CrossRef]

Demus, D.; Goodby, J.; Gray, G.W.; Spiess, H.-W.; Vill, V. (Eds.) *Handbook of Liquid Crystals*; WILEY-VCH Verlag GmbH: Weinheim, Germany, 1998.

Tomczyk, W.; Marzec, M.; Juszyńska-Gałązka, E.; Węgłowska, D. Mesomorphic and physicochemical properties of liquid crystal mixture composed of chiral molecules with perfluorinated terminal chains. *J. Mol. Struc.* **2017**, *1130*, 503–510. [CrossRef]

Novotná, V.; Hamplová, V.; Podoliak, N.; Kašpar, M.; Glogarová, M.; Pociecha, D.; Górecka, E. Chiral liquid crystalline compounds with a re-entrant SmA* phase. *J. Mater. Chem.* **2011**, *21*, 14807–14814. [CrossRef]

Gleeson, H.F.; Wang, Y.; Watson, S.; Sahagun-Sanchez, D.; Goodby, J.W.; Hird, M.; Petrenko, A.; Osipov, M.A. On the temperature dependence of the tilt and spontaneous polarisation in high tilt antiferroelectric liquid crystals. *J. Mater. Chem.* **2004**, *14*, 1480–1485. [CrossRef]

Żurowska, M.; Dąbrowski, R.; Dziaduszek, J.; Czupryński, K.; Skrzypek, K.; Filipowicz, M. Synthesis and Mesomorphic Properties of Chiral Esters Comprising Partially Fluorinated Alkoxyalkoxy Terminal Chains and a 1-methylheptyl Chiral Moiety. *Mol. Cryst. Liq. Cryst.* **2008**, *495*, 145/[497]–157/[509]. [CrossRef]

Frisch, M.J.; Trucks, G.W.; Schlegel, H.B.; Scuseria, G.E.; Robb, M.A.; Cheeseman, J.R.; Scalmani, G.; Barone, V.; Petersson, G.A.; Nakatsuji, H.; et al. (Eds.) *Gaussian 09, Revision A.02*; Gaussian, Inc.: Wallingford, CT, USA, 2016.

21. Weigend, F.; Ahlrichs, R. Balanced basis sets of split valence, triple zeta valence and quadruple zeta valence quality for H Design and assessment of accuracy. *Phys. Chem. Chem. Phys.* **2005**, *7*, 3297–3305. [CrossRef]
22. Becke, A.D. Density-functional exchange-energy approximation with correct asymptotic behaviour. *Phys. Rev. A* **19** 3098–3100. [CrossRef]
23. Lee, C.; Yang, W.; Parr, R.G. Development of the Colle-Salvetti correlation-energy formula into a functional of the electron d *Phys. Rev. B* **1988**, *37*, 785–789. [CrossRef] [PubMed]
24. Grimme, S.; Antony, J.; Ehrlich, S.; Kries, H. A consistent and accurate ab initio parametrization of density functional disp correction (DFT-D) for the 94 elements H-Pu. *J. Chem. Phys.* **2010**, *132*, 154104. [CrossRef] [PubMed]
25. Grimme, S.; Ehrlich, S.; Goerigk, L. Effect of the damping function in dispersion corrected density functional theory. *J. C Chem.* **2011**, *32*, 1456–1465. [CrossRef] [PubMed]
26. Hanwell, M.D.; Curtis, D.E.; Lonie, D.C.; Vandermeersch, T.; Zurek, E.; Hutchison, G.R. Avogadro: An advanced ser chemical editor, visualization, and analysis platform. *J. Cheminf.* **2012**, *4*, 17. [CrossRef] [PubMed]
27. Rüger, R.; Yakovlev, A.; Philipsen, P.; Borini, S.; Melix, P.; Oliveira, A.F.; Franchini, M.; van Vuren, T.; Soini, T.; de Re et al. *AMS DFTB 2024.1*; SCM, Theoretical Chemistry, Vrije Universiteit: Amsterdam, The Netherlands, 2024. Available https://www.scm.com (accessed on 17 June 2024).
28. Grimme, S.; Bannwarth, C.; Shushkov, P. A Robust and Accurate Tight-Binding Quantum Chemical Method for Stru Vibrational Frequencies, and Noncovalent Interactions of Large Molecular Systems Parametrized for All spd-Block Elemen 1–86). *J. Chem. Theor. Comp.* **2017**, *13*, 1989–2009. [CrossRef] [PubMed]
29. Wróbel, P.; Kubisiak, P.; Eilmes, A. MeTFSI (Me = Li, Na) Solvation in Ethylene Carbonate and Fluorinated Ethylene Carbor Molecular Dynamics Study. *J. Phys. Chem. B* **2021**, *125*, 1248–1258. [CrossRef] [PubMed]
30. Wróbel, P.; Eilmes, A. Effects of Me–Solvent Interactions on the Structure and Infrared Spectra of MeTFSI (Me = Li, Na) Soluti Carbonate Solvents—A Test of the GFN2-xTB Approach in Molecular Dynamics Simulations. *Molecules* **2023**, *28*, 6736. [Cro [PubMed]
31. Roisnel, T.; Rodriguez-Carvajal, J. WinPLOTR: A Windows tool for powder diffraction patterns analysis. *Mater. Sci. Forum 378–381*, 118–123. [CrossRef]
32. Zapata Trujillo, J.C.; McKemmish, L.K. Model Chemistry Recommendations for Scaled Harmonic Frequency Calculati Benchmark Study. *J. Phys. Chem. A* **2023**, *127*, 1715–1735. [CrossRef] [PubMed]
33. Origin Help. 15.4.1. Additional Information of R-Square. Available online: https://www.originlab.com/doc/en/Origin-Details_of_R_square (accessed on 17 June 2024).
34. Novotný, J.; Bazzi, S.; Marek, R.; Kozelka, J. Lone-pair-π interactions: Analysis of the physical origin and biological implic *Phys. Chem. Chem. Phys.* **2016**, *18*, 19472–19481. [CrossRef]
35. Hirst, L.S.; Watson, S.J.; Gleeson, H.F.; Cluzeau, P.; Barois, P.; Pindak, R.; Pitney, J.; Cady, A.; Johnson, P.M.; Huang, C.C Interlayer structures of the chiral smectic liquid crystal phases revealed by resonant X-ray scattering. *Phys. Rev. E* **2002**, *65*, C [CrossRef] [PubMed]
36. Hong, S.H.; Verduzco, R.; Williams, J.C.; Twieg, R.J.; DiMasi, E.; Pindak, R.; Jákli, A.; Gleeson, J.T.; Sprunt, S. Short-range s order in bent-core nematic liquid crystals. *Soft Matter* **2010**, *6*, 4819–4827. [CrossRef]
37. Tonge, P.J.; Fausto, R.; Carey, P.R. FTIR studies of hydrogen bonding between α,β-unsaturated esters and alcohols. *J. Mol.* **1996**, *379*, 135–142. [CrossRef]
38. Žagar, E.; Grdadolnik, J. An infrared spectroscopic study of H-bond network in hyperbranched polyester polyol. *J. Mol.* **2003**, *658*, 143–152. [CrossRef]
39. Liu, Y.; Zhao, J.; Peng, Y.; Luo, J.; Cao, L.; Liu, X. Comparative Study on the Properties of Epoxy Derived from Aromat Heteroaromatic Compounds: The Role of Hydrogen Bonding. *Ind. Eng. Chem. Res.* **2020**, *59*, 1914–1924. [CrossRef]
40. Ossowska-Chruściel, M.D.; Karczmarzyk, Z.; Chruściel, J. The Polymorphism Of 4-n-Pentylphenyl-4′-n-Butyloxythio-Ben (4OS5) In The Crystalline State. *Mol. Cryst. Liq. Cryst.* **2002**, *382*, 37–52. [CrossRef]
41. Chruściel, J.; Pniewska, B.; Ossowska-Chruściel, M.D. The Crystal and Molecular Structure of 4-Pentylphenyl-4′-Pentioxythi (5S5). *Mol. Cryst. Liq. Cryst.* **1995**, *258*, 325–331. [CrossRef]

crystals

MDPI

e

alysing the Photo-Physical Properties of Liquid Crystals

n Hobbs, Johan Mattsson and Mamatha Nagaraj *

School of Physics and Astronomy, University of Leeds, Leeds LS2 9BW, UK
* Correspondence: m.nagaraj@leeds.ac.uk

Abstract: Intrinsically fluorescent liquid crystals are highly sought after for a variety of applications. Most of the measurements of photo-physical properties of liquid crystals are made in dilute solutions, which is mainly due to the relative ease of both these measurements and the interpretation of data. The fluorescence spectra depend on a number of parameters including the concentration in liquid crystal solutions, the device geometry, and the mesophase in which the spectra have been measured. Working with neat, or concentrated, liquid crystal samples adds experimental complexities such as the inner filter effect (IFE), which affects the collection of data, interpretation of the results, and accuracy of the conclusions. In this paper, we present a systematic study of the photo-physical properties of both a model reference material, Nile red, and a nematic liquid crystal, 4-cyano-4′-pentylbiphenyl (5CB). The influence on the emission spectra of an increasing solute concentration is investigated and discussed. Moreover, a detailed investigation of the influence of the used device geometry, as well as the choice of appropriate data fitting methodologies, are presented.

Keywords: emission; fluorescence; inner filter effect; liquid crystals; liquid crystal devices; photo-physics

1. Introduction

Thermotropic liquid crystals (LCs) have been used in many applications, such as the now ubiquitous LC displays (LCDs), as the LC director (bulk molecular orientation) can be controlled by external manipulation which, in turn, produces unique changes to the optical behaviour. In LCDs, by far the most successful application of LCs to date [1], an electric field is used to reorient the director in order to obtain optical contrast [2]. Similarly, intrinsically fluorescent LCs are desirable for many applications as their emission properties could be electrically controlled by reorientation of the LC director.

Fluorescence, in general, is the process of spontaneous emission of light by electronically excited species of an organic or inorganic material. To achieve fluorescence, various methods of electronic excitation can be used, including: light in photo-luminescence [3], an electric field in electro-luminescence [4], and ultrasound in sono-luminescence [5]. Applications of intrinsically fluorescent LCs are widespread and include polarised lasers [6,7], fluorescent LCD materials to replace the back-light component found in LCDs [8,9] and improve viewing angles [10], anisotropic organic light-emitting diodes [11,12], fluorescent LC gels for photonic applications [13,14], ink-jet print dyes [15], and 1D semiconductors [16].

The fluorescence emission from LCs has been studied previously. A number of these studies investigate cyano-biphenyl (CB)-based LCs. Through these studies, it has been identified that CB-based LCs form 'excimers'. An excimer is a transient complex formed between the same type of molecules, where one is in the ground state, while the other in the excited state [17]. Excimers are observed in CB materials both in solutions at higher concentrations (>100 mM) and in the neat form [18–23]. While LCs are often highly fluorescent in dilute solution, once they are concentrated, either due to the nature of the particular LC phase, or simply due to the increased solution concentration, excimer formation either red-shifts and alters the spectrum, or it quenches the emission entirely [24]. In neat LCs, excimers have been identified both in their LC and isotropic phases [19,21,22].

ı: Hobbs, J.; Mattsson, J.;
, M. Analysing the
hysical Properties of Liquid
. *Crystals* **2024**, *14*, 362.
/doi.org/10.3390/
)40362

ıic Editor: Serguei
:h Palto

d: 15 March 2024
: 27 March 2024
ːd: 29 March 2024
ed: 11 April 2024

These excimers are typically found to be long-lived, and they have been suggested related to the anti-parallel pair formation that occurs in CB-based materials [23,25,2

Understanding the photo-physical properties of LCs in solution, i.e., as single mol isolated from neighbouring LC molecules, is an important precursor to understandi photo-physical properties of neat LCs. As high-quality UV-Vis and photolumines spectrometers are relatively commonplace, making photo-physical measurements on solution is relatively straightforward. However, interpretation of the data can be mor plex, and it is not clear that conclusions based on measurements of LCs in dilute sol apply to LCs used in the neat form. Furthermore, often when the fluorescence behavi studied for high concentration solutions and for neat LCs, either the contributions fro so-called inner-filter effect (IFE) (details in Section 2.2) are ignored, or not enough d are provided in the description of the experimental geometry to accurately interpre understand the presented results. In CB-based LCs, often the IFE, and its contribut measured emission and excitation spectra, has not been considered and discusse note that the shift in emission wavelength caused by the excimer formation is often enough in CB-based LCs that correct conclusions have often been reached in spite c oversight even though this is not always true.

To address these issues, this paper provides a systematic investigation of som photo-physical properties of a cyanobiphenyl LC, 5CB, which is often used in the lite as a 'standard' nematic LC for such investigations. The photo-physical effects are fi vestigated using a standard fluorophore, Nile red, which does not show excimer form unlike the cyanobiphenyl LCs. The results from Nile red are compared to correspo results from 5CB. The paper establishes the effect of Nile red and 5CB concentratio solution) on the emission properties, and it explains how the IFE can alter the obs spectral concentration dependencies seen in both materials. The investigation of the sion and excitation spectra of Nile red also shows how such spectra can distort a concentrations in sample geometries such as cuvettes. Even though Nile red do undergo any changes to the ground state at the concentrations studied here, spectral c tions to the emission and excitation spectra of Nile red are still observed even at rela low concentrations. We further evaluate the role of the measurement device geo on the measured emission spectra, using results from front-facing (defined in de Section 2.2) measurements on LC cells, which significantly minimise spectral alter due to the IFE. Finally, the influence of the applied analysis procedures such as f routines of the emission spectra, on the final results, are discussed in detail.

2. Theory

2.1. The Effect of Concentration

The quantum yield of a fluorophore, Φ_F, is defined as the number of emitted ph relative to the number of absorbed photons. Mathematically, this can be written as

$$\Phi_F = \int_0^\infty F_\lambda(\lambda_F) d\lambda_F,$$

where $F_\lambda(\lambda_F)$ represents the probability that an absorbed photon is emitted at a sp wavelength, λ_F [3]. The fluorescence intensity, I_F, is thus simply proportional to F and the number of photons absorbed by the fluorophore, $I_A(\lambda_E)$. This means th fluorescence intensity of a fluorophore can be written as

$$I_F(\lambda_E, \lambda_F) = kF_\lambda(\lambda_F) I_A(\lambda_E),$$

where λ_E and λ_F denote the wavelength of excitation and emission, respectively, and constant which includes various experimental contributions, such as the bandwidth monochromators and the optical configuration [3].

$I_A(\lambda)$ can be written as:

$$I_A(\lambda_E) = I_0(\lambda_E) - I_T(\lambda_E), \tag{3}$$

where $I_T(\lambda)$ is the intensity transmitted through the sample. From the Beer–Lambert law, $I_T(\lambda_E)$ can be written as:

$$I_T(\lambda_E) = I_0(\lambda_E)\exp(-2.3\varepsilon(\lambda_E)lc), \tag{4}$$

where l is the path length through the sample, epsilon(symbol) is the molar absorption coefficient and c is the concentration of the fluorophore in solution. All three of the Equations (2)–(4) can be combined to give the following:

$$I_F(\lambda_E, \lambda_F) = kF_\lambda(\lambda_F)I_0(\lambda_E)[1 - \exp(-2.3\varepsilon(\lambda_E)lc)]. \tag{5}$$

The exponential can then be expanded to an infinite series for which the higher-order terms become negligible at low concentrations, allowing for further simplification:

$$I_F(\lambda_E, \lambda_F) = 2.3kF_\lambda(\lambda_F)I_0(\lambda_E)A(\lambda_E), \tag{6}$$

where $A(\lambda_E) = \varepsilon(\lambda_E)lc$ is the absorption at the excitation wavelength λ_E (i.e., the Beer–Lambert law [27]). The final step is to integrate over the whole emission spectrum, which yields the following:

$$I_F(\lambda_E) = 2.3k\Phi_F I_0(\lambda_E)A(\lambda_E). \tag{7}$$

This expression gives the fluorescence intensity as a function of parameters that are relatively simple to measure, except for k. Equation (7) shows that for lower concentrations, the fluorescence intensity of emission is directly proportional to the fluorophore concentration. However, expansion of the exponential concentration dependence ($I_F(\lambda_E) \propto e^c$) to a linear behaviour ($I_F(\lambda_E) \propto c$) is only valid for low concentrations, and at some concentration, all of the incoming light will be absorbed by the fluorophore, leading to an upper-limit/saturation in fluorescence intensity. Equations (2) and (3) can be combined and then integrated over all emission wavelengths to give [28]:

$$I_F(\lambda_E, c) = k\Phi_F[I_0(\lambda_E) - I_T(\lambda_E, c)]. \tag{8}$$

A clear implication of this is that when all the light is absorbed by the sample, i.e., $I_0(\lambda_F) >> I_T(\lambda_F)$, the effect of concentration on the emitted intensity is lost. By measuring the fluorescence intensity for increasing concentrations e.g., of LCs in solution, this behaviour is often not observed due to another effect called the inner filter effect [28].

2.2. The Inner Filter Effect (IFE)

Liquid crystals are generally optically very dense in their LC state due to high amounts of absorption and scattering, which makes them difficult to study in neat form using traditional fluorescence spectroscopy methods. The IFE is suggested to be the root cause for this [28,29]. The IFE has been suggested to be present at some level in almost all published emission measurements [30]. However, despite its almost universal appearance, it is often ignored in the literature, which can lead to incorrect and misleading conclusions; the IFE can both dramatically quench emission intensity and induce significant spectral distortions [28,29].

There are two key types of IFE: primary and secondary IFE. Primary IFE is a natural consequence of the sample absorbing light. As light travels through the sample, individual photons are absorbed before being emitted in a random direction, resulting in fewer photons reaching the measurement volume. If the concentration of absorbing species is high enough, almost no light will reach the measurement volume and the measured fluorescence intensity will be significantly reduced. Secondary IFE, sometimes known as secondary absorption, is where a photon emitted by the fluorophore is absorbed by another molecule and is emitted in a random direction before it reaches the detector. This will induce signi spectral distortions due to lost photons. The smaller the overlap between the emissio absorption, or excitation peaks (the Stoke's shift), the stronger the effect of secondar becomes [29].

In the most commonly used experimental geometry for measuring fluorescence tra, the sample is contained in a cuvette placed along the beam path, and the da collected at a right angle (shown in Figure 1a). As both primary and secondary significantly influence experiments in this geometry, many suggestions to correct for effects have been proposed [28,29,31,32]. However, these correction methods often re complicated experimental setups, significant time for data collection, and do not pr notable reduction in errors. Therefore, it has been suggested that front-face measure (such as the one shown in Figure 1b) offer significant advantages over these other niques; the front-face measurement geometry avoids the excitation and emission passing through the entire sample [33], which makes correcting for IFE simpler. recently been shown that it is extremely difficult to completely eliminate the IFE fro measurements, and the IFE can only be minimised by use of front-face measureme thin-film samples [29]. Thus, importantly, the thin sandwich cells typically used to many properties of LCs provide an ideal geometry for front-facing photo-physical st

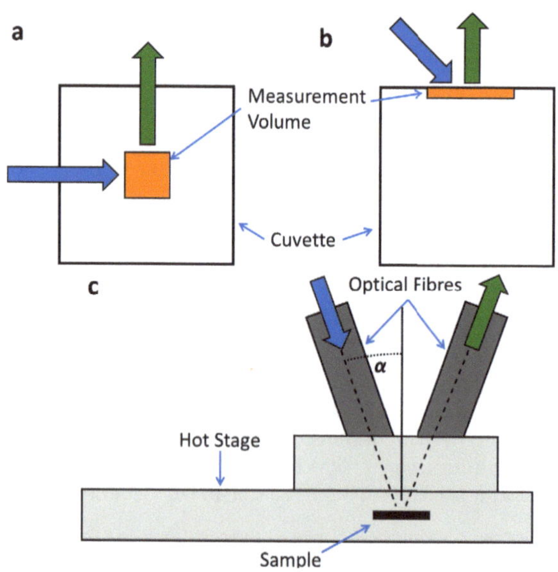

Figure 1. Schematic diagram of two commonly used sample geometries for fluorescence me ments. (**a**) Shows the most commonly found 90° geometry and (**b**) shows the alternative front- geometry, in cuvettes. (**c**) Schematic representation of the alternative geometry of measur compatible with 'standard' LC cells, which allows for measurement of the emission proper LCs as function of temperature, surface conditions etc. As the angle α is 20°, this can be consi as a front-face measurement [34]. In all figures, blue arrows indicates the direction of exci green arrow indicates the direction of observed emission and the orange area indicates the sa measurement volume.

3. Materials and Methods

All materials, including solvents, used in this study were purchased from the mercial supplier Sigma-Aldrich and used without further purification. Care was to only use spectroscopic grade solvents of the highest purity (>99.5% as rated b manufacturer) since trace impurities are often fluorescent and so would be detecta the measured emission and excitation spectra. The chemical structures of 9-(Diethylamino)-5H-benzo[a]phenoxazin-5-one (Nile red) and the liquid crystal used in this study, 4′-pentyl-[1,1′-biphenyl]-4-carbonitrile (5CB), are shown in Figure 2. The LC exhibits a nematic phase at room temperature and a nematic to isotropic phase transition at 35.5 °C. Both Nile red and 5CB are easily soluble at relatively high concentrations using standard solvents such as methanol.

The fluorescence experiments were carried out using an Edinburgh Instruments FLS1000 photoluminescence spectrometer equipped with dual monochromators on both the excitation and emission sections of the fluorimeter. An ozone-free xenon arc lamp was used to excite the samples. The slit bandgaps, which control the bandwidth of the emission and excitation, were set to values between 0.5 and 1 nm depending on the intensity of the sample. The values were kept consistent for each individual sample in order to allow for a direct comparison of intensities. For experiments involving a cuvette geometry, quartz cuvettes with a 1 cm path length were used for the solutions. A N-K02exd TE sample holder was used to hold the cuvettes within the spectrometer. All measurements using cuvettes were conducted using the right-angle geometry shown in Figure 1a.

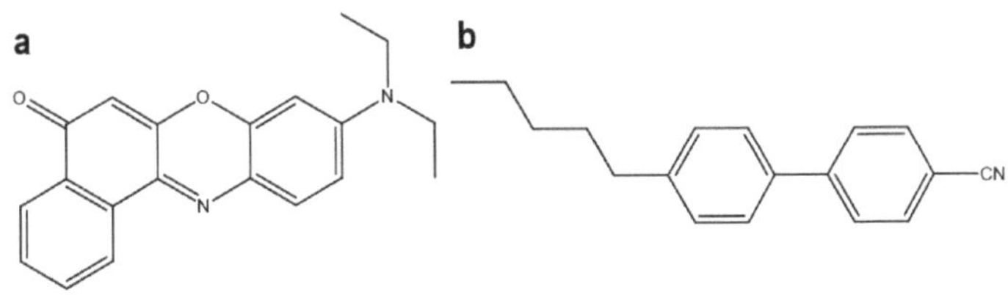

Figure 2. Chemical structures of (**a**) Nile red and (**b**) 5CB, the fluorescent materials used in this study.

In experiments where LC cells were used, the light was directed via optical fibres to a temperature-controlled hotstage (T95, Linkam Instruments). The hotstage was mounted in a reflection geometry, such that the excitation and subsequent collected emission light were normal to the LC director for samples in a planar aligned nematic LC cell. A schematic representation of this experimental geometry is provided in Figure 1c. The LC cells/devices used in this study were purchased from commercial suppliers (AWAT, Poland), and the cells were filled in the nematic phase of 5CB. Planar alignment is achieved using rubbed polyimides with the two substrates assembled in an anti-parallel arrangement. The alignment quality and phase transition temperatures in the cells were checked using polarised optical microscopy (POM) before being used. The glass substrates (with a glass thickness of ~1 mm) cause an attenuation of the excitation beam and a slight distortion of the emission spectra at wavelengths below 350 nm. Beam attenuation reduces the overall emission intensity, and this will be wavelength independent. The soda-lime glass shows reduced transparency below 350 nm (≃20% at 320 nm), which in turn reduces the intensity of the emission spectra below 350 nm. These effects are consistent across all experiments conducted in LC cells and do not change as a function of temperature, cell gap or concentration; therefore, these effects could be ignored. Cells fabricated using quartz glass could be used instead if the investigated material requires access to wavelengths below 350 nm. For the materials used in this study, this was not necessary. All emission and excitation measurements were corrected for the optical efficiency of the optical components within the spectrometer, and for a variation in bulb intensity across the spectrum using the stock correction files provided by the spectrometer manufacturer, which includes the optical efficiency of the optical fibres.

4. Results and Discussion

4.1. Nile Red—A Non Liquid Crystalline Fluorophore

Nile red is a commonly used dye [35,36] that can form H-aggregates (these are side-to-side parallel aggregates [37]) while in aqueous solution, which quenches the emission [38].

The dye does not undergo any changes in its photo-physical behaviour for the conc tions studied here in non-aqueous solutions such as methanol [36,39]. Typical exci and emission spectra for three different concentrations of Nile red in methanol, rec at 640 nm and 555 nm, respectively, are shown in Figure 3. From these graphs, it is that Nile red in 100 μM solution has a lower emission intensity than in the 10 μM sol The normalised spectra also demonstrate that as the concentration of Nile red incr the shape of the peak changes on the lower wavelength side.

Equation (7) shows a linear relationship between intensity and concentration th it should be noted that this equation is specifically only valid for low concentratio discussed in Section 2.1, and the results here could be interpreted as a violation c linear dependence. However, the lack of a linear relationship between intensit concentration, and the variation in peak shape, can in fact be attributed to the IFE. A primary IFE acts to absorb the excitation beam before it reaches the measurement vc the observed reduction in emission intensity can be unrelated to any changes in ma properties, such as quantum yield or intermolecular interactions. Similarly, the dist in the emission shape profile is due to the secondary IFE, where the light that is em from the measurement volume is re-absorbed by Nile red molecules situated bet the measurement volume and the detector. These photons are then re-emitted in ra directions, which means that the chance of them being captured by the detector i and many of these photons are effectively lost [40].

The excitation spectra of Nile red also show effects of IFEs: both the lack of a concentration dependence of the peak intensity, and shape changes of the emission are observed; the latter is most clearly observed for the 100 μM solution, where the emission peak splits into two. The latter behaviour could be interpreted as a physical c in the ground state of the sample; however, we know for Nile red that this is not like these concentrations in this solvent. The more likely cause is primary IFE, where Ni absorbs strongly at the investigated concentrations, thus preventing the excitation bean reaching the measurement volume at its full strength, which reduces the emission int Outside of the peak center, Nile red is less absorbent, and more incoming light can thus the measurement volume, resulting in artificial peaks at 480 nm and 605 nm.

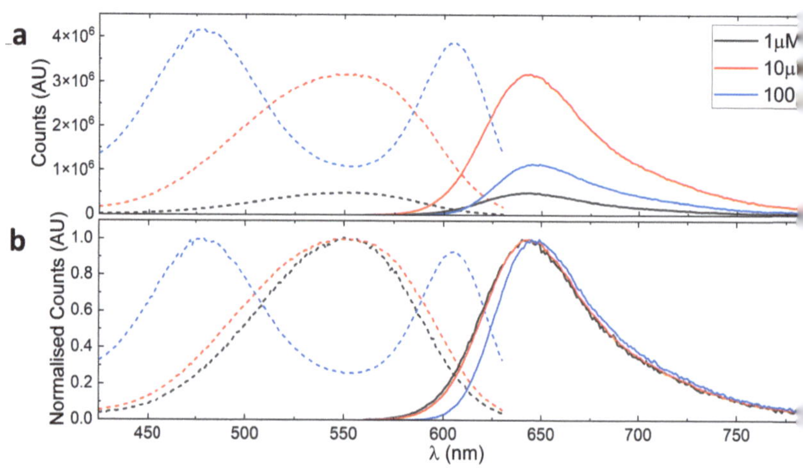

Figure 3. Excitation (dashed lines) and emission (solid lines) spectra of Nile red at various c trations in methanol. The excitation and emission spectra were recorded at 640 nm and 5 respectively. Shown on (**a**) is the raw spectra while on the (**b**) is the normalised spectra. The flat of the excitation spectra of 10 μM and the subsequent splitting of the spectra into two peaks at 1 are due to the primary IFE. The normalised emission spectra show the effects of secondary IFE blue edge where there is some spectral distortion.

The peak positions shown in the excitation spectra are thus determined from a balance between the increased intensity of the excitation source reaching the measurement volume and the reduced absorption coefficient of Nile red at these wavelengths. Spectral distortions due to IFE can be seen even at the relatively low concentrations of 10 μM where a slight increase in peak width can be seen when compared to the peak of the 1 μM solution.

Table 1 provides the intensity counts for the Nile red excitation peaks observed at different concentrations. The table shows that the intensity for the 550 nm emission peak at a concentration of 10 μM is 6.4 times the intensity of the 1 μM solution, even though a 10 times increase is expected based on a linear intensity vs. concentration. These non-linearities are reduced at the edges of the peak, where the absorption is reduced and hence also the effect of primary IFE. Excitation measurements at 450 nm, well away from the peak absorption, show the best agreement with the linear concentration vs. intensity dependence, even though the correlation is still not accurately following a linear behaviour.

Table 1. List of intensity counts taken from the raw excitation spectra of Figure 3 for specific wavelengths for the three different concentrations of Nile red in methanol. The factors in brackets indicate the multiplication factor of the counts from their respective lower concentrations.

c (μM)	450 nm	500 nm	550 nm	600 nm
1	43,000	254,100	491,500	169,000
10	419,400 (9.7×)	2,005,200 (7.9×)	3,161,100 (6.4×)	1,429,100 (8.5×)
100	2,665,800 (6.4×)	3,267,200 (1.6×)	1,077,500 (0.34×)	3,743,100 (2.6×)

Overall, the investigation of the emission and excitation spectra of Nile red shows the limitations of these types of measurements performed in cuvettes, especially at relatively high fluorophore concentrations. Even though Nile red does not undergo any changes to its ground state for the concentrations studied here [36,39], spectral distortions to the emission and excitation spectra are observed for concentrations even as low as 10 μM. While the concentration at which one begins to observe spectral distortions due to IFE changes from sample to sample (as demonstrated in the next section for 5CB), as shown from the work presented here, 10 μM is a commonly used dilute concentration. The presented results also show that the spectra, even for concentrations as low as 10 μM, are slightly distorted, which is a fact that must be considered when analysing the photo-physical measurements of liquid crystals.

4.2. 5CB—Measurements Made in Cuvettes

Figure 4 shows the emission and excitation spectra of 5CB measured at 280 nm and 340 nm, respectively. Since 5CB is a nematic LC at room temperature, it is easily dispersed even at high concentrations using standard lab solvents such as methanol or acetone. This allows measurements of emission spectra up to concentrations as high as 100 mM. In the excitation spectra, the same splitting of the excitation peak that was seen for Nile red occurs also for 5CB at a concentration of 100 μM. As the concentration was increased further, the double peak increasingly separates until it becomes a single peak, which then continuously moves to longer wavelengths.

Similar to the behaviour of Nile red, the emission spectrum of 5CB does not follow a linear relationship over the investigated concentrations as we would expect from the simplified expression in Equation (7), although it should be noted that this equation is specifically only valid for low concentrations as discussed in Section 2.1. Spectral distortions are clearly visible, and the blue edge of the emission spectra red-shifts for increasing concentrations. 5CB has a smaller emission and excitation peak overlap compared to Nile red and is thus less affected by secondary IFE: concentrations as high as 1 mM are required for red-shifting to become visible, while for Nile red, red-shifting effects became visible at 10 μM. As mentioned before, we note that both the peak shape, the peak width, and the

Stoke's shift affects the overlap between the emission and excitation peaks of a
material and thus how much a particular material is affected by the IFE.

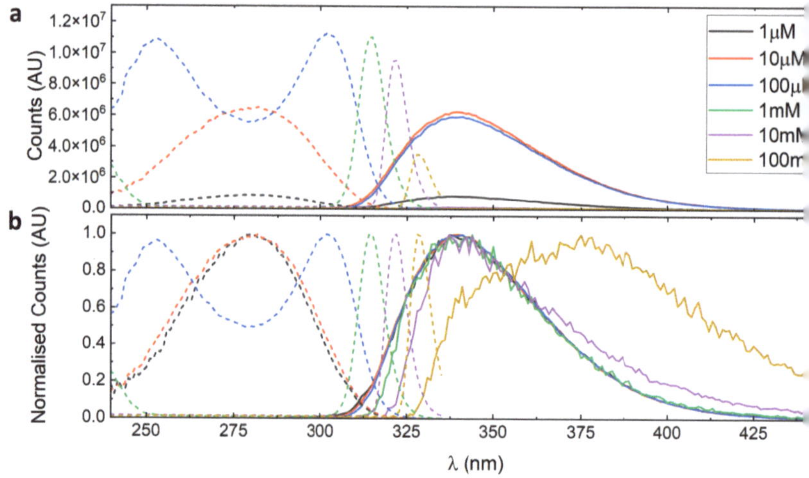

Figure 4. Excitation (dashed lines) emission (solid lines) spectra of 5CB in various concentrati methanol recorded at 340 nm and 280 nm, respectively. The raw spectra are shown in (**a**), ar normalised spectra is shown in (**b**).

An additional effect that is present in 5CB, but does not occur in Nile red, is the f tion and subsequent emission from excimers. The presence of excimer emission in l crystals has been proposed based on the emission measurements of highly concent solutions of LCs in isotropic solvents [19–23]. For example, the calamitic LC, 5CB, diss at low concentration (1.7 μM) in isopropyl alcohol (IPA) showed fluorescence emi at 335 nm, while at high concentrations (0.5 M), it instead showed a new emission at 380 nm but with a shoulder at 335 nm [21]. This red shift was attributed to ex formation between anti-parallel (AP) pairs of 5CB molecules. These AP pairs are known as 'dimers'. Moreover, it has been shown that excimer formation in all nCB nOCBs is related to the same anti-parallel (AP) pair formation that is often stated 'stabiliser' for the LC phase [23,25,26]. However, although excimer emission for the and nOCB series of LCs has been well studied [19–23,41–44], correct consideration to the IFE may have affected those results has not always been given.

In Figure 4, for the high concentration samples (10 mM and 100 mM), an inc in the relative intensity can be seen near 400 nm. While some of this effect will be to the secondary IFE, IFE-induced spectral distortions affect the lower wavelength of the peaks more strongly and would be unlikely to be the cause of the big grow emission at 375 nm for the 100 mM solutions. The main cause of this significant red is more likely to be due to excimer formation, which for 5CB tends to happen in h concentrated solutions. For the 100 mM sample, 5CB now emits more strongly at 37 than at the original peak position of 330 nm, which suggests that excimers appear to b dominant emitting species. However, without being able to fully deconvolute the IF excimer induced red-shifts, it is difficult to conclude this for certain. Therefore, both b understanding excimer emission, and being able to effectively separate it from the pri and secondary IFE, is key to properly understanding the photo-physical properties o liquid crystals.

4.3. 5CB—Front-Facing Measurements

Measurements of 5CB in methanol at concentrations of 100 mM and comparisons literature where excimer formation has been demonstrated in solutions of 5CB in IPA demonstrate that excimers are present and that they influence the measured spectra. It is also clear that above a certain system-dependent concentration, the IFE is very strong in these optically dense systems, and this effect will be exacerbated for neat materials as the nematic phase will have considerably higher scattering due to director fluctuations and defect formation.

To minimise the IFE effects, measurements were performed in a front-facing geometry using LC cells, as discussed in Section 3. As described in Figure 1c, fibre optics are used both for the excitation light and for collecting the emission from the sample; the LC sample was contained in a LC cell and placed on a hot-stage connected to a temperature controller. The LC cells contained alignment layers on both the top and bottom substrates. These layers, and the rest of the LC cell assembly, induce a stable and weak spectral contribution; this contribution can be accurately removed by subtraction from the recorded data. This process is shown in Figure 5a. For 5CB, a relatively highly fluorescent fluorophore, the empty cell is not particularly significant; however, for very thin samples or samples with a lower fluorescence output, this step can become extremely important.

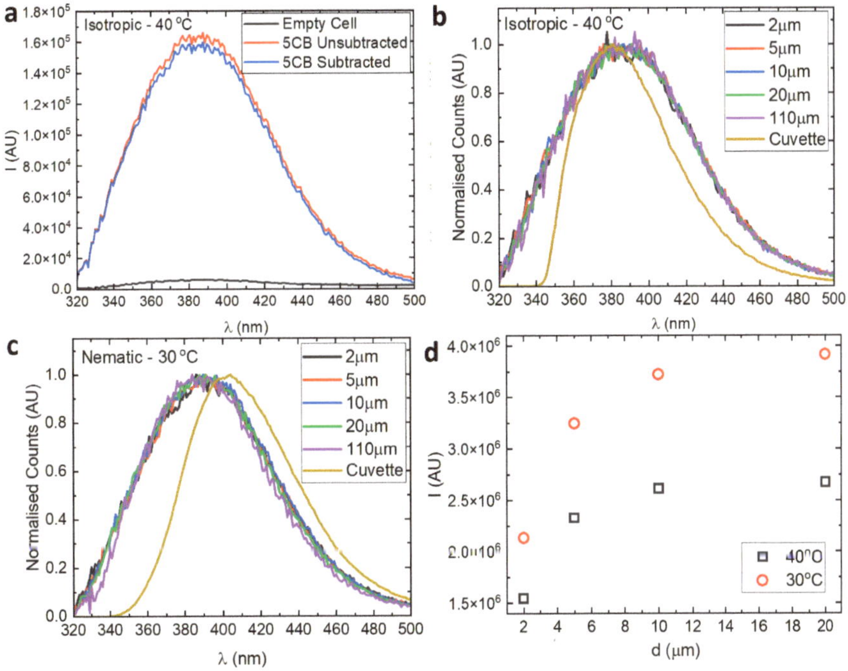

Figure 5. (**a**) Demonstration of the background subtraction process for 5CB at 40 °C. Emission spectra of neat 5CB liquid crystal at (**b**) 40 °C (isotropic phase) and (**c**) 30 °C (nematic phase). The results were obtained using front-face measurements in LC cells with various cell gaps. This is compared with the result obtained from neat 5CB in a cuvette of 1cm path length using the 90° angle method (orange solid line). (**d**) shows the cell gap dependence of emitted intensity, obtained by integrating the area under the emission spectra, for isotropic (black squares) and nematic phases (red circles).

Figure 5b,c show the emission spectra of neat 5CB in its isotropic and nematic phases respectively, which were obtained using front-facing measurements in both LC cells and in a cuvette using the 90° angle method, respectively. The figure contains results from LC cells of different cell gaps, varying from 2 to 110 μm. These cells thus correspond to different optical path lengths for the excitation and emission beams. In both the isotropic and nematic LC phases, the IFE effects are minimised in the front-facing geometry compared with results from cuvettes. Therefore, the spectra presented here are highly accurate emission spectra of 5CB without experimental artefacts. A comparison of measurements from cell cuvettes also demonstrates the magnitude of the IFE-induced spectral distortions i (both in the nematic and isotropic phases).

Overall, spectra for all of the investigated cell gaps, for 5CB in both its isotropi nematic phase, show good agreement. This demonstrates that the resulting spectra r largely unchanged even though thinner cells show significant surface anchoring e whereas for thicker cells (e.g., 110 μm), the anchoring effects are much less releva 30 °C, in the nematic phase, a good agreement between different cells is observed for facing measurements. The spectra from the 110 μm cell show a slight change compa the rest, which could be due an imperfect director alignment in the cell; this will in scattering effects and slightly change the spectra. However, these distortions are neg when compared to the significant differences between the spectra from cells and cu where different peak positions, and shapes, are observed for cells vs. cuvettes.

The comparison of results from cells and cuvettes shows the significance of c consideration, and minimisation of the IFEs, in photo-physical studies. The effects could also be used to understand the penetration depth of incoming light in optically materials. We know that primary IFE attenuates the excitation light, which implies th sample fluorescence only has contributions from before this sample-characteristic d

Figure 5c shows a plot of the overall intensity, defined as the numerically integ area of the peak, as a function of the cell gap, for a 5CB LC in both its isotropic and ne phases. It is clear that by 20 μm, the emitted intensity is significantly saturated, sugg that the measurement can only study the sample to a depth of around 20 μm.

4.4. 5CB—Fitting the Emission Spectra

Once the emission spectra have been measured, a quantitative analysis can b ducted. Often, Gaussian peaks are used to fit fluorescence data [42,45–49], although times other peak types [50] are being used. The fitting of peaks to the fluorescence s can be an important step and needs to be carried out correctly to reach correct conclu on the physical origin of these spectra.

Spectrometers record photon counts per wavelength; however, processes that broaden emission spectra act on the energy scale, not on a wavelength scale. Further inhomogeneous broadening effects such as Doppler broadening [51] and the solvent effect, where the variations in local environment leads to a statistical distribution [52], result in the emission shape being Gaussian on an energy-based scale [53–55]. The length is inversely proportional to energy; therefore, the y-axis conversion from int as a function of wavelength to intensity as a function of energy is not as simple x-axis conversion. Instead, a Jacobian conversion is required to convert the y-axis measured emission spectra. The conversion is needed before applying a fitting proc using Gaussians [56].

Due to the inverse relationship between wavelength and energy, measurements e spaced on a wavelength scale will not be evenly spaced on an energy-scale. For ple, a properly converted constant signal on a wavelength scale will not be consta the energy scale; instead, the counts per unit energy is higher at lower energy/h wavelength, resulting in a curved behaviour. The conversion has an even more signi effect on a Gaussian peak, and incorrectly fitting Gaussians on a wavelength-sca thus lead to erroneous results. Figure 6 shows how three Gaussians, differing only i peak positions, change shape when converted from wavenumber to wavelength s Peak 1 (black; left in Figure 6a) and right in Figure 6b) is significantly stretched in w and shrunk in height compared to peak 3 (blue; right in Figure 6a and left in Figu Note that all the Gaussians in the wavelength scale plot (Figure 6b) have been norm relative to the tallest Gaussian, which has also been normalised to 1.

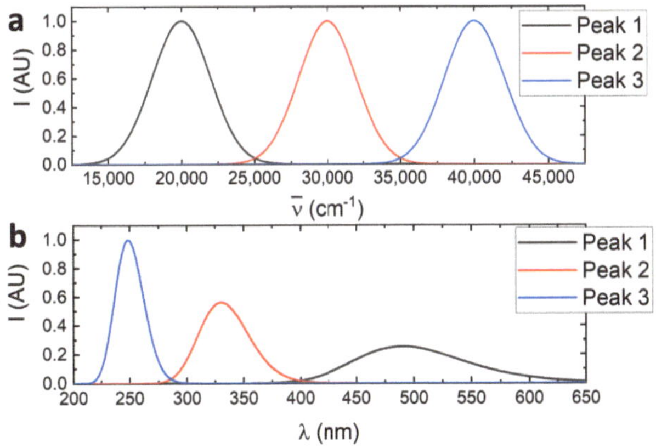

Figure 6. (**a**) Arbitrary Gaussian peaks with the same height and width parameters evenly spaced on a wavenumber scale. (**b**) The same arbitrary Gaussian peaks correctly converted to the wavelength scale showing how Gaussian-shaped spectra are asymmetrically stretched on the wavelength scale vs. the wavenumber.

The Jacobian conversion is defined as

$$f(E) = f(\lambda)\frac{d\lambda}{dE}, \tag{9}$$

where E is the energy and λ is the wavelength, although E can be any quantity that is proportional to the energy. For the wavenumber-scale, this means that

$$f(\overline{\nu}) = -f(\lambda)\lambda^2, \tag{10}$$

where $\overline{\nu}$ is the wavenumber and

$$\overline{\nu} = \frac{1}{\lambda}. \tag{11}$$

Once the data have been correctly converted, Gaussians can be fitted to obtain details about the spectra. A good general approach is to first perform a fit using only a single Gaussian, which is followed by checking the resulting fit residuals. If the fit residuals show only random noise, the Gaussian fit can be accepted. However, if the residual demonstrates a trend beyond random noise, an approach using a sum of multiple Gaussians can be attempted. The investigation of the residuals is repeated until a sufficiently good fit is obtained.

Figure 7 shows the emission spectrum of 5CB in its isotropic phase at 60 °C. The graphs show the results of fitting (a) one and (b) two Gaussians to the experimental spectra. For the single Gaussian fit, the residual contains a significant structure clearly not resulting from random noise. This is also clear from Figure 7a, which further confirms the inadequate quality of the fit. By adding a second Gaussian, the structure in the residuals is significantly decreased, and a good agreement between the fit and the data are observed, as shown in Figure 7b. For 5CB, the dual Gaussians correspond to 'monomer' emission and 'excimer' emission for the high wavenumber (lower wavelength) and low wavenumber (high wavelength) peaks, respectively. For the cuvette measurements, only a small emission shoulder at 350 nm (28,600 cm^{-1}) is observed for the monomer contribution (Figure 4 100 mM), suggesting that the excimer emission is the dominant contribution. However, due to the dominance of IFE, this is difficult to conclude with certainty. Using front-facing measurements, the IFE can be significantly reduced, and hence it can be concluded with certainty that the excimer emission is the dominant contribution even at 60 °C, which is 25 °C above the transition to the nematic phase. Since excimer formation has been linked to AP pair formation in cyanobiphenyl liquid crystals [23], from front-facing experimental results, it is clear that the AP pair formation occurs far from the nematic phase, wh in agreement with the measurements made using dielectric spectroscopy on these crystals [57].

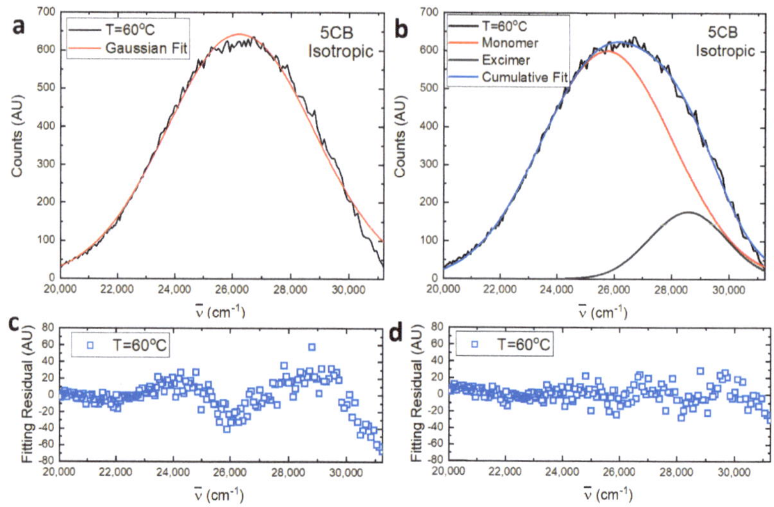

Figure 7. Results from fitting (**a**) a single Gaussian peak and (**b**) a double Gaussian peak emission spectra of 5CB in the isotropic phase using front-facing measurements. Figures (**c**,**d**) the fitting residuals. Studying the fitting residuals here shows how a single Gaussian do adequately describe the emission spectra of 5CB at this temperature.

The approach discussed here can be applied to the emission spectra of materi general, and even though the interpretation of the physical/molecular origin of each Gaussian is not necessarily trivial, performed correctly, this simple analysis can allow deeper understanding of the photo-physical properties of materials [58].

5. Conclusions

Fluorescent LCs are highly desirable for a wide variety of applications. How measuring, understanding, and interpreting their photo-physical properties can be di due to secondary effects such as the IFE. Both highly concentrated LC solutions, and LCs, are optically highly dense and even though IFEs are present in almost all p physical measurements, they are exacerbated in optically dense media. This paper pro a systematic investigation of how IFEs can manifest in the fluorescence data. We illu the effects for two classes of materials, using (i) the fluorophore, Nile red, which do show any spectral changes due to aggregation/collective behaviour for the solvents here, and (ii) the liquid crystal, 5CB, which shows excimer formation at high concentra and in neat solutions. The influences of solute concentration and optical path leng the results and the interpretations are discussed in detail. We demonstrate that by front-face measurements, spectral distortions can be minimised, resulting in highly acc emission spectra. Once, accurate spectra are obtained, the Jacobian wavelength-to-e conversion should be used for correct descriptions of the peak behaviour; this proce is illustrated using a sum of two Gaussians to describe the monomer and the ex emission contributions, respectively, for 5CB.

Author Contributions: Conceptualisation, J.H. and M.N.; methodology, J.H.; software, J.H.; validation, J.H., J.M. and M.N.; formal analysis, J.H.; investigation, J.H., J.M. and M.N.; resources, M.N.; data curation, J.H.; writing—original draft preparation, J.H. and M.N.; writing—review and editing, J.H., J.M. and M.N.; visualisation J.H.; supervision, J.M. and M.N.; project administration, J.M. and M.N.; funding acquisition, J.M. and M.N. All authors have read and agreed to the published version of the manuscript.

Funding: This research was funded by the Engineering and Physical Sciences Research Council (EPSRC) funded Centre for Doctoral Training in Soft Matter and Functional Interfaces (grant EP/L015536/1).

Data Availability Statement: The data in this paper are available in the Leeds Data Repository (https://doi.org/10.5518/1500).

Conflicts of Interest: The authors declare no conflicts of interest. The funders had no role in the design of the study; in the collection, analyses, or interpretation of data; in the writing of the manuscript; or in the decision to publish the results.

ences

ones, J.C. The fiftieth anniversary of the liquid crystal display. *Liq. Cryst. Today* **2018**, *27*, 44–70. [CrossRef]

Klasen-Memmer, M.; Hirschmann, H. Nematic Liquid Crystals for Display Applications. In *Handbook of Liquid Crystals: Volume 3*, 2nd ed.; Goodby, J., Collings, P., Kato, T., Tschierske, C., Gleeson, H., Raynes, P., Vill, V., Eds.; John Wiley & Sons, Ltd.: Hoboken, NJ, USA, 2014. [CrossRef]

Valeur, B.; Berberan-Santos, M. Characteristics of Fluorescence Emission. In *Molecular Fluorescence*, 2nd ed.; John Wiley & Sons, Ltd.: Hoboken, NJ, USA, 2012; pp. 53–74.

Mitschke, U.; Bäuerle, P. The electroluminescence of organic materials. *J. Mater. Chem.* **2000**, *10*, 1471–1507. [CrossRef]

Brenner, M.P.; Hilgenfeldt, S.; Lohse, D. Single-bubble sonoluminescence. *Rev. Mod. Phys.* **2002**, *74*, 425–484. [CrossRef]

Bertolotti, M.; Sansoni, G.; Scudieri, F. Dye laser emission in liquid crystal hosts. *Appl. Opt.* **1979**, *18*, 528–531. [CrossRef]

Mysliwiec, J.; Szukalska, A.; Szukalski, A.; Sznitko, L. Liquid crystal lasers: The last decade and the future. *Nanophotonics* **2021**, *10*, 2309–2346. [CrossRef]

Grell, M.; Bradley, D.D.C.; Inbasekaran, M.; Woo, E.P. A glass-forming conjugated main-chain liquid crystal polymer for polarized electroluminescence applications. *Adv. Mater.* **1997**, *9*, 798–802. [CrossRef]

Grell, M.; Bradley, D.D.C. Polarized Luminescence from Oriented Molecular Materials. *Adv. Mater.* **1999**, *11*, 895–905. [CrossRef]

van Ewyk, R.; O'Connor, I.; Mosley, A.; Cuddy, A.; Hilsum, C.; Blackburn, C.; Griffiths, J.; Jones, F. Anisotropic fluorophors for liquid crystal displays. *Displays* **1986**, *7*, 155–160. [CrossRef]

O'Neill, M.; Kelly, S. Liquid Crystals for Charge Transport, Luminescence, and Photonics. *Adv. Mater.* **2003**, *15*, 1135–1146. [CrossRef]

De, J.; Abdul Haseeb, M.M.; Yadav, R.A.K.; Gupta, S.P.; Bala, I.; Chawla, P.; Kesavan, K.K.; Jou, J.H.; Pal, S.K. AIE-active mechanoluminescent discotic liquid crystals for applications in OLEDs and bio-imaging. *Chem. Commun.* **2020**, *56*, 14279–14282. [CrossRef]

Tong, X.; Zhao, Y.; An, B.K.; Park, S.Y. Fluorescent Liquid-Crystal Gels with Electrically Switchable Photoluminescence. *Adv. Funct. Mater.* **2006**, *16*, 1799–1804. [CrossRef]

Bobrovsky, A.; Shibaev, V.; Hamplová, V.; Novotna, V.; Kašpar, M. Photochromic and fluorescent LC gels based on a bent-shaped azobenzene-containing gelator. *RSC Adv.* **2015**, *5*, 56891–56895. [CrossRef]

Zhang, L.; Cui, Y.; Wang, Q.; Zhou, H.; Wang, H.; Li, Y.; Yang, Z.; Cao, H.; Wang, D.; He, W. Spatial Patterning of Fluorescent Liquid Crystal Ink Based on Inkjet Printing. *Molecules* **2022**, *27*, 5536. [CrossRef]

Sergeyev, S.; Pisula, W.; Geerts, Y.H. Discotic liquid crystals: A new generation of organic semiconductors. *Chem. Soc. Rev.* **2007**, *36*, 1902–1929. [CrossRef]

Birks, J.B. Excimers. *Rep. Prog. Phys.* **1975**, *38*, 903. [CrossRef]

David, C.; Baeyens-volant, D. Absorption and Fluorescence Spectra of 4-Cyanobiphenyl and 4′-Alkyl- or 4′-Alkoxy-Substituted Liquid Crystalline Derivatives. *Mol. Cryst. Liq. Cryst.* **1980**, *59*, 181–196. [CrossRef]

Subramanian, R.; Patterson, L.; Levanon, H. Luminescence behavior as a probe for phase transitions and excimer formation in liquid crystals: Dodecylcyanobiphenyl. *Chem. Phys. Lett.* **1982**, *93*, 578–581. [CrossRef]

Tamai, N.; Yamazaki, I.; Masuhara, H.; Mataga, N. Picosecond time-resolved fluorescence spectra of a liquid crystal: Fluorescence behavior related to phase transitions in cyanooctyloxybiphenyl. *Chem. Phys. Lett.* **1984**, *104*, 485–488. [CrossRef]

Ikeda, T.; Kurihara, S.; Tazuke, S. Excimer formation kinetics in liquid-crystalline alkylcyanobiphenyls. *J. Phys. Chem.* **1990**, *94*, 6550–6555. [CrossRef]

Ikeda, T.; Kurihara, S.; Tazuke, S. Persistence of ordering in 4-n-pentyl-4′-cyanobiphenyl above the nematic-isotropic transition as detected by picosecond time-resolved fluorescence spectroscopy. *Liq. Cryst.* **1990**, *7*, 749–752. [CrossRef]

23. Klock, A.; Rettig, W.; Hofkens, J.; van Damme, M.; De Schryver, F. Excited state relaxation channels of liquid-crys cyanobiphenyls and a ring-bridged model compound. Comparison of bulk and dilute solution properties. *J. Photochem. Ph A Chem.* **1995**, *85*, 11–21. [CrossRef]
24. Voskuhl, J.; Giese, M. Mesogens with aggregation-induced emission properties: Materials with a bright future. *Aggregat* 3, e124. [CrossRef]
25. Goodby, J.W.; Davis, E.J.; Mandle, R.J.; Cowling, S.J. Nano-Segregation and Directed Self-Assembly in the Formation of Fun Liquid Crystals. *Isr. J. Chem.* **2012**, *52*, 863–880. [CrossRef]
26. Dabrowski, R. From the discovery of the partially bilayer smectic A phase to blue phases in polar liquid crystals. *Liq. Crys* *42*, 783–818. [CrossRef]
27. Mayerhöfer, T.G.; Pahlow, S.; Popp, J. The Bouguer-Beer-Lambert Law: Shining Light on the Obscure. *ChemPhysChe* *21*, 2029–2046. [CrossRef]
28. Kimball, J.; Chavez, J.; Ceresa, L.; Kitchner, E.; Nurekeyev, Z.; Doan, H.; Szabelski, M.; Borejdo, J.; Gryczynski, I.; Gryczy On the origin and correction for inner filter effects in fluorescence Part I: Primary inner filter effect - the proper appro sample absorbance correction. *Methods Appl. Fluoresc.* **2020**, *8*, 033002. [CrossRef] [PubMed]
29. Ceresa, L.; Kimball, J.; Chavez, J.; Kitchner, E.; Nurekeyev, Z.; Doan, H.; Borejdo, J.; Gryczynski, I.; Gryczynski, Z. On the and correction for inner filter effects in fluorescence. Part II: Secondary inner filter effect-the proper use of front-face config for highly absorbing and scattering samples. *Methods Appl. Fluoresc.* **2021**, *9*, 035005. [CrossRef] [PubMed]
30. Bevilacqua, M.; Rinnan, A.; Lund, M.N. Investigating challenges with scattering and inner filter effects in front-face fluore by PARAFAC. *J. Chemom.* **2020**, *34*, e3286. [CrossRef]
31. Fonin, A.V.; Sulatskaya, A.I.; Kuznetsova, I.M.; Turoverov, K.K. Fluorescence of Dyes in Solutions with High Absorbance Filter Effect Correction. *PLoS ONE* **2014**, *9*, 0103878. [CrossRef]
32. Wang, T.; Zeng, L.H.; Li, D.L. A review on the methods for correcting the fluorescence inner-filter effect of fluorescence spe *Appl. Spectrosc. Rev.* **2017**, *52*, 883–908. [CrossRef]
33. Valeur, B.; Berberan-Santos, M. Steady-State Spectrofluorometry. In *Molecular Fluorescence*, 2nd ed.; John Wiley & Son Hoboken, NJ, USA, 2012; pp. 263–283. [CrossRef]
34. Gryczynski, Z.K.; Gryczynski, I. Steady-State Fluorescence: Applications. In *Practical Fluorescence Spectroscopy*; CRC Boca Raton, FL, USA, 2019; pp. 315–401.
35. Jessop, P.G.; Jessop, D.A.; Fu, D.; Phan, L. Solvatochromic parameters for solvents of interest in green chemistry. *Green* **2012**, *14*, 1245–1259. [CrossRef]
36. Martinez, V.; Henary, M. Nile Red and Nile Blue: Applications and Syntheses of Structural Analogues. *Chem. A Eur. J* *22*, 13764–13782. [CrossRef] [PubMed]
37. Hestand, N.J.; Spano, F.C. Expanded Theory of H- and J-Molecular Aggregates: The Effects of Vibronic Coupling and Inter ular Charge Transfer. *Chem. Rev.* **2018**, *118*, 7069–7163. [CrossRef]
38. Ray, A.; Das, S.; Chattopadhyay, N. Aggregation of Nile Red in Water: Prevention through Encapsulation in β-Cyclodextrir *Omega* **2019**, *4*, 15–24. [CrossRef] [PubMed]
39. Dutta, A.K.; Kamada, K.; Ohta, K. Spectroscopic studies of nile red in organic solvents and polymers. *J. Photochem. Photo Chem.* **1996**, *93*, 57–64. [CrossRef]
40. Kuśba, J.; Grajek, H.; Gryczynski, I. Secondary emission influenced fluorescence decay of a homogeneous fluorophore so *Methods Appl. Fluoresc.* **2013**, *2*, 015001. [CrossRef]
41. Abe, K.; Usami, A.; Ishida, K.; Fukushima, Y.; Shigenari, T. Dielectric and fluorescence study on phase transitions in liquid 5CB and 8CB. *J. Korean Phys. Soc.* **2005**, *46*, 220–223.
42. Bezrodna, T.; Melnyk, V.; Vorobjev, V.; Puchkovska, G. Low-temperature photoluminescence of 5CB liquid crystal. *J. Lumin* *130*, 1134–1141. [CrossRef]
43. Klishevich, G.V.; Kurmei, N.D.; Melnik, V.I.; Tereshchenko, A.G. Temperature Dependence of the Luminescence Spectra of Liquid Crystal and its Phase Transitions. *J. Appl. Spectrosc.* **2018**, *85*, 904–908. [CrossRef]
44. Oladepo, S.A. Temperature-dependent fluorescence emission of 4-cyano-4′-pentylbiphenyl and 4-cyano-4′-hexylbiphenyl crystals and their bulk phase transitions. *J. Mol. Liq.* **2021**, *323*, 114590. [CrossRef]
45. Subhash, N.; Mohanan, C. Curve-fit analysis of chlorophyll fluorescence spectra: Application to nutrient stress detec sunflower. *Remote Sens. Environ.* **1997**, *60*, 347–356. [CrossRef]
46. Allabergenov, B.; Chung, S.H.; Jeong, S.M.; Kim, S.; Choi, B. Enhanced blue photoluminescence realized by copper dif doping of ZnO thin films. *Opt. Mater. Express* **2013**, *3*, 1733–1741. [CrossRef]
47. Meira, M.; Quintella, C.M.; de O. Ribeiro, E.M.; Silva, W.L. Gaussian fit to the fluorescence spectra for determinat adulteration to diesel by addition of residual oil. *Biomass Convers. Biorefinery* **2015**, *5*, 295–297. [CrossRef]
48. Ashenfelter, B.A.; Desireddy, A.; Yau, S.H.; Goodson, T.I.; Bigioni, T.P. Fluorescence from Molecular Silver Nanoparticles. *J Chem. C* **2015**, *119*, 20728–20734. [CrossRef]
49. Papagiorgis, P.; Stavrinadis, A.; Othonos, A.; Konstantatos, G.; Itskos, G. The Influence of Doping on the Optoelectronic Pro of PbS Colloidal Quantum Dot Solids. *Sci. Rep.* **2016**, *6*, 18735. [CrossRef]
50. Bacalum, M.; Zorilă, B.; Radu, M. Fluorescence spectra decomposition by asymmetric functions: Laurdan spectrum re *Anal. Biochem.* **2013**, *440*, 123–129. [CrossRef] [PubMed]

Struve, W. Spectral Lineshapes and Oscillator Strengths. In *Fundamentals of Molecular Spectroscopy*; Wiley: Hoboken, NJ, USA, 1989; pp. 267–282.
Wehry, E.L. Effects of Molecular Environment on Fluorescence and Phosphorescence. In *Practical Fluorescence*, 2nd ed.; Guilbault, G., Ed.; Modern Monographs in Analytical Chemistry, Taylor & Francis: Abingdon, UK, 1990; pp. 127–184.
Valeur, B.; Berberan-Santos, M. Environmental Effects on Fluorescence Emission. In *Molecular Fluorescence*, 2nd ed.; John Wiley & Sons, Ltd.: Hoboken, NJ, USA, 2012; pp. 109–140.
Kador, L. Stochastic theory of inhomogeneous spectroscopic line shapes reinvestigated. *J. Chem. Phys.* **1991**, *95*, 5574–5581. [CrossRef]
Nemkovich, N.A.; Rubinov, A.N.; Tomin, V.I. Inhomogeneous Broadening of Electronic Spectra of Dye Molecules in Solutions. In *Topics in Fluorescence Spectroscopy: Principles*; Lakowicz, J.R., Ed.; Springer: Berlin/Heidelberg, Germany, 2002; pp. 367–428. [CrossRef]
Mooney, J.; Kambhampati, P. Get the Basics Right: Jacobian Conversion of Wavelength and Energy Scales for Quantitative Analysis of Emission Spectra. *J. Phys. Chem. Lett.* **2013**, *4*, 3316–3318. Corrected in *J. Phys. Chem. Lett.* **2014**, *5*, 3497–3497. [CrossRef]
Dalmolen, L.G.P.; Picken, S.J.; de Jong, A.F.; de Jeu, W.H. The order parameters $\langle P_2 \rangle$ and $\langle P_4 \rangle$ in nematic p-alkyl-p′-cyanobiphenyls: Polarized Raman measurements and the influence of molecular association. *J. Phys.* **1985**, *46*, 1443–1449. [CrossRef]
Dimitriev, O.P.; Piryatinski, Y.P.; Slominskii, Y.L. Excimer Emission in J-Aggregates. *J. Phys. Chem. Lett.* **2018**, *9*, 2138–2143. [CrossRef] [PubMed]

crystals

Article

Machine Learning Analysis of Umbilic Defect Annihilation i Nematic Liquid Crystals in the Presence of Nanoparticles

Ingo Dierking *, Adam Moyle, Gabriele Maria Cepparulo, Katherine Skingle, Laura Hernández and Juhan Rai

Department of Physics and Astronomy, University of Manchester, Oxford Road, Manchester M139PL, UK
adam.moyle@student.manchester.ac.uk (A.M.); gabriele_cepparulo@yahoo.it (G.M.C.);
katherineskingle@gmail.com (K.S.); laura.hrndez@proton.me (L.H.); juhan.raidal@student.manchester.ac.u
* Correspondence: ingo.dierking@manchester.ac.uk

Abstract: Machine learning-based image recognition is employed to investigate the a lation dynamics of umbilic defects induced in systems of nematic liquid crystals doped nanoparticles. A machine learning methodology based on a YOLO algorithm is tr and optimized to identify defects of strength s = ±1 and determine their trajectories d the annihilation process of umbilics of opposite sign. Universal scaling laws descr the distance between two defects as a function of time to annihilation are determined average scaling exponents α are calculated for an ensemble of events. It is observe the defect annihilation scaling exponents deviate from the theoretically predicted of α = 1/2 when nanoparticles of varying size and concentration are introduced system. Scaling laws of the form $D \sim t^{\alpha}$ do not yield the typical square-root law nor observed, but the experiments suggest a decrease in the exponent to saturation val approximately α = 0.38 ± 0.01 as the size, particle concentration, and mass concentrati the nanoparticles is increased. Interestingly, the defect density itself is not affected, v implies that the nanoparticles do not act as defect formation sites.

Keywords: liquid crystal; nematic; defect annihilation; topological defect; umbilic; uni scaling law; machine learning

Academic Editors: Oleg D. Lavrentovich and Maxim V. Gorkunov

Received: 27 January 2025
Revised: 20 February 2025
Accepted: 21 February 2025
Published: 24 February 2025

Citation: Dierking, I.; Moyle, A.; Cepparulo, G.M.; Skingle, K.; Hernández, L.; Raidal, J. Machine Learning Analysis of Umbilic Defect Annihilation in Nematic Liquid Crystals in the Presence of Nanoparticles. *Crystals* **2025**, *15*, 214. https://doi.org/10.3390/cryst15030214

1. Introduction

The coarsening dynamics of nematic liquid crystal topological defects have studied by a number of authors ever since an analogy to cosmological defects was fo lated [1]. Investigations have been carried out both experimentally and via simula and theory. Experimental studies have included string defects [1–4] and defect loops but were mainly concentrated on the annihilation of defect pairs of opposite sign equal strength [6–8], as they are found in the schlieren textures of the nematic phase a temperature or pressure quench from the isotropic liquid. These defects can h strength of s = ±1/2 or s = ±1: they are singularities in the nematic director field an illustrated in Figure 1, both as a texture observed experimentally via polarizing micros and schematically. In an ideal, infinite sample, the total sum over all defects will be to zero.

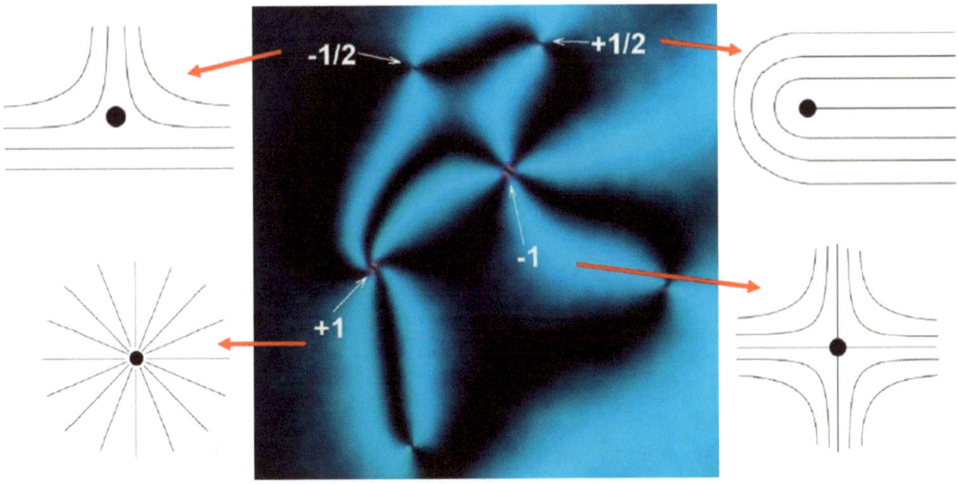

Figure 1. Nematic schlieren texture obtained after a temperature quench from the isotropic liquid between crossed polarizers, illustrating the four singular defects with strengths of s = ±1/2 and s = ±1, also shown schematically (reproduced from ref. [9] with permission).

These types of topological defects arise in many physical systems, such as in cosmology [10], super-fluids [11], solid-state materials [12], active matter [13], and biological [14] and living systems [15]. It was found that liquid crystals are simple model systems to study the mechanisms and dynamics of symmetry breaking through phase transitions, during which topological defects are formed [16].

As liquid crystals are elastic fluids, the defect dynamics can be described by continuum theory, provided that the defect motion is overdamped and adiabatic [17]. Previous experimental work has shown that the dynamics of defect annihilation comprise a complex combination of elastic attractions, viscous drag forces, backflow effects, director configurations, and cell confinement [18]. In this analysis, however, only elastic attractions and viscous drag forces are considered, and elasticity is treated in the one-constant approximation $K_{11} = K_{22} = K_{33} = K$, i.e., splay-, twist-, and bend-elastic constants are taken as equal in magnitude.

Considering two defects of strength s_1 and s_2 separated by distance D, the attractive elastic force is given as follows:

$$F_{elast} = 2\pi s_1 s_2 \frac{K}{D} \tag{1}$$

The drag force for a defect of strength s with velocity v in a film with no flow is given as follows [19]:

$$F_{drag} = \pi\gamma s^2 v \, ln\left(\frac{3.6}{E_r}\right) \tag{2}$$

where γ is the rotational viscosity and E_r is the Eriksen number, equal to the ratio of viscous to elastic forces.

Equating the elastic and drag forces yields a terminal velocity of defects as follows:

$$v = \pm \frac{2K}{\gamma \ln\left(\frac{3.6}{E_r}\right) D} \tag{3}$$

Integration gives the following:

$$\int_{D}^{0} D'dD' = \pm \int_{t}^{t_0} \frac{2K}{\gamma \ln\left(\frac{3.6}{E_r}\right)} dt\prime$$

and thus,

$$D^2(t) = \mp \frac{2K}{\gamma \ln\left(\frac{3.6}{E_r}\right)} (t - t_0)$$

The distance between two annihilating defects therefore follows a simple scalin of the following form:

$$D(t) \sim (t_0 - t)^{\alpha}$$

where t_0 is the time of annihilation and α is the scaling exponent; which takes a va 1/2 according to the simple considerations above. A more explicit theoretical treatme well as computer simulations [20–24], support the scaling law of $D(t)$~$(t_0 - t)^{1/2}$, wh indeed also found experimentally [6–8].

The symmetry-breaking phase transition from isotropic to liquid crystal is induced by temperature quenches, as in the case of the nematic phase [6], but al smectic C [25] or nematic polymers [26,27]. Equivalent is the use of another inte variable of state: pressure quenches to induce the symmetry breaking phase transitio the formation of defects [1,4–6]. We here utilize a somewhat different mechanism to i defects in the form of the application of electric fields to a dielectrically negative ne liquid crystal subjected to homeotropic boundary conditions—the Freedericksz tran as it is commonly called. The director switches from a homeotropic to planar orient while the substrates provide no in-plane directionality and thus cause a large numl defects to be formed. These defects are called umbilics, and they are not singulari the director field [28–30]. Nevertheless, they represent very localized regions of e where the director reorients parallel to the electric field, while the in-plane comp of the director rotates through $\pm 2\pi$. This resembles closely the s = ±1 defects, e that the parallel alignment of the core matches smoothly onto the homeotropic su anchoring instead of ending in singular points. The scaling laws of defect annihilatic equivalent to those of the schlieren defects discussed above, which has been demons in a number of experiments [7,8,31,32] with a varying range of external parameters as applied electric field amplitude and frequency, cell gap, and temperature, as w nematic LC material. The advantage of using this methodology to induce defects lies fact that a much cleaner system of defects is obtained with only s = +1 and s = −1 d being present.

In all cases of experimental, theoretical, and simulation work for various defect ating methods as well as different liquid crystal materials and external applied cond the above scaling law $D(t)$~$(t_0 - t)^{1/2}$ was confirmed to hold and is thus seen to be u sal. In this paper, we investigated the defect annihilation and its corresponding sc exponent for a nematic liquid crystal with dispersed nanospheres at varying concentr and sizes. The experimental results appear to indicate a deviation from universality, would be of much interest to see theoretical and simulation work investigating this pro further to shed further light on this issue.

2. Experimental

2.1. Materials and Experimental Data

The nematic liquid crystal used in this investigation was N-(4-Methoxybenzyli 4-butylaniline (MBBA) from Synthon Chemicals, Bitterfeld-Wolfen, Germany, whic combined with varying amounts of barium titanate nanoparticles of differing sizes between 45 and 500 nm (Nanografi, Jena, Germany). Both materials were used as received. The liquid crystal MBBA has a negative dielectric anisotropy of approximately $\Delta\varepsilon \approx -0.75$ [33]. Dispersions were made by mixing the nanospheres into MBBA with isopropanol at (i) varying mass concentrations from 0.01 to 0.4% for a constant nanosphere diameter of 280 nm (thus varying the particle concentration from 0.00006 to 0.003%, assuming spherical particles of uniform density), (ii) varying nanosphere size from 45 to 500 nm at a constant particle concentration of 0.0002% (thus varying the mass density from 0.00015 to 0.2%), and (iii) varying nanosphere size from 45 to 500 nm at constant mass concentration of 0.1% (thus varying the particle concentration from 0.15 to 0.0001%). After solvent evaporation and sonification of the dispersions for 15 min, the materials were filled into ITO-coated Hele–Shaw cells of 15 μm thickness via capillary action in the nematic state at room temperature.

Defects were induced through the application of electric fields well above the Freedericksz threshold at a voltage amplitude of 20 V peak-to-peak with a frequency of 50 kHz (Agilent 33220A function generator and in-house built power amplifier). Defect formation and annihilation were followed using polarizing microscopy (Leica Optipol) in combination with a digital video camera (UI-3360CP-C-HQ, uEye Gigabit Ethernet) at a frame rate of 30 frames per second and resolution of 2048 × 1088 pixels, corresponding to 1546 × 821 μm. The transient defect formation process lasted about 3 s, and the following annihilation process was recorded for at least 3 min (180 s). Figure 2 shows an exemplary time series during defect annihilation of a MBBA/370 nm particle system. Many defect annihilation events can be observed over time, while the defect density decreases.

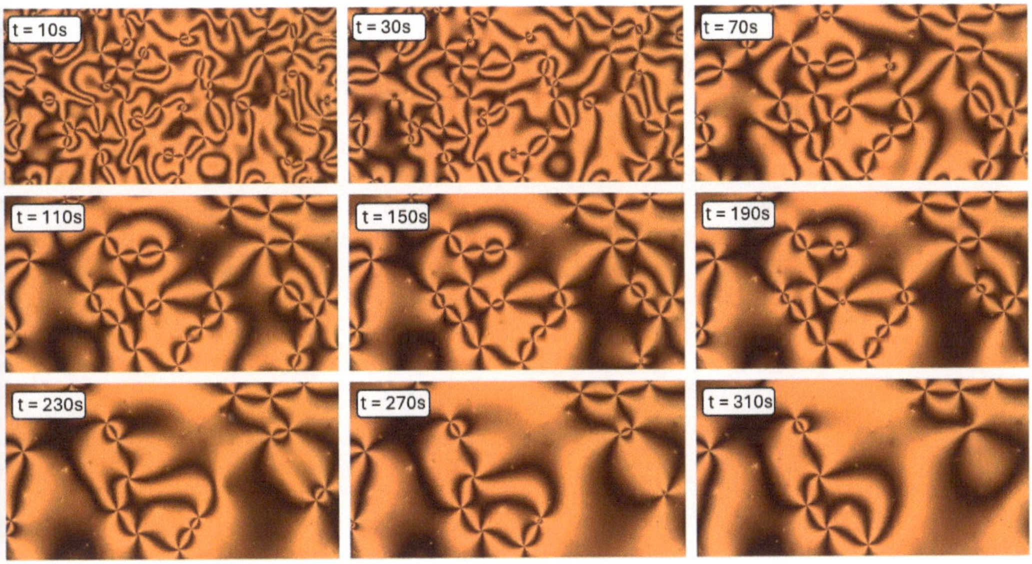

Figure 2. Annihilation of umbilic defects, exemplarily depicted as a time series for a sample of MBBA with 370 nm nanoparticles at a mass concentration of 0.1%.

Time series of frames extracted from video sequences therefore offer a large number of defect pairs that can be analyzed simultaneously, if the individual defects can be recognized automatically. To this end, a machine learning algorithm was utilized.

2.2. Computer Vision and Image Recognition

To detect the defects in the videos taken, a machine learning architecture c YOLOv5 (by Ultralytics) was used. YOLOv5, which is short for "You Only Look version 5", is an object detection algorithm developed using PyTorch 2.1.2 that en fast custom object detection. In general, the YOLOv5 algorithms consist of three part backbone, the neck, and the head. The backbone of the YOLOv5 model is a convolu neural network that aggregates and forms image features at different granularitie it detects different objects in images provided. The neck is a series of layers to mi combine image features to pass them forward to prediction, and the head consumes fea from the neck, classifies the detected objects, and encases them in bounding boxes architecture comes with four prebuilt models, with near-identical methods of oper differing only in the size of the models themselves: YOLOv5s (small), YOLOv5m (med YOLOv5l (large), YOLOv5x (extra-large). The small YOLOv5 model consists of 283 n convolutional layers, while the extra-large model consists of 604 layers. YOLOv5 object detection algorithm that is well known due to its fast object detection and accuracy [34]. From testing the various model sizes that are available, it was conclude YOLOv5s was sufficient to produce a reasonable precision for our anticipated applic while maintaining fast inference times. The YOLOv5 models are pretrained on different commercially available datasets, such as COCO—Common Objects in Co However, since topological defects are not commonly found objects such as cats and are, the models needed to be manually trained on a custom dataset, as discussed bel

To train YOLOv5 to recognize the topological defects, a labelled dataset of ne 4-fold defects was assembled. This was done using an online tool called Roboflow, v enables drawing bounding boxes around objects in images and labelling each cla objects. A total of 380 images were processed via the online platform Roboflow in to use the Roboflow Annotate tool; this allowed each defect to be classified and man labelled with a bounding box. Within Roboflow, the training images were then augm and preprocessed to provide the model with more data; a random selection of in was rotated, flipped, or divided into four. To achieve maximal accuracy in the dete of the defect positions, care was taken during the labelling process. The bounding b were made as small as possible without generating oversight of defects, and the d "singularity" was centered within each bounding box. Testing the models produc YOLOv5 on different versions of the labelled dataset over 100 epochs demonstrate the tighter bounding boxes were more suitable for purpose within this investigation

To quantify the tests, precision and recall were calculated, which are perform metrics defined as follows:

$$precision = \frac{relevant\ retrieved\ instances}{all\ retrieved\ instances} = \frac{tp}{tp + fp}$$

and

$$recall = \frac{relevant\ retrieved\ instances}{all\ relevant\ intances} = \frac{tp}{tp + fn},$$

where *tp* are true positives, *fp* are false positives, and *fn* are false negatives. A h precision implies that an image recognition algorithm returns more relevant results irrelevant ones. A higher recall means that an algorithm returns most of the relevant re

Loss scores are also used to evaluate the success of an object detection mode For example, in YOLOv5, the box loss measures how successfully bounding boxes l the centers and areas of objects. Class, or CLS, loss measures to what degree a m misclassifies objects, which is irrelevant here, as we only have one class. Thus, class l equal to zero. Object loss gives a score of how probable it is that an object resides wi predicted bounding box [36]. An exemplary set of metrics is depicted in Figure 3 over a training period of 100 epochs.

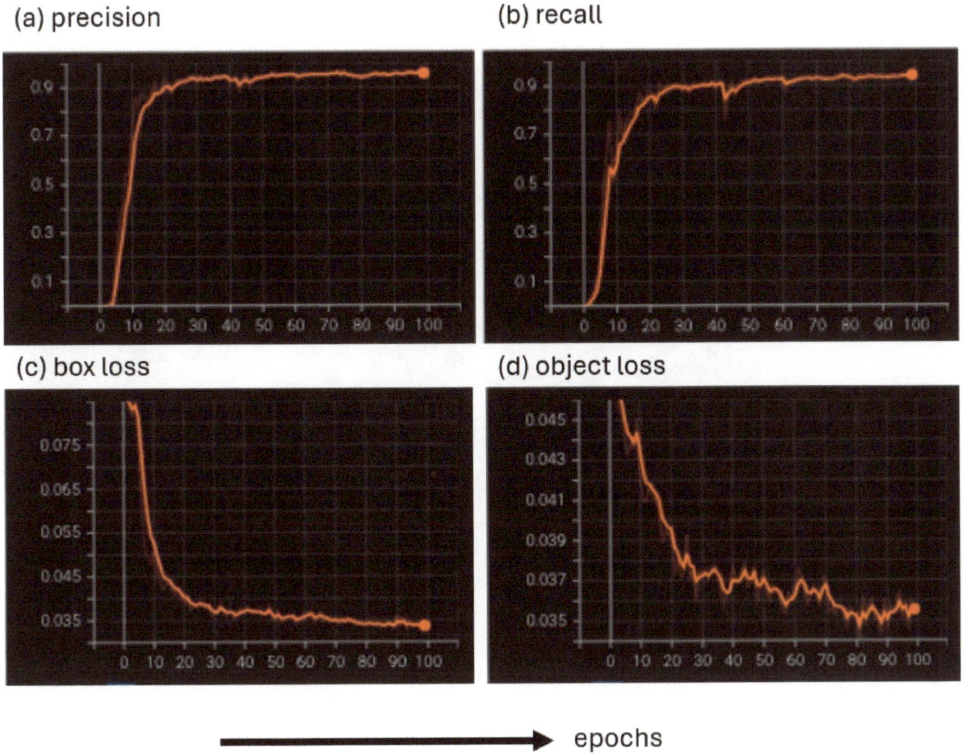

Figure 3. Exemplary demonstration of the metrics to optimize the object recognition algorithm over 100 epochs: (**a**) precision, (**b**) recall, (**c**) box loss, and (**d**) object loss.

The so-called 'best' weights were extracted, which optimized precision vs. recall and were then used to analyze videos of defects. These videos were sped up by three times to reduce runtime, which was adequate given the large number of frames per second. They were also cropped to remove the sections prior to and including defect formation, concentrating on defect annihilation.

DeepSORT (Simple Online and Realtime Tracking with a Deep association metric) is an object tracking software [37]. A notebook in Google Colab was used with DeepSORT in order to track the defects in the videos, analyzing them on a per-frame basis and producing further videos showing bounding boxes on the defects with confidence levels, as depicted in Figure 4. The confidence level here is of the order of 0.65, which is sufficient, as only one class of defects is studied. For other cases, this could be improved through the use of larger training datasets.

By observing the video with tracked defects, appropriate defects within an annihilating pair were selected, and the distance $D(t)$ between annihilating defect pairs was calculated as a function of time.

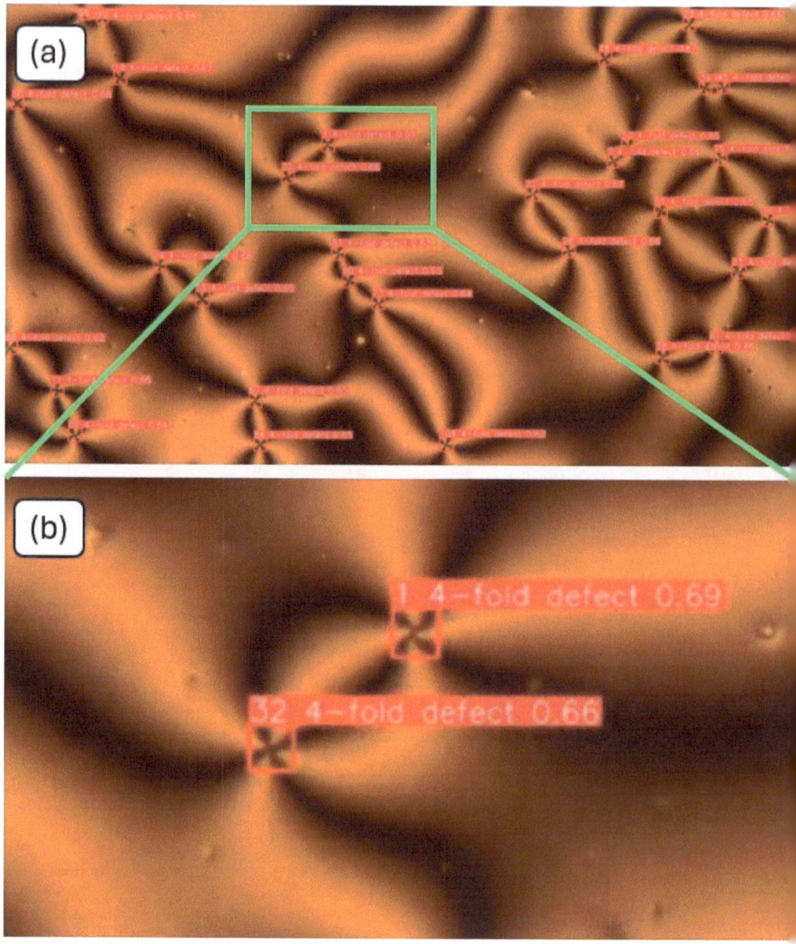

Figure 4. (**a**) Test data illustrating that the trained algorithm not only recognizes the defect
excellent accuracy but (**b**) also places the defect "singularity" in the center of the bounding box,
guarantees highly accurate (x,y)-positions of the tracked defects. The information for each de
the identification number, the class (here, only 4-fold defects), and the confidence level.

3. Experimental Results and Discussion

Firstly, it should be mentioned that for consecutive measurement series, defec
not formed in the same locations for each experiment. Nevertheless, this does not i
that there are no nanoparticles at the center of the defects because the former are gen
attracted to defects and interfaces. Given the size of the nanoparticles employed, this c
be verified using optical microscopy. At first, the defect density, i.e., the number of d
per unit area, was investigated for varying nanoparticle sizes at a constant concentrati
well as varying concentrations at a constant particle size. Some exemplary texture seri
depicted in Figure 5 for varying the particle size from 45 nm to 500 nm at a concentr
of 0.2% (part (a)) and for varying the nanoparticle concentration from 0.01% to 0.4
a constant particle size of 280 nm (part (b)). In each case, the frame at time t = 3
selected for comparison. It is interesting to note that the defect density is almost co
and independent of nanoparticle size and nanoparticle mass concentration.

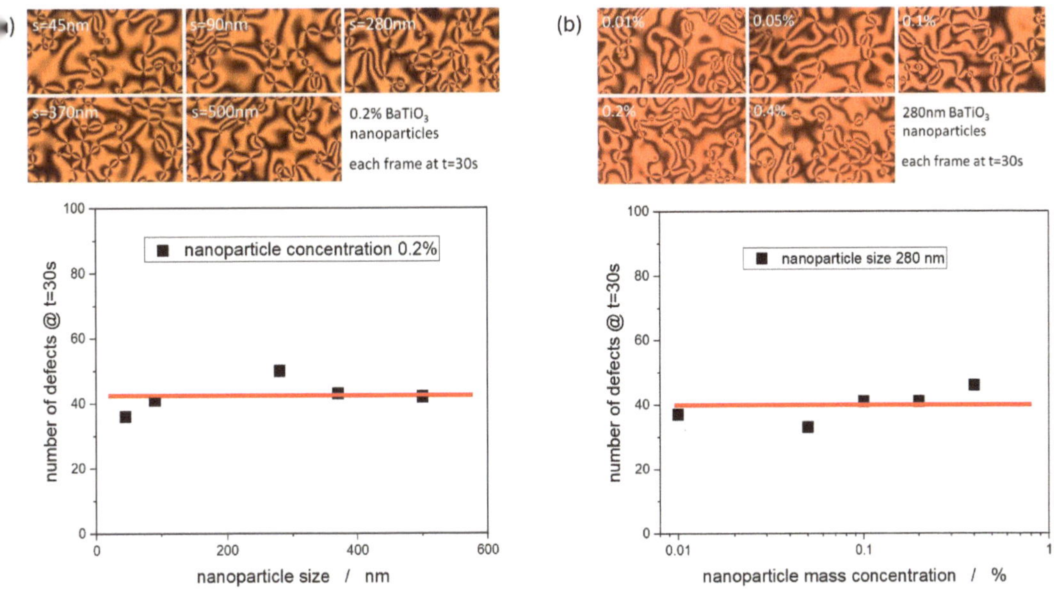

Figure 5. Textures at t = 30 s and corresponding number of defects observed (**a**) as a function of nanoparticle size at a concentration of 0.1% and (**b**) nanoparticle mass concentration for a particle size of 280 nm.

This implies that the addition of nanoparticles neither appears to influence the defect formation nor the attractive force between the defects. At the same time, the defect density $\rho(d,c,t) = \rho(t)$, with d being the particle size and c being the particle concentration, and in this particular case analyzed above, $\rho(t = 30\text{ s}) = (33 \pm 4)\text{ mm}^{-2}$.

Figure 6 is an exemplary demonstration of defect distance acquisition and subsequent determination of the defect annihilation exponent α from the obtained position data acquired via machine learning as a function of time. In Figure 6a, the original data $D(t)$ are depicted, which are clearly described by a nonlinear relationship $D(t)\sim(t_0 - t)^{\alpha}$. The annihilation exponent can then be calculated from a log–log representation, which exhibits a linear scaling regime after some time following the application of the electric field. At early times, the relation is not suitable for analysis because of the convolution of two simultaneous processes: defect formation and defect annihilation.

It should be pointed out that this linear scaling regime is observed for all of the subsequent measurement series, but care needs to be taken to avoid the regime of overlap between formation and annihilation of defects. This way, we could analyze between 6 to 40 defect pairs for each of the individual nematic nanoparticle dispersions. Outliers, such as defects that become trapped with dust contamination within the liquid crystal, were removed from the dataset considering the Chauvenet criterion, which defines an acceptable scattering of values around the mean value from N measurements. This is a widely accepted and soft-touch approach to identify outliers, even for small datasets.

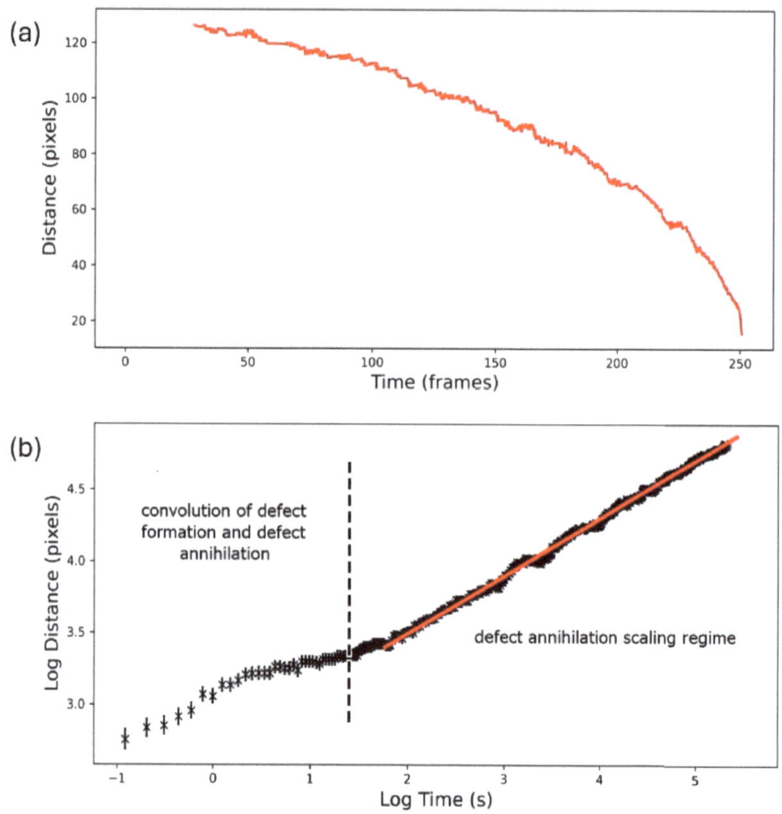

Figure 6. (**a**) Original data of $D(t)$, determined via machine learning, and (**b**) log–log represen for the determination of the annihilation exponent α from the slope of the linear scaling regim

3.1. Variation of Mass Concentration at a Constant Particle Size

With this, we can proceed to discuss the annihilation exponents as a functi nanoparticle mass concentration at constant particle size of 280 nm, which is close t middle of the investigated size range. The results are shown in Figure 7, with the red being a guide to the eye.

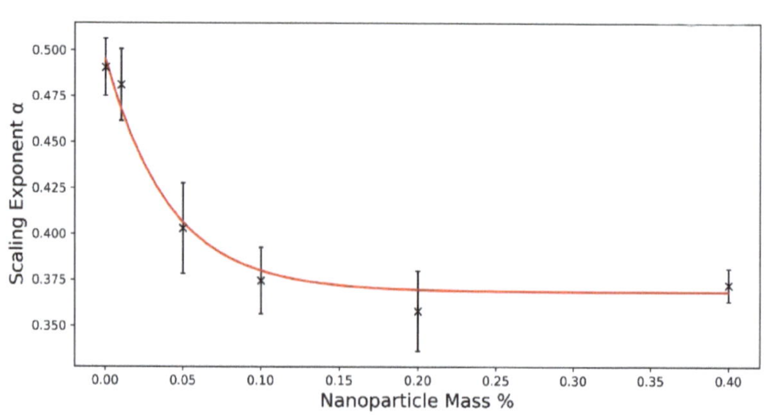

Figure 7. Scaling exponent α as a function of nanoparticle mass concentration at a constant pa size of 280 nm.

At first, one can notice that for the neat liquid crystal, it is found within the limits of error that $\alpha = 0.49 \pm 0.02$; thus, $\alpha = 1/2$, as expected from theoretical predictions and simulations [20–24] and confirmed by a range of experiments [6–8,31]. For increasing mass concentrations, the annihilation exponent decreases before saturating at approximately $\alpha = 0.37$, which is clearly smaller than $\alpha = 1/2$.

3.2. Variation of Particle Size at a Constant Particle Concentration

In the next series of experiments, the nanoparticle concentration was kept constant at 0.0002% and the particle size was varied between 45 nm and 500 nm, thus also varying the mass concentration (Figure 8).

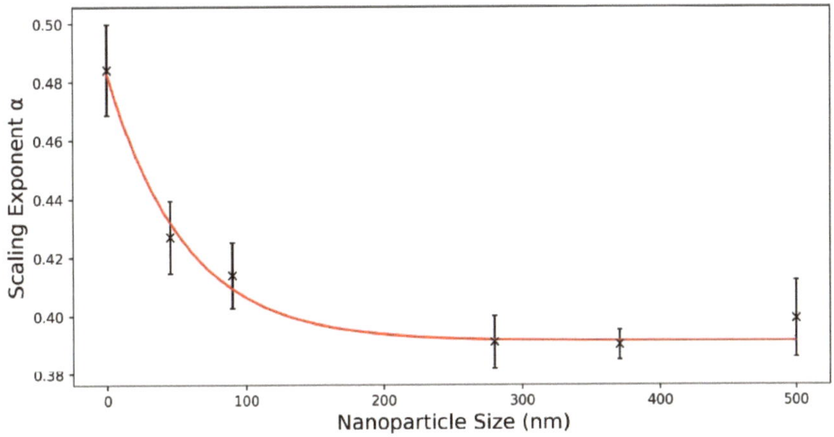

Figure 8. Scaling exponent α for variations in particle size at a constant nanoparticle concentration of 0.0002%.

Within the limits of error, the neat liquid crystal shows a value of the annihilation exponent near 1/2, $\alpha = 0.483 \pm 0.007$, as expected. Again, the for the particle-laden systems, the annihilation exponent decreases, approaching a value of approximately $\alpha = 0.39$, clearly lower than the predicted exponent.

3.3. Variation of Nanoparticle Size at a Constant Mass Concentration

In the last experimental series, the nanoparticle size was varied between 45 and 500 nm at constant mass concentration of 0.1%, thus also varying the particle concentration. The results are shown in Figure 9.

The exponent of the neat liquid crystal once again follows the predicted behavior, showing a value of $\alpha = 0.49 \pm 0.02$, which reproduces the predicted annihilation exponent of 1/2. The exponent decreases and saturates at a value of about $\alpha = 0.37$.

The experiments in general suggest that neat liquid crystals, independent of their molecular species and of applied external conditions such as confining cell gap, applied electric field amplitude and frequency, or temperature, away from the vicinity of phase transitions, follow the predicted behavior based on theory and computer simulations. Nevertheless, nanoparticle-doped nematics appear to produce deviations from the accepted behavior, not with respect to the number density of defects observed when the nanoparticle size or concentration is varied, but rather with respect to the fundamental dynamics of defect annihilation, changing respective scaling exponents.

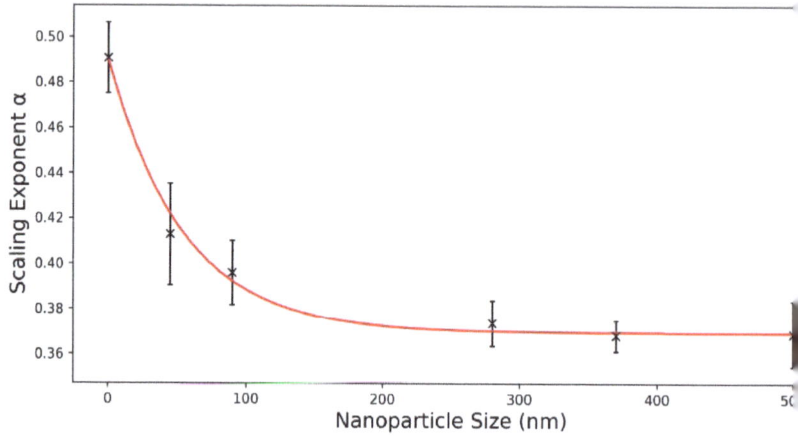

Figure 9. Defect annihilation scaling exponent α for variations in particle size at a constan concentration of 0.1%.

A possible explanation could be the accumulation of particles at the defects as th to annihilation proceeds. It is known that nanoparticles accumulate within defect re and at interfaces. Even though the particles do not appear to be acting as defect nucl centers, they can be collected within the defect region while the defect moves throu liquid crystal–nanoparticle dispersion and also through diffusion and attractive forces can vary the defect velocity, as discussed below in relation to Figure 10, and thus cl the drag force F_{drag} (Equation (2)) with time. A number of factors would play int scenario: (i) particle concentration, because more particles would be accumulated for l concentrations within the same time, and (ii) particle size, because the larger the pa the smaller the velocity caused by a constant drag. The velocity v (Equation (3)) w then become a function of time t, parametrized via particle concentration and pa size, which would change the obtained scaling exponent in Equation (6). It is jus to assume that there is a certain maximum number of particles of a given size wl defect can accommodate, explaining the saturation of the exponent α for increasing pa concentration, size, and thus overall mass. It would be of interest to verify such beh through computer simulations.

It is known based on theory and simulations [38,39] that attracting defects d the annihilation approach often travel in an S-shaped trajectory. A similar behav also observed in the experiments presented here, with an exemplary trajectory sho Figure 10, which was determined via the same machine learning algorithm as out above. This shows that, ideally, machine learning in the future will facilitate investig of this phenomenon in detail and provide insight into the fundamental mechanism c type of dynamic behavior.

However, it is also known from experiments and simulations [32] of undoped l crystals that in general, the $s = +1$ defect has a speed about twice that of the $s = -1$ c $v_{(+1)} = 2v_{(-1)}$, which is attributed to backflow effects [32,40]. The obtained data in Figu on the other hand, suggest very equal speeds for both defects of opposite sign, indic a reduction of $v(_{+1})$ compared to a pure liquid crystal without nanoparticles. Pos the addition of nanoparticles prevents or diminishes backflow effects and modifie viscous drag force, thus causing deviations in the generally observed behavior d defect annihilation and its scaling compared to non-doped systems.

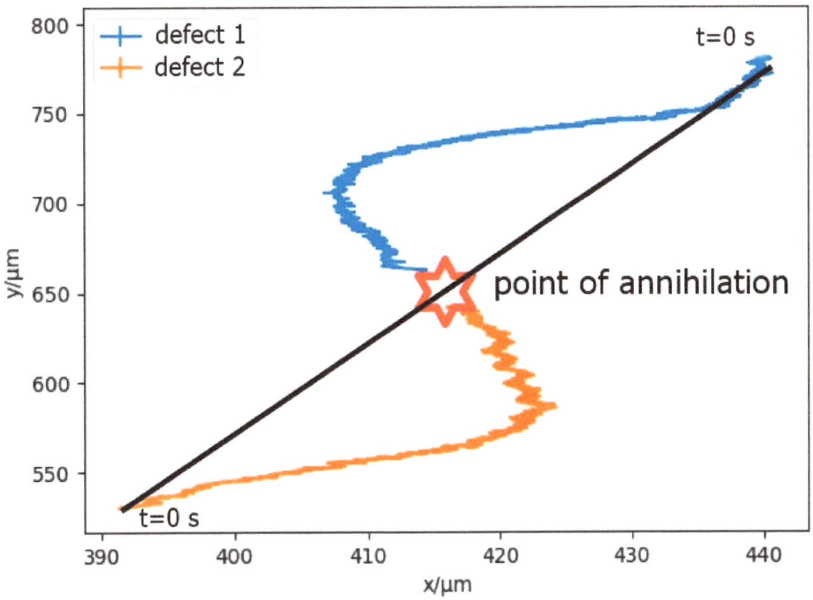

Figure 10. S-shaped trajectory of two defects approaching the point of annihilation. Particle size is 280 nm at a mass concentration of 0.1%.

4. Conclusions

By training and optimizing an image recognition algorithm based on machine learning, we were able to investigate the defect annihilation dynamics for an ensemble of annihilation events of umbilic defects of strength $s = \pm 1$ in detail. While the defect density is not affected by the addition of nanoparticles of various sizes and concentrations, the annihilation exponent is clearly affected when compared to the neat nematic liquid crystal system. The undoped nematic follows the predicted scaling law of square root functionality with exponent $\alpha = 1/2$, while the addition of nanoparticles decreases the exponent by approximately 25% to saturate at about $\alpha = 0.38 \pm 0.01$ for increasing particle size and increasing particle concentration. A qualitative explanation is proposed. The experimental behavior suggests that for dispersions, the scaling of topological defect annihilation might be altered, despite the fact that changes in external conditions, such as liquid crystal material, applied electric field amplitude or frequency, temperature, and confining cell gap have been shown to not affect the respective scaling law.

Furthermore, the S-shaped defect trajectories, which have been predicted theoretically and via simulations, were experimentally verified, which have been proposed theoretically and via simulation. Nevertheless, the previously observed speed anisotropy of defects of opposite sign, which is attributed to backflow effects, could not be observed in nanoparticle-doped systems. One possible explanation is that the dispersed nanoparticles diminish backflow, which might also be the reason for the modified annihilation scaling laws. It would be of fundamental interest to further investigate this issue theoretically or with computer simulations.

Author Contributions: Conceptualization, I.D.; initial methodology, J.R.; improved methodology, A.M., G.M.C., K.S. and L.H.; investigation, K.S., L.H., A.M. and G.M.C.; resources, I.D.; writing—original draft preparation, I.D.; review and editing, I.D. and K.S.; supervision, I.D. All authors have read and agreed to the published version of the manuscript.

Funding: This research received no external funding.

Data Availability Statement: The original contributions presented in this study are included article. Further inquiries can be directed to the corresponding author.

Conflicts of Interest: The authors declare no conflict of interest.

References

1. Chuang, I.; Durrer, R.; Turok, N.; Yurke, B. Cosmology in the laboratory: Defect dynamics in liquid crystals. *Science* **1991**, 25 [CrossRef]
2. Orihara, H.; Ishibashi, Y. Dynamics of Disclinations in Twisted Nematics Quenched below the Clearing Point. *J. Phys. S* **1986**, *55*, 2151–2156. [CrossRef]
3. Orihara, H.; Nagaya, T.; Ishibashi, Y. Collective Motion of an Assembly of Disclinations in the Non-Orthogonally T Nematics Quenched below the Clearing Point. *Phys. Soc. Jpn.* **1987**, *56*, 3086–3091. [CrossRef]
4. Chuang, I.; Turok, N.; Yurke, B. Late-time coarsening dynamics in a nematic liquid crystal. *Phys. Rev. Lett.* **1991**, *66* [CrossRef]
5. Chuang, I.; Yurke, B.; Pargellis, A.N.; Turok, N. Coarsening dynamics in uniaxial nematic liquid crystals. *Phys. Rev.* *47*, 3343. [CrossRef]
6. Pargellis, A.N.; Turok, N.; Yurke, B. Monopole-antimonopole annihilation in a nematic liquid crystal. *Phys. Rev. Lett.* *67*, 1570. [CrossRef] [PubMed]
7. Nagaya, T.; Hotta, H.; Orihara, H.; Ishibashi, Y. Observation of Annihilation Process of Disclinations Emerging from Domains. *J. Phys. Soc. Jpn.* **1991**, *60*, 1572–1578. [CrossRef]
8. Nagaya, T.; Hotta, H.; Orihara, H.; Ishibashi, Y. Experimental Study of the Coarsening Dynamics of +1 and -1 Disclinat *Phys. Soc. Jpn.* **1992**, *61*, 3511–3517. [CrossRef]
9. Dierking, I.; Archer, P. Imaging liquid crystal defects. *RSC Adv.* **2013**, *3*, 26433. [CrossRef]
10. Hindmarsh, M.B.; Kibble, T.W.B. Cosmic strings. *Rep. Prog. Phys.* **1995**, *58*, 411. [CrossRef]
11. Zurek, W.H. Cosmological experiments in superfluid helium? *Nature* **1985**, *317*, 505–508. [CrossRef]
12. Zurek, W.H. Cosmological experiments in condensed matter system. *Phys. Rep.* **1996**, *276*, 177–221. [CrossRef]
13. Shankar, S.; Souslov, A.; Bowick, M.J.; Marchetti, M.C.; Vitelli, V. Topological active matter. *Nat. Rev. Phys.* **2022**, *4*, 38 [CrossRef]
14. Ardaševa, A.; Doostmohammadi, A. Topological defects in biological matter. *Nat. Rev. Phys.* **2022**, *4*, 354–356. [CrossRef]
15. Fardin, M.-A.; Ladoux, B. Living proof of effective defects. *Nat. Phys.* **2021**, *17*, 164–173. [CrossRef]
16. Kralj, M.; Kralj, M.; Kralj, S. Topological Defects in Nematic Liquid Crystals: Laboratory of Fundamental Physics. *Phys. Solidi A* **2021**, *218*, 2000752. [CrossRef]
17. Harth, K.; Stannarius, R. Topological Point Defects of Liquid Crystals in Quasi-Two-Dimensional Geometries. *Front. Phys* *8*, 112. [CrossRef]
18. Shen, Y.; Dierking, I. Annihilation dynamics of topological defects induced by microparticles in nematic liquid crystal *Matter* **2019**, *15*, 8749–8757. [CrossRef] [PubMed]
19. Imura, H.; Okano, K. Friction coefficient for a moving disinclination in a nematic liquid crystal. *Phys. Lett. A* **1973**, *42*, 40 [CrossRef]
20. Toyoki, H. Pair annihilation of point like topological defects in the ordering process of quenched systems. *Phys. Rev. A* *42*, 911. [CrossRef]
21. Mondello, M.; Goldenfeld, N. Scaling and vortex dynamics after the quench of a system with a continuous symmetry. *Phy A* **1990**, *42*, 5865. [CrossRef]
22. Toyoki, H. Cell dynamics simulation for the phase ordering of nematic liquid crystals. *Phys. Rev. E* **1993**, *47*, 2558. [Cro [PubMed]
23. Jang, W.G.; Ginzburg, V.V.; Muzny, C.D.; Clark, N.A. Annihilation rate and scaling in a two-dimensional system of ch particles. *Phys. Rev. E* **1995**, *51*, 411. [CrossRef] [PubMed]
24. Liu, C.; Muthukumar, M. Annihilation kinetics of liquid crystal defects. *J. Chem. Phys.* **1997**, *106*, 7822–7828. [CrossRef]
25. Pargellis, A.N.; Finn, P.; Goodby, J.W.; Panizza, P.; Yurke, B.; Cladis, P.E. Defect dynamics and coarsening dynamics in sme films. *Phys. Rev. A* **1992**, *46*, 7765. [CrossRef]
26. Ding, D.-K.; Thomas, E.L. Structures of Point Integer Disclinations and Their Annihilation Behavior in Thermotropic L Crystal Polyesters. *Mol. Cryst. Liq. Cryst.* **1994**, *241*, 103–117. [CrossRef]
27. Wang, W.; Shiwaku, T.; Hashimoto, T. Experimental study of dynamics of topological defects in nematic polymer liquid cr *J. Chem. Phys.* **1998**, *108*, 1614–1625. [CrossRef]
28. Rapini, A. Umbilics: Static properties and shear-induced displacements. *J. Phys.* **1973**, *34*, 629–633. [CrossRef]

Saupe, A. Disclinations and Properties of the Director field in Nematic and Cholesteric Liquid Crystals. *Mol. Cryst. Liq. Cryst.* 1973, *21*, 211–238. [CrossRef]

Meyer, R.B. Point Disclinations at a Nematic-lsotropic Liquid Interface. *Mol. Cryst. Liq. Cryst.* **1972**, *16*, 355–369. [CrossRef]

Dierking, I.; Marshall, O.; Wright, J.; Bulleid, N. Annihilation dynamics of umbilical defects in nematic liquid crystals under applied electric fields. *Phys. Rev. E* **2005**, *71*, 061709. [CrossRef]

Dierking, I.; Ravnik, M.; Lark, E.; Healey, J.; Alexander, G.P.; Yeomans, J.M. Anisotropy in the annihilation dynamics of umbilic defects in nematic liquid crystals. *Phys. Rev. E* **2012**, *85*, 021703. [CrossRef] [PubMed]

Beigmohammadi, M.; Sadigh, M.K.; Poursamad, J. Dielectric anisotropy changes in MBBA liquid crystal doped with barium titanate by a new method. *Sci. Rep.* **2024**, *14*, 5756. [CrossRef] [PubMed]

Nelson, J.; Solawetz, J. Yolov5 is Here. 2020. Available online: https://blog.roboflow.com/yolov5-is-here/ (accessed on 5 January 2025).

Redmon, J.; Divvala, S.; Girshick, R.; Farhadi, A. You only look once: Unified, realtime object detection. *arXiv* **2015**, arXiv:1506.02640.

Patel, M. Detailed Information on Different Versions of YOLO. Available online: https://towardsai.net/ (accessed on 5 January 2025).

Wojke, N.; Bewley, A.; Paulus, D. Simple online and real-time tracking with a deep association metric. In Proceedings of the 2017 IEEE International Conference on Image Processing (ICIP), Beijing, China, 17–20 September 2017. [CrossRef]

Missaoui, A.; Harth, K.; Salamon, P.; Stannarius, R. Annihilation of point defect pairs in freely suspended liquid-crystal films. *Phys Rev Res.* **2020**, *2*, 013080. [CrossRef]

Tang, X.; Selinger, J.V. Annihilation trajectory of defects in smectic-C films. *Phys. Rev. E* **2020**, *102*, 012702. [CrossRef] [PubMed]

Missaoui, A.; Lacaze, E.; Eremin, A.; Stannarius, R. Observation of Backflow during the Annihilation of Topological Defects in Freely Suspended Smectic Films. *Crystals* **2021**, *11*, 430. [CrossRef]

crystals

Article

Polarization Coupling between Ferroelectric Liquids and Ferroelectric Solids: Effects of the Fringing Field Profile

Stefano Marni [1], Raouf Barboza [1], Ayomide S. Oluwajoba [1], Riccardo Zamboni [2,3] and Liana Lucchetti [1,*]

[1] Dipartimento SIMAU, Università Politecnica delle Marche, Via Brecce Bianche, 60131 Ancona, Italy; s.marni@univpm.it (S.M.); r.barboza@univpm.it (R.B.); s1101521@studenti.univpm.it (A.S.O.)
[2] Institute of Applied Physics, University of Münster, Correnstr. 2/4, 48149 Münster, Germany; riccardo.zamboni@uni-muenster.de
[3] Dipartimento di Fisica e Astronomia G. Galilei, Universitá di Padova, 35131 Padova, Italy
* Correspondence: l.lucchetti@univpm.it

Abstract: Recent experiments devoted to characterizing the behavior of sessile ferroelectric droplets on ferroelectric solid substrates have shown the existence of a droplet electromech Rayleigh-like instability. The instability is induced by the bulk polarization of the ferroelectri which couples to the polarization of the underlying substrate through its fringing field and solic interface coupling. With the aim of characterizing this phenomenon, namely the coupling betwe polarizations of a fluid and a solid material, we studied the behavior of ferroelectric liquid d confined between two solid substrates, arranged in different configurations, realized to ge fringing fields with different profiles. The results show that the features of the droplets instabi indeed affected by the specific fringing field shape in a way dominated by the minimization electrostatic energy associated with the bulk polarization of the ferroelectric fluid.

Keywords: ferroelectric nematic liquid crystals; lithium niobate; fringing field; polarization co

Citation: Marni, S.; Barboza, R.; Oluwajoba, A.S.; Zamboni, R.; Lucchetti, L. Polarization Coupling between Ferroelectric Liquids and Ferroelectric Solids: Effects of the Fringing Field Profile. *Crystals* **2024**, *14*, 425. https://doi.org/10.3390 cryst14050425

Academic Editor: Ingo Dierking

Received: 10 April 2024
Revised: 25 April 2024
Accepted: 27 April 2024
Published: 29 April 2024

1. Introduction

The discovery of the ferroelectric nematic phase [1–5] was a groundbreaking since the combination of fluidity in ferroelectric fluid and its polar coupling to electric allow the observation of a whole new world of phenomena, which are rapidly becc the focus of an increasing number of scientists, as demonstrated by the large num peer review articles already published on the subject [6–16].

In this scenario, we recently performed experiments aimed at characterizing t havior of sessile ferroelectric nematic liquid crystal droplets into contact with a ferroe solid substrate [8]. We observed that upon entering the ferroelectric nematic phase, dr experience an electromechanical instability that manifests itself through the sudde tion of fluid jets, which branch into smaller streams and eventually form secondar droplets [8]. This behavior resembles the instability predicted by Lord Rayleigh in 18 charged conductive liquid droplets that are above the critical charge-to-volume rati Since ferroelectric nematic droplets are neutral, in our case this process occurs in t sence of free charges, but the required charging within the droplet arises from the int polarization of the ferroelectric liquid crystal via its contact with the ferroelectric sub The coupling between the polarization in the solid and fluid materials is mediated b fringing field generated by the pyroelectric charging of the substrate. This polariz coupling induces the accumulation of surface charges on the droplet–air interface. A droplet polarization grows by cooling the material, the local accumulation of polariz charges produces repulsive forces that overcome the surface tension. As this condit met, the instability turns to an explosive runaway process, since the flow of the ferroe liquid crystal in the arising jet induces the orientational order of the nematic director the jet direction. This transports polarization charge to the tip, thereby increasing the electrostatic repulsion.

The observed polarization-induced droplet instability crucially depends on the properties of the ferroelectric nematic phase and on the combination of polarization and fluidity unique to this system. Ferroelectric nematic liquid crystals offer for the first time the opportunity of studying the coupling between the polarization of a fluid and a solid ferroelectric, which understanding might provide the basis for novel electro-hydromechanical applications.

In this work, we aim at characterizing in detail the effects of the fringing field created by the ferroelectric substrate on the droplets' instability. To this purpose, we did not work with sessile droplets but confined them between two ferroelectric substrates arranged in two different ways, to generate fringing fields with different profiles. Specifically, we realized sandwich cells where the substrates' polarization vectors are parallel or antiparallel to each other, so to develop opposite or equal surface electrostatic charges at the interfaces with the liquid crystal droplet.

Noteworthy, conventional nematic liquid crystals in combination with ferroelectric substrates have been studied in several different configurations see for example [18,19]. The consequence of substrate charging on the liquid crystal's average molecular orientation has been demonstrated; however, no effect comparable to those observed with ferroelectric nematic droplets has been reported.

2. Materials and Methods

The ferroelectric liquid crystal used in this work is 4-[(4-nitrophenoxy)carbonyl]phenyl 2,4-dimethoxybenzoate (RM734). This compound was synthesized as described in [1], and its structure and phase diagram have already been reported [1]. The ferroelectric nematic phase (N_F) appears here through a weakly fist order phase transition upon cooling from the conventional higher temperature nematic (N) phase and is stable in the temperature range 133–80 °C [1]. The value of the spontaneous polarization **P** of RM734 exceeds 6 $\mu C/cm^2$ at the lowest temperature in the N_F phase; moreover, **P** is locally collinear to the molecular director **n**, defining the average orientation of the molecular axis, and is either parallel or antiparallel to it [1].

As ferroelectric solid substrates, we used 900 μm thick z-cut undoped lithium niobate (LN) crystals (PI-Kem). Although the bulk spontaneous polarization of LN crystals along the [0001] z-axis is of the order of 70 $\mu C/cm^2$, because of very efficient compensation mechanisms at the z-cut surfaces, the surface charge at equilibrium is only of the order of about 10^{-2} $\mu C/cm^2$ [20]. When, however, temperature variations are induced in the crystal, the surface charge of LN can significantly increase thanks to the pyroelectric effect [21–23], a transient phenomenon observable during and shortly after the temperature variation, and due to the slow free charge relaxation in LN. The pyroelectric coefficient of undoped LN is known to be of the order of 10^{-4} C/m^2K at room temperature [24], increases by one order of magnitude around 100 °C [25]. Given the temperature used in our experiments, dictated by the RM734 phase diagram, we can thus expect an induced surface charge density of the order of 1 $\mu C/cm^2$, for temperature variations of a few degrees, ramped in a short time compared to the LN charge relaxation. To match the conditions of the previous work on sessile droplets [8], LN crystals were used as bare substrates, with no coating applied.

To prepare the samples used in this work, a RM734 droplet was deposited on a LN substrate at a temperature T = 200 °C, which corresponds to the liquid crystal isotropic phase, and then covered by a second LN slab, previously heated at the same T. RM734 droplets were realized following two steps. First, a small amount of RM734 powder was deposited at room temperature on a clean glass slide and heated to 150 °C until it melted. To create the initial droplets, a cold stainless needle was dipped into the melt and retracted, causing the droplet on its tip to solidify upon contact with the surrounding air. To increase the size of the RM734 "pearl", rapid (to avoid re-melting) successive dipping is performed. Then, the pearl was remelted into a droplet on the heated LN substrate. The two substrates were then stuck together by means of two 100 μm thick stripes of Kapton tape, whic define the cell thickness. The liquid crystal droplet confined between two solid subs assumes the form of a capillary bridge (Figure 1a). The entire cell was then transf to a small, closed oven, suitable for the optical microscope, and cooled down to th phase. In order to perform a systematic analysis of the instability events, we decr the cell temperature by steps of 5 °C ramped in 60 s each, resulting in a cooling r 0.08 °C/s. Due to the structure of the oven used, the temperature variation is the sa the entire cell, which guarantees the same amount of pyroelectric charging on both th confining crystals.

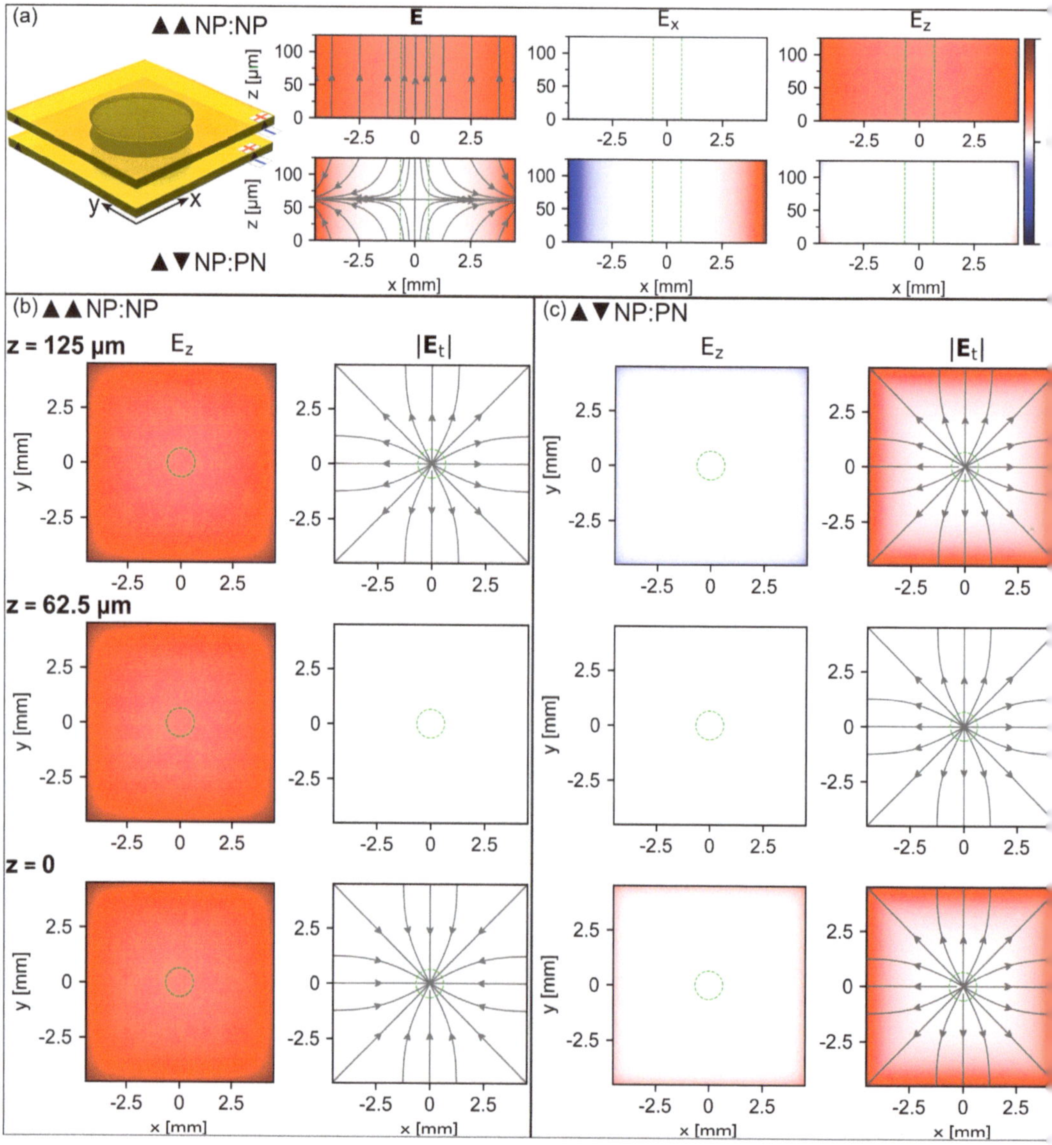

Figure 1. Sketch of the liquid crystal cells used in this work and fringing field profile and value two configurations. (**a**) The left-hand side shows the sketch of an RM734 droplet confined be two LN slabs, assuming the form of a capillary bridge. The two LN substrates in the sketch

parallel polarization vectors and develop opposite charges at the interface with the liquid crystal, which corresponds to the *np*/*np* configuration. Right-hand side reports the normalized fringing field with its z and x-components at y = 0. The top line refers to *np*/*np*, the bottom line to *np*/*pn*. (**b**) Vertical ($\mathbf{E}_z$) and in-plane ($\mathbf{E}_t$) components of the fringing field at different positions along the cell thickness d, for the *np*/*pn* configuration. (**c**) Same as panel b, for the *np*/*pn* configuration. The maximum values of the fringing field are $|\mathbf{E}_{ff}|_{max}$ = 843 V/cm for *np*/*np* and $|\mathbf{E}_{ff}|_{max}$ = 321 V/cm for *np*/*pn*.

The liquid crystal cells realized with two LN substrates are of two different kinds. Specifically, LN crystals were arranged so as to expose equally or oppositely charged surfaces at the interfaces with the liquid crystal, as sketched in Figure 1a. The two configurations will be referred to as *np*/*pn* and *np*/*np*, where *n* and *p* stand for negative and positive, respectively. Noteworthy, the large thickness of the cells (100 µm) makes the droplets' volume high enough to balance the additional friction due to confinement and the much lower thermal gradients on the LN substrates resulting from the closed arrangement.

The details of the electromechanical droplets' instability have been analyzed by POM observations and videos recorded by a CCD camera operating at 25 frames per second.

3. Results and Discussion

Due to the finite size of LN crystals, their pyroelectric charging gives rise to an external fringing field. This field is a fraction *f* of the internal field $\sigma_{LN}/\varepsilon_0$, with *f* depending on the crystal shape and size, and of the order of $f \approx 10^{-3}$ in our experimental conditions [8]. The fringing fields generated by the LN slabs combine in different ways in *np*/*np* and *np*/*pn* cells, which results in different profiles and values of the total field present within the region between the substrates. The total fringing field $\mathbf{E}_{ff}$ in the two cells, in the absence of the liquid crystalline fluid bridge, is reported in Figure 1. Figure 1a shows the normalized field E (intensity and lines) and its x- and z-components on the xz plane at y = 0 for the two configurations (top: *np*/*np*; bottom: *np*/*pn*). The *np*/*pn* arrangement gives rise to a total fringing field, which is mainly vertical and quite uniform along the cell thickness. The x-component is different from zero, although very weak, only close to the substrate's corners. On the contrary, the *np*/*pn* configuration produces a total fringing field with a very weak vertical component, different from zero only at the corners, and an x-component that is uniform along the thickness, and increases from the center to the edges of the LN plates. The vertical and in-plane field components are reported in panels b (*np*/*np*) and c (*np*/*pn*) at three different positions along the cell thickness d: z = 0, z = d/2 and z = d. For the in-plane component $\mathbf{E}_t$, the field lines are also shown. They are radially distributed on the xy planes. The absolute values of the field are also different in the two cells, being about 8.4×10^5 V/m for the *np*/*np* cells and 3.2×10^5 V/m for the *np*/*pn* ones (maximum values).

The specific profile of $\mathbf{E}_{ff}$ affects the features of the electromechanical instability of the N_F fluid bridge. In particular, (i) the number of instability events observed within the temperature window corresponding to the N_F phase, (ii) the number of ejected fluid jets for each of these events and (iii) the temperature at which these events start, depend on the specific substrates' arrangement. This is shown in Figure 2a, where the number N of observed fluid jets is reported as a function of the instability temperature T_i, defined as the temperature at which the instability event takes place. Different colors correspond to different kinds of cells. It is evident that the instability starts at higher temperatures in *np*/*np* cells, which thus exhibit the highest number of instability events. Moreover, these events are, on average, characterized by the highest number of ejected jets. Since a higher temperature corresponds to a lower value of the N_F polarization **P** [1], the results in Figure 2a indicate that in *np*/*np* cells, the charging threshold is reached for lower values of **P** than in the other kind of cells.

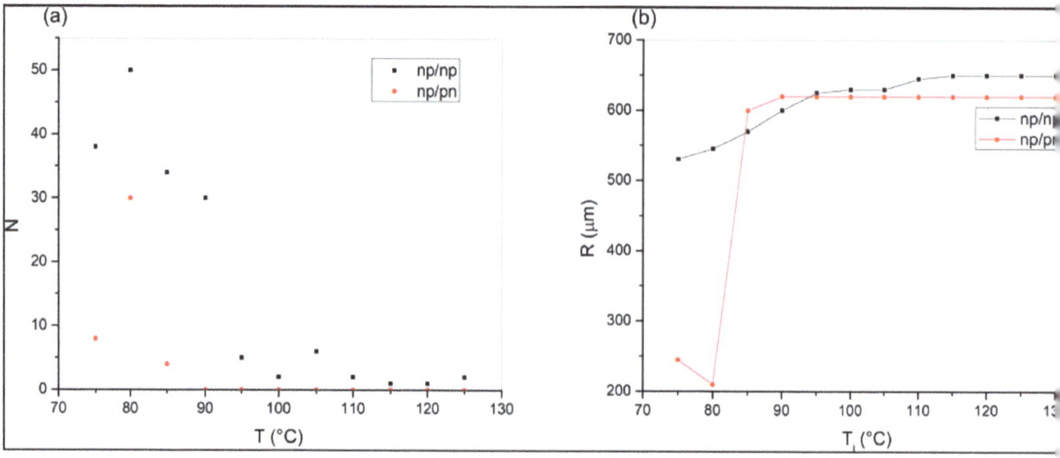

Figure 2. (**a**) Number of fluid jets ejected for each instability event for *np*/*np* (black square *np*/*pn* (red circles) cells, as a function of the instability temperature T_i; (**b**) average radius N_F fluid bridge measured from the cells' top view, after each instability event for *np*/*np* squares) and *np*/*pn* (red circles) cells, as a function of the instability temperature T_i. Full lines a guides for the eyes. Data shown in the figure are related to two specific cells. However, the measurements have been repeated several times on different samples and gave similar result

The *np*/*np* configuration, with its mainly vertical and uniform fringing field, is si to the geometry used by Mathe et al. in [26]. In their study, a ferroelectric fluid b was confined between two conductive glasses and an external electric field was ap perpendicular to the cell substrates. The instability caused by this field was interp as a kind of labyrinthine instability, already observed both in magnetic fluid expos a magnetic field orthogonal to the bounding plates and in dielectric fluids [27–2 both cases, the presence of a gap separating the fluid from the magnetic poles or fro electrodes is necessary for the instability to take place. Such a gap allows the presenc field component parallel to the plates. In our case, the gap is intrinsically present sin is itself an insulator.

The presence of the ferroelectric fluid bridge changes the fringing field prof ported in Figure 1. In *np*/*np* cells, this change consists in the appearance of an addi component of the field parallel to the bounding surfaces. It is known that the bu larization of the N_F liquid crystal spontaneously self organizes to minimize the in and external electric fields. Generally, **P** will end up parallel to the interfaces to the accumulation of surface charge $\sigma = \mathbf{P}.\mathbf{u}$ (where **u** is the unit vector perpendicu the surfaces). Additionally, **P** will adopt bend deformations, which do not produce charge, thus preventing nonzero $\nabla \cdot \mathbf{P}$ as much as compatible with geometric const In the presence of the fringing field, which in this geometry is mainly normal to the LN interface planes, the ferroelectric nematic becomes polarized. This happens through a reorientation of **P** by an angle such that it deposits polarization charge on the fluid b surfaces, canceling the internal field, a peculiarity of the N_F phase referred to as ' superscreening" [8,13]. This process leads to a mismatch between the field inside the b and the one outside. Indeed, in the first case, the potential difference is virtually negli being due to the field in the thin gap between the LN-charged surface and the liquid c interfacial layer; in the second case, the potential difference is determined by the fi air. This mismatch generates an additional in-plane component of the field, whic be arbitrarily large, depending on the thickness of the layer along which the pot difference changes from zero to the value $d\sigma_{LN}/\varepsilon_0$. The additional field component c jet ejection from the charge accumulation sites. The result is the occurrence of se instability events, characterized by the ejection of a large number of jets from different portions of the capillary bridge at the interfaces with the bounding plates (Figure 3a–d).

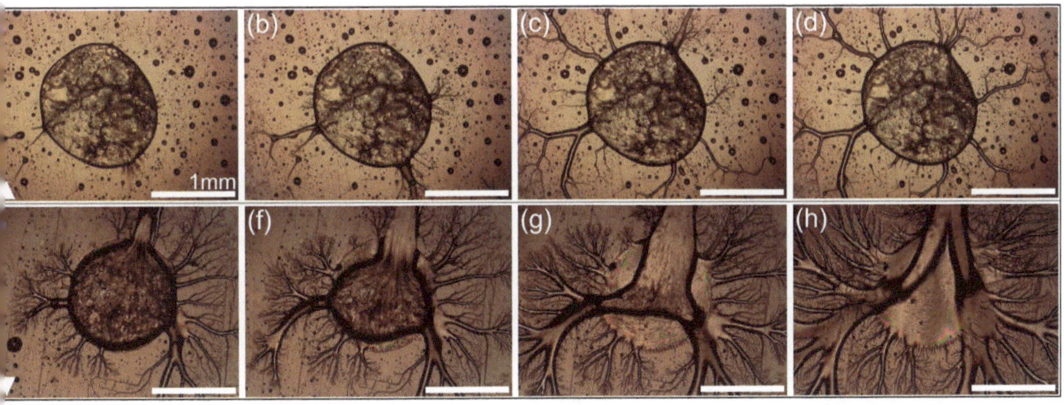

Figure 3. Examples of the observed electromechanical instability in N_F fluid bridges formed by confining an RM734 droplet between two LN crystals. (**a**,**d**) Sequence of frames showing the evolution of the instability in the *np*/*np* configuration. The onset of instability consists in the formation of several small spikes protruding from different regions of the bridge rim (**a**). As the instability proceeds, some jets grow (**b**) and new ones appear (**c**,**d**). Some of the jets are clearly ejected from locations on different planes, namely the two LN/N_F interface planes. (**e**,**h**) Sequence of frames showing the evolution of the instability in the *np*/*pn* configuration. In this case, jets rapidly grow in diameter (**e**,**f**), encompassing large portions of the droplet rim and volume on both interface planes (**g**,**h**). Note that the small droplets already present in the images (**a**,**e**) are due to fragmentation of the main droplet during sample preparation. Frames are not consecutive.

An additional feature of *np*/*np* cells is the temperature at which the instability starts, which is higher than that in *np*/*pn* cells, a behavior that we ascribe to the higher value of $\mathbf{E}_{ff}$ components that characterize this specific substrates' arrangement.

In *np*/*pn* cells, the fringing field has a lower absolute value and is mostly in plane, while the vertical component is weak and practically negligible at the center of the cell. **P** and $\mathbf{E}_{ff}$ are thus both parallel to the LN/N_F interface planes, and the required in-plane component of the field is present from the beginning due to the specific LN substrate arrangement. In these conditions, the polarization charges that cancel the internal field are generated at the N_F/air lateral interfaces and are most probably due to twist distortions compatible with **P** being parallel to the two bounding surfaces. Again, this creates an additional electric field that, in this case, is comparable in magnitude to the original value of the fringing field. The threshold charging that gives rise to the electromechanical instability is reached here at temperatures lower than those in the *np*/*np* cells, corresponding to higher values of **P**, and on average produce the ejection of a lower number of fluid jets. We understand this phenomenon to be a result of the weaker fringing field that characterizes this configuration. Noteworthy, experiments on N_F sessile droplets deposited on glass surfaces with patterned electrodes showed that jet ejection preferentially occurs in regions where the fringing field is in the vertical direction [26]. This is an additional indication that the in-plane electric field component, arising in these conditions due to the N_F superscreening, is higher than in any other configuration.

Once jets are formed in *np*/*pn* samples, they are, however, more disruptive compared to the other kind of cell. This is shown in Figure 2b, where the variation of the average droplet radius after each instability event is reported as a function of T_i. This parameter exhibits a decreasing trend for both configurations, but such a decreasing trend is different in the two situations, being smooth for the *np/np* cells and very steep for the *np/pn* ones.

In this latter case, the average radius decreases by more than 60% after the first insta events, indicating an extremely explosive and disruptive phenomenon.

We understand this behavior to be due to the radial shape of $\mathbf{E}_{ff}$ in the *np/p* arrangement. Indeed, jets are polar fluid tubes carrying polarization charges on thei which thus keep on moving in the direction of the field, accelerated by the field itsel polar nature of the ejected jets is well visualized in Figure 4a, where jets parallel polarizer appear dark, while those at 45 deg. exhibit the maximum brightness, sho that the liquid crystal director **n** is along the jet axis. Since the polarization vector is l collinear to the optical axis **n**, Figure 4a indicates that **P** is also along the jet's axis, direction of flow. Noteworthy, POM analysis of the nematic director in conditions invo thick cells with LN substrates is cumbersome and made difficult by the formatio thick fluid bridge and by the birefringence of the substrates. For this reason, the ima Figure 4a is related to a RM734 N_F sessile droplet lying on a LN substrate. The resul however, extendable to the double substrate configuration.

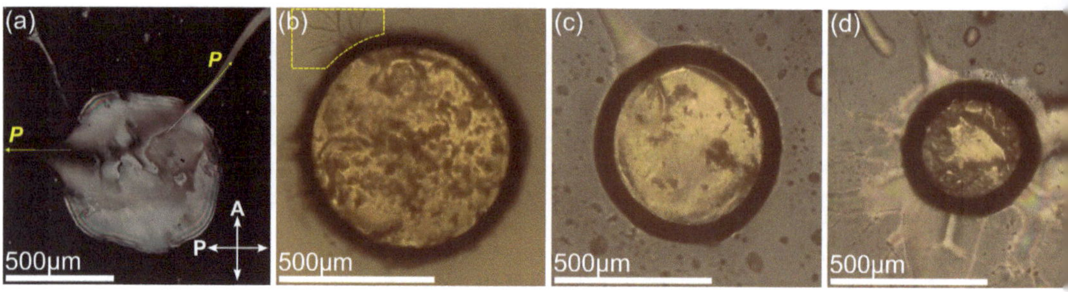

Figure 4. (**a**) RM734 N_F droplet on a LN substrate observed under a crossed polarizer. Jets p to either polarizer or analyzer appear dark, while those at 45 deg. exhibit the maximum brig This indicates that **n** is along the jets' axis, which also defines the direction of the polarization **P**. Note that the attribution of the specific direction of **P** (parallel or antiparallel to **n**) is here arb (**b**,**c**) evolution of the electromechanical instability of the N_F fluid bridge in case of a 150 μm *np/pn* cell. (**b**) Onset of the instability consisting in the formation of small spikes protruding several portions of the bridge rim. The yellow dashed line identifies one very thin jet that bra on one of the bounding substrates; (**c**) as the instability proceeds, some of the small spikes r or disrupt by forming tiny droplets, leaving only one large, main jet that continues to grow; (is followed by the ejection of a large portion of the whole fluid mass from all around the b perimeter. Frames are not consecutive.

In the *np/pn* configuration, the in-plane field that accelerates the jet tips depen the specific substrates' arrangement and remains approximately unperturbed duri motion. On the contrary, in the *np/np* case, the in-plane component of the field is gene by N_F superscreening and depends on the position of the polarization charges. As it changes both value and direction as soon as jets protrude from the fluid bridge. I case, there is not a "constant" radial field that moves the charged expelled fluid al fixed direction, and jets do not show the collective disruptive motion observed in n cells. The observed disruptive instability in *np/pn* cells might additionally be due acceleration experienced by the jets' tips, which may generate large distortions o fluid bridge portions close to the ejection sites. This, in turn, causes additional c accumulation in a sort of self-sustained effect.

An example of the instability of *np/pn* cells is reported in Figure 3e–h.

The evolution of the electromechanical instability of N_F fluid bridges is shown in detail in Figure 4b–d. In these images, the gap between the LN crystals is larger th Figure 3 (150 vs. 100 μm), which results in a more evident meniscus (thicker dark around the "droplet"). Figure 4b shows the appearance of small spikes that protrude several portions of the bridge rim at the onset of the instability event. This is follow the formation of a single main jet (Figure 4c) and by the ejection of a large portion of the total fluid mass (Figure 4d). These events are accompanied by the decrease of the droplet diameter, which is quantified in Figure 2b for both kinds of cells. This specific sequence of frames is related to the *np*/*pn* configuration. Noteworthy, large variations of the droplet texture can be observed by comparing the three images, which suggests that instability induces severe rearrangements of the N_F liquid crystal polarization. This reasonably gives rise to the additional accumulation of bound charges which, in turn, stimulates additional fluid ejection as a strategy to decrease electrostatic energy.

Noteworthy, measuring the position of the jet tips from the initial frames after the ejection, we noticed that the average jet speed for equal values of T_i is higher in *np*/*pn* cells compared to the others, in agreement with the notion that fluid motion is faster in this configuration.

In conclusion, we studied the behavior of ferroelectric liquid bridges confined between two solid ferroelectric substrates arranged in different configurations realized so that, once pyroelectrically charged, they generate fringing fields of different values and shapes. Our observations highlighted that the features of the liquid crystal instability are affected by the specific fringing field profile in a way dominated by the minimization of the electrostatic energy associated with the bulk polarization of the ferroelectric fluid.

Our results show that the electromechanical instability of ferroelectric droplets confined between two ferroelectric solid substrates can be controlled in terms of instability, temperature, number of ejected jets and violence by acting on the substrates' arrangement. This might allow the realization of novel electro-hydrodynamic applications based on the electrostatic instability of polar liquids.

Author Contributions: Conceptualization, L.L.; methodology, L.L., S.M., R.Z. and R.B.; software, R.B.; validation, L.L. and S.M.; formal analysis, S.M. and A.S.O.; investigation, S.M.; data curation, S.M.; writing—original draft preparation, L.L. and R.Z.; writing—review and editing, L.L. and R.B.; supervision, L.L. and S.M. All authors have read and agreed to the published version of the manuscript.

Funding: L.L. and R.B. acknowledge the European Union—Next Generation EU, project code: ECS00000041; project title: Innovation, digitalization, and sustainability for the diffused economy in Central Italy—VITALITY.

Data Availability Statement: The raw data supporting the conclusions of this article will be made available by the authors on request.

Acknowledgments: L.L. and S.M. are thankful to Tommaso Bellini for useful discussions.

Conflicts of Interest: The authors declare no conflict of interest.

rences

Chen, X.; Korblova, E.; Dong, D.; Wei, X.; Shao, R.; Radzihovsky, L.; Glaser, M.A.; Maclennan, J.E.; Bedrov, D.; Walba, D.M.; et al. First-principles experimental demonstration of ferroelectricity in a thermotropic nematic liquid crystal: Polar domains and striking electro-optics. *Proc. Natl. Acad. Sci. USA* **2020**, *117*, 14021. [CrossRef] [PubMed]

Mandle, R.J.; Cowling, S.J.; Goodby, J.W. A nematic-to-nematic transformation exhibited by a rod-like liquid crystal. *Phys. Chem. Chem. Phys.* **2017**, *19*, 11429–11435. [CrossRef] [PubMed]

Nishikawa, H.; Shiroshita, K.; Higuchi, H.; Okumura, Y.; Haseba, Y.; Yamamoto, S.I.; Sago, K.; Kikuchi, H. A fluid liquid-crystal material with highly polar order. *Adv. Mater.* **2017**, *29*, 1702354. [CrossRef] [PubMed]

Mandle, R.J.; Cowling, S.J.; Goodby, J.W. Rational design of rod-like liquid crystals exhibiting two nematic phases. *Chemistry* **2017**, *23*, 14554–14562. [CrossRef] [PubMed]

Mertelj, A.; Cmok, L.; Sebastián, N.; Mandle, R.J.; Parker, R.R.; Whitwood, A.C.; Goodby, J.W.; Čopič, M. Splay nematic phase. *Phys. Rev. X* **2018**, *8*, 41025. [CrossRef]

Lavrenovich, O.D. Ferroelectric nematic liquid crystal, a century in waiting. *Proc. Natl. Acad. Sci. USA* **2020**, *117*, 14629–14631. [CrossRef] [PubMed]

Chen, X.; Korblova, E.; Glaser, M.A.; Maclennan, J.E.; Walba, D.M.; Clark, N.A. Polar in-plane surface orientation of a ferroelectric nematic liquid crystal: Polar monodomains and twisted state electro-optics. *Proc. Natl. Acad. Sci. USA* **2021**, *118*, e2104092118. [CrossRef] [PubMed]

8. Barboza, R.; Marni, S.; Ciciulla, F.; Mir, F.A.; Nava, G.; Caimi, F.; Zaltron, A.; Clark, N.A.; Bellini, T.; Lucchetti, L. Explosive e static instability of ferroelectric liquid droplets on ferroelectric solid surfaces. *Proc. Natl. Acad. Sci. USA* **2022**, *119*, e2207 [CrossRef]
9. Sebastian, N.; Čopič, M.; Mertelj, A. Ferroelectric nematic liquid-crystalline phases. *Phys. Rev. E* **2022**, *106*, 021001. [Cross
10. Zavvou, E.; Klasen-Memmer, M.; Manabe, A.; Bremer, M.; Eremin, A. Polarisation-driven magneto-optical and nonlinear- behaviour of a room-temperature ferroelectric nematic phase. *Soft Matter* **2022**, *18*, 8804. [CrossRef] [PubMed]
11. Máthé, M.T.; Perera, K.; Buka, Á.; Salamon, P.; Jákli, A. Fluid ferroelectric filaments. *Adv. Sci.* **2023**, *9*, 2305950. [Cro [PubMed]
12. Marni, S.; Nava, G.; Barboza, R.; Bellini, T.; Lucchetti, L. Walking ferroelectric liquid droplets with light. *Adv.* **2023**, *35*, 2212067. [CrossRef] [PubMed]
13. Caimi, F.; Nava, G.; Fuschetto, S.; Lucchetti, L.; Paiè, P.; Osellame, R.; Chen, X.; Clark, N.A.; Glaser, M.A.; Bellini, T. superscreening and polarization following in confined ferroelectric nematics. *Nat. Phys.* **2023**, *19*, 1658–1666. [CrossRef]
14. Máthé, M.T.; Himel, M.S.H.; Adaka, A.; Gleeson, J.T.; Sprunt, S.; Salamon, P.; Jákli, A. Liquid Piezoelectric Materials: Electromechanical Effect in Fluid Ferroelectric Nematic Liquid Crystals. *Adv. Funct. Mater.* **2024**, *34*, 2314158. [CrossRef]
15. Sebastián, N.; Lovšin, M.; Berteloot, B.; Osterman, N.; Petelin, A.; Mandle, R.J.; Aya, S.; Huang, M.; Drevenšek-Olenik, I.; Ne et al. Polarization patterning in ferroelectric nematic liquids via flexoelectric coupling. *Nat. Commun.* **2023**, *14*, 3029. [Cro
16. Basnet, B.; Rajabi, M.; Wang, H.; Kumari, P.; Thapa, K.; Paul, S.; Lavrentovich, M.O.; Lavrentovich, O.D. Soliton walls pai polar surface interactions in a ferroelectric nematic liquid crystal. *Nat. Commun.* **2022**, *13*, 3932. [CrossRef] [PubMed]
17. Rayleigh, L. XX. On the equilibrium of liquid conducting masses charged with electricity. *Phil. Mag.* **1882**, *14*, 184–186. [Cro
18. Habibpourmoghadam, A.; Lucchetti, L.; Evans, D.; Reshetnyak, V.; Omairat, F.; Schafforz, S.L.; Lorenz, A. Laser-induced er patterns in a N* liquid crystal on an iron doped lithium niobate surface. *Opt. Express* **2017**, *25*, 26148. [CrossRef] [PubMe
19. Carns, J.L.; Cook, G.; Saleh, M.A.; Serak, S.V.; Tabiryan, N.; Evans, D.R. Self-activated liquid-crystal cells with photov substrates. *Opt. Lett.* **2006**, *31*, 993. [CrossRef] [PubMed]
20. Sanna, S.; Schmidt, W.G. $LiNbO_3$ surfaces from a microscopic perspective. *J. Phys. Condens. Matter* **2017**, *29*, 413001. [Cro [PubMed]
21. Kostritskii, S.M.; Sevostyanov, O.G.; Aillerie, M.; Bourson, P. Suppression of photorefractive damage with aid of stead temperature gradient in nominally pure LiNbO3 crystals. *J. Appl. Phys.* **2008**, *104*, 114104. [CrossRef]
22. Kostritskii, S.M.; Aillerie, M.; Sevostyanov, O.G. Self-compensation of optical damage in reduced nominally pure $LiNbO_3$ c *J. Appl. Phys.* **2010**, *107*, 123526. [CrossRef]
23. Ferraro, P.; Grilli, S.; Miccio, L.; Vespini, V. Wettability patterning of lithium niobate substrate by modulating pyroelectric e form microarray of sessile droplets. *Appl. Phys. Lett.* **2008**, *92*, 213107. [CrossRef]
24. Byer, R.L.; Roundy, C.B. Pyroelectric coefficient direct measurement technique and application to a nsec response time de *Ferroelectrics* **2011**, *43*, 333–338.
25. Gebre, T.; Batra, A.K.; Guggilla, P.; Aggarwal, M.D.; Lal, R.B. Pyroelectric properties of pure and doped lithium niobate c for infrared sensors. *Ferroelectr. Lett. Sect.* **2010**, *31*, 131–139. [CrossRef]
26. Máthé, M.T.; Farkas, B.; Péter, L.; Buka, Á.; Jákli, A.; Salamon, P. Electric field-induced interfacial instability in a ferroe nematic liquid crystal. *Sci. Rep.* **2023**, *13*, 6981. [CrossRef] [PubMed]
27. Rosensweig, R.E.; Zahn, M.; Shumovich, R.J. Labyrinthine instability in magnetic and dielectric fluids. *Magn. Magn.* **1983**, *39*, 127–132. [CrossRef]
28. Zahn, M.; Shumovich, R. Labyrinthine instability in dielectric fluids. *IEEE Trans. Ind. Appl.* **1985**, *21*, 53–61. [CrossRef]
29. Igonin, M.; Cebers, A. Labyrinthine instability of miscible magnetic fluids. *Phys. Fluids* **2003**, *15*, 1734. [CrossRef]

crystals

e

cal Orientation Transitions to a Lying Helix State in Negative lectric Anisotropy Cholesteric Liquid Crystal

V. Simdyankin, Artur R. Geivandov *, Irina V. Kasyanova and Serguei P. Palto

Shubnikov Institute of Crystallography, Kurchatovsky Complex "Crystallography and Photonics", National Research Centre "Kurchatovsky Institute", Leninskii pr.59, 119333 Moscow, Russia; serguei.palto@gmail.com (S.P.P.)
* Correspondence: ageivandov@yandex.ru

Abstract: Orientation transitions in a cholesteric liquid crystal (CLC) layer with negative dielectric anisotropy, under the influence of a non-uniform spatially periodic electric field created using a planar system of interdigitated electrodes, were studied experimentally and numerically. In the interelectrode space, transitions are observed from a planar Grandjean texture, with the helix axis perpendicular to the layer plane, to states with a lying helix, when the helix axis is parallel to the layer plane and perpendicular to the electrode stripes. It was found that the relaxation time of the induced state in the Grandjean zones, corresponding to two or more half-turns of the helix, significantly exceeded the relaxation time for the first Grandjean zone with one half-turn. An analysis of experimentally observed and numerically simulated textures shows that slow relaxation to the initial state in the second Grandjean zone, as well as in higher-order zones, is associated with the formation of local topologically equivalent states. In these states, the helix has a reduced integer number of helix half-turns throughout the layer thickness or unwound into the planar alignment state.

Keywords: cholesteric liquid crystal; lying helix; negative dielectric anisotropy; numerical simulation

n: Simdyankin, I.V.; dov, A.R.; Kasyanova, I.V.; .P. Local Orientation ons to a Lying Helix State in ve Dielectric Anisotropy eric Liquid Crystal. *Crystals* 4, 891. https:// /10.3390/cryst14100891

nic Editor: Ingo Dierking

ed: 13 September 2024
l: 8 October 2024
ed: 11 October 2024
ed: 13 October 2024

1. Introduction

Cholesteric liquid crystals (CLCs) stand out among other types of LCs because of their ability to spontaneously form a helical director distribution [1]. This feature allows us to consider these liquid crystal structures as one-dimensional photonic crystals with a forbidden photonic stopband [2], which determines their demand for numerous photonic applications, such as tunable optical filters, mirrors, beam deflectors [3–5], and, especially, laser elements [6,7]. Lasing was studied by many authors in a typical geometry of a Grandjean texture with a helical axis directed along the normal to the layer plane, under planar alignment conditions at the confining substrates [8–10]. Interesting results on lasing were obtained in Cano–Grandjean wedge geometry [11,12]. However, there are limitations related to the longitudinal pumping of CLC-based lasers in planar Grandjean textures [13], which can be avoided by means of the employment of so-called "lying helix" (LH) geometry, with the axis of the cholesteric helix lying in the plane of the layer [13–15].

The problem of forming a stable and defect-free LH has been actively discussed since the late 1990s [16]. Over several decades, numerous approaches have been proposed to obtain LH, including the use of periodic anchoring conditions [16], surface relief formed by laser lithography [17], cooling of CLCs from an isotropic phase in an electric field [18,19], or employing photo- or thermally cross-linkable polymers to stabilize LC texture by forming polymer networks [20], microchannels [14], grafts at the confining surfaces [21], and light stimuli [22,23]. It should be noted that the very fact of obtaining LH in most cases was confirmed only indirectly based on the optical texture features observed in a polarizing microscope. In addition, the intrinsic thermodynamic instability of planar helical textures under typical boundary conditions [24] significantly complicates their formation.

We previously showed that a thermodynamically stable LH state can be obt in an electric field under periodic boundary conditions with binary modulation easy axis direction, induced in the alignment layer with an ion beam treatment [photoalignment [25]. To note, in [15,25], one of the planarly aligned cell surface patterned with homeotropic-alignment stripes, making the tilt angle periodically cl in binary mode (planar homeotropic) along the in-plane axis. For such a transitio CLC natural pitch p_0 must be less than the period of binary modulation of the bou conditions. It is important that, in this case, the helix is strongly deformed by the Also, the LH state disappears when the field is turned off. In the context of this worl also essential that, in the aforementioned studies, CLC has a positive dielectric aniso and the electric field vector is normal to the CLC layer that is directed along the axis initial helix. In the case of planar electric field geometry, when the electric field ve is perpendicular to the helical axis, such an orientation transition does not occur in with positive anisotropy. The helix either becomes strongly deformed in pulsed maintaining the original pitch [26,27], or, in static electric fields, is unwound wi formation of a defective texture [28].

In [29], the behavior of CLCs with both positive and negative anisotropy in the p geometry of a spatially periodic electric field was investigated. Using the fluores method, the authors visualized a fundamentally different field-induced change initial helical director distribution in the interelectrode area, depending on the sign dielectric anisotropy. In particular, for CLCs with negative dielectric anisotropy, the au reported an orientation transition with the appearance of a periodic texture near one surfaces and interpreted this as a rotation of the axis of the cholesteric helix by 90°. result seems to be very important and was confirmed by the results of the current st

A few other studies in this area illustrate highly inhomogeneous textures observ an electric field, which is apparently due to the significant thickness of the layers stud comparison with the natural pitch of the helix [30–34].

In this work, we studied in detail the behavior of CLCs with negative diel anisotropy in a planar electric field geometry for thicknesses comparable to the helix These studies were conducted for the first four Grandjean zones. There are two parts t The first section briefly presents optical observations of electric field-induced orient transitions using a polarizing optical microscope and the features of relaxation induced textures after turning off the field in the region of thicknesses corresponding second and fourth Grandjean zones. In the second section, the results of optical observa and features of orientation transitions for the four Grandjean zones are discussed in considering the data obtained by numerical simulation.

2. Materials and Methods

An experimental CLC cell is composed of two glass substrates, one of which b system of opaque (chromium) interdigitated electrodes (IDEs) with a period $L = 1$ The width of the electrode strip was $w = 5$ µm and the distance between the electrode $l = 10$ µm. Glass without electrodes was used as the opposite substrate. The surfa both substrates provided a planar orientation of the CLC in the y-direction, as sho Figure 1. To obtain a planar alignment, standard technology was used to form poly layers on the surface of the substrates, followed by rubbing them with a soft cloth pa to the electrode strips. In the assembled cell, the rubbing directions on the two subs are opposite. To study the orientation transitions depending on the thickness of the layer, we used a wedge-shaped cell. The thickness of the CLC layer d varied from 0 to owing to the use of a Teflon gasket on only one edge of the cell.

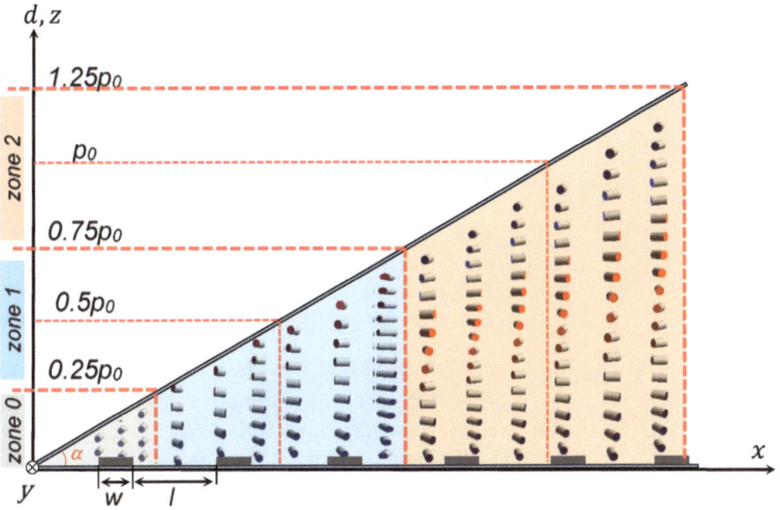

Figure 1. Schematic of experimental wedge-shaped LC cell. An IDE system was formed on the bottom substrate; the width of the electrode strips was w = 5 μm, with a gap between them of l = 10 μm. Thick dashed lines separate adjacent Grandjean zones, with the number of half-turns of the helix differing by one. At the centers of each Grandjean zone, with the exception of the zero zone, the number of half-turns of the helix was a multiple of $p_0/2$. In a real cell, the angle α is small, and the width of each zone is several millimeters.

To prepare CLCs, we used as a matrix a nematic mixture that was developed with the following characteristics: main refractive indices $n_\perp$ = 1.49, $n_\parallel$ = 1.58, low-frequency dielectric constant $\varepsilon_\perp$ = 8.2, dielectric anisotropy $\Delta\varepsilon = \varepsilon_\parallel - \varepsilon_\perp = -4.15$, and transition temperature to the isotropic state T_{iso} = 79 °C. To impart chirality, an optically active additive 1,4:3,6-Dianhydro-D-sorbitol-2,5-bis(4-hexyloxybenzoate) with a twisting capacity of 49 μm^{-1} was used. An additive weight concentration of 0.7 wt. % ensured the natural pitch of the cholesteric helix $p_0 \cong 2.3$ μm, which was measured using a wedge-shaped cell with a known thickness at the centers of the resulting Grandjean zones.

In the absence of an electric field, the CLC molecules were oriented planarly, and the helix axis was perpendicular to the surface with the electrodes. In a wedge-shaped cell, where the thickness varies along the x-axis, the pitch of the cholesteric helix cannot change monotonically. Instead, the formation of Grandjean zones occurs, in which the number of half-turns of the helix is fixed and changes strictly by one when moving to the neighboring zone. The helical pitch p in each Grandjean zone varies monotonically only within the zone, for example, in the 2nd zone from $p_0 - 0.25p_0$ to $p_0 + 0.25p_0$. At the center of each zone, $2p/p_0 = m$, where m is an integer corresponding to the number of the Grandjean zone. In the zero zone, the helical pitch p tends to infinity (the helix is untwisted), and a uniform planar orientation is realized.

Numerical simulations were performed using the *LCDTDK v.4.0* software created by one of the authors (S.P.P.). Since the calculation for the experimental geometric parameters of the IDE (L = 15 μm) requires significant computing resources, the calculations were performed for the values of the geometric parameters reduced several times. The period of the IDE array was reduced to L = 3 μm with a distance between the electrodes of 2 μm. The natural pitch of the cholesteric helix p_0 and the thickness of the simulated layer also decreased compared to the experimental samples. In the simulation, the minimum layer thickness was 1 μm, and the maximum was 1.5 μm. The number of half-turns per layer thickness varied in accordance with the number of the Grandjean zone. It should be noted that in the simulation, the natural pitch p_0 was chosen such that the helix in a given zone was stressed (for example, in the case of the 2nd zone, 1.2 $p_0 = d$; Figure 1). In this case, it was possible to obtain the best agreement with the experiment. Since the optical aniso of CLC, determined by the difference in the main refractive indices, is quite small experiment ($n_\perp$ = 1.49; $n_\|$ = 1.58), then with a significant change in the thicknes accordingly, the optical retardation $(n_\| - n_\perp)d$, the direct comparison of optical pa for simulated and experimental textures remained possible up to the 4th Grandjean

3. Results

Observations of texture changes in an electric field showed sharp differences be the first and second Grandjean zones. In contrast to the first zone, in the second after turning off the electric field, the induced texture remains for a long time. Simila between the second and fourth zones were also revealed. For this reason, we begi discussion with the results of observations in the second and fourth zones.

3.1. Induced Transitions in Second and Fourth Grandjean Zones

Figure 2 shows photographs of field-induced CLC textures in the second (a, b fourth (c, d) Grandjean zones at electrical voltages of 50 V and 30 V (rectangular altern voltage at a frequency of 5 KHz was used), as well as at different time intervals after turning off the electrical voltage. The voltage of about 30 V is close to the tran threshold, so the field-induced texture is not well established. At U = 50 V, we dea the well-established optical texture. It is noteworthy that before the application of ele voltage, the texture in each zone appeared to be homogeneous in the crossed polar which is typical for Grandjean textures. In an electric field in the interelectrode s alternating bands of different intensities are observed, oriented along the electrodes number of induced bands depends on the number of the Grandjean zone. For examp the zone with m = 2, two bright stripes are observed, separated between themselves ar electrodes by stripes of lower intensity of transmitted light, as shown in Figure 2a. turning off the electrical voltage, the induced periodic texture persists for a long time form of two bright stripes on a deep black background, as shown in Figure 2b.

In the zone with m = 4, the resulting texture is more defective, as seen in Figure 2c the electric field, both one bright band and several bright bands were observed. How after the electric field was turned off, seven bright stripes could be observed in the electrode space. Among these bands, the two brightest stand out, located symmetr relative to the middle of the interelectrode space.

As can be seen from the photo in Figure 2b,d, the resulting periodic textures rela slowly. Even for the m = 2 zone, where the layer thickness is about 2.5 μm, the per texture in the interelectrode space is maintained for several seconds after the electric fi turned off. In the zone with m = 4, the relaxation time increases by tens of times. Thus turning off the field, the relaxation time τ cannot be estimated from the well-known re

$$\tau = \frac{\gamma d^2}{\pi^2 K},$$

which for the thickness d = 2.5 μm, rotational viscosity γ = 0.1 Pa·s, and effective e coefficient K = 10 pN gives the value τ = 6 ms. The periodic texture, for example, i zone with m = 2 persists for 60 s, and in the zone with m = 4 does not disappear after 10 min, as seen in Figure 2d. Thus, in the second and fourth Grandjean zone appearance of a planar periodic texture can be considered as a field-induced transitio long-living metastable state.

In the first Grandjean zone, a very rapid relaxation of the induced texture was obs with a characteristic time that approximately coincides with estimate (1). The reaso such rapid relaxation in the first zone is explained below, taking into account the num simulation results.

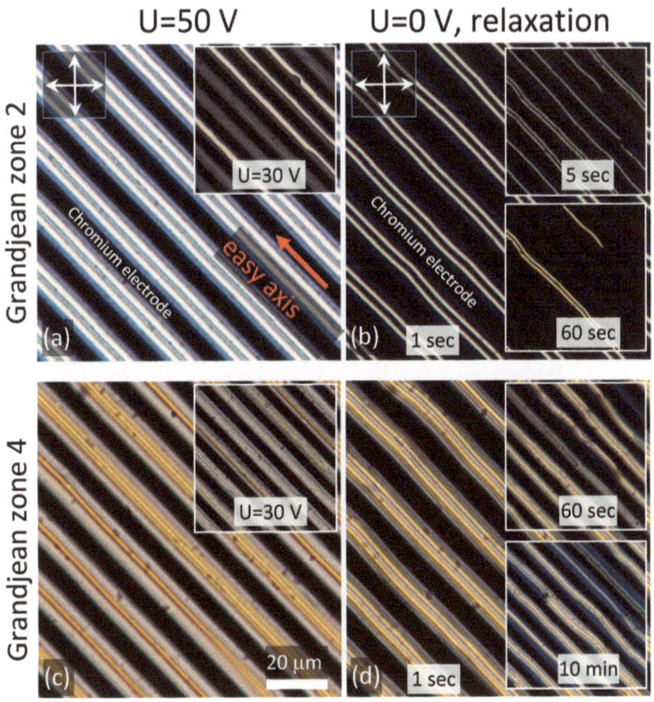

Figure 2. Photographs of LC cell textures in crossed polarizers in the Grandjean zones m = 2 (**a**,**b**) and m = 4 (**c**,**d**) in an electric field (**a**,**c**) and after turning it off (**b**,**d**) at the time intervals indicated in the figure. The natural pitch of the helix is p_0~2.3 μm, the IDE period is L = 15 μm, and the frequency of the applied electric voltage U is f = 5 KHz. White arrows indicate the direction of the axes of the crossed polarizers.

3.2. Simulation of Orientation Transitions and Discussion

The numerical modeling is grounded on the equations of the continuum theory of liquid crystals for finding the spatial–temporal distributions of the LC director, as well as Maxwell's equations for calculating the distributions of the low-frequency electric and optical fields. The solution to the problem is implemented in the *LCD TDK* software package (developed by S.P.P.).

In particular, the three-dimensional dynamics of the liquid crystal director $\mathbf{n} = (n_x, n_y, n_z)$ is described by the following equations:

$$-\frac{\partial(F+g)}{\partial n_i}+\frac{d}{dx}\left(\frac{\partial(F+g)}{\frac{\partial n_i}{\partial x}}\right)+\frac{d}{dx}\left(\frac{\partial(F+g)}{\frac{\partial n_i}{\partial y}}\right)+\frac{d}{dx}\left(\frac{\partial(F+g)}{\frac{\partial n_i}{\partial z}}\right)=\gamma\frac{dn_i}{dt}, \quad i\in\{x,y,z\},\quad g\equiv\frac{1}{2}\mu\left(1-\textstyle\sum_i n_i^2\right)=0, \tag{2}$$

$$n_x^2+n_y^2+n_z^2=1, \tag{3}$$

wherein F stands for free energy density that is described as follows:

$$F=\frac{1}{2}\left\{K_1(\nabla\cdot\mathbf{n})^2+K_2(\mathbf{n}\cdot(\nabla\times\mathbf{n})-q_0)^2+K_3(\mathbf{n}\times(\nabla\times\mathbf{n}))^2\right\}-\mathbf{E}\mathbf{P}_f-\frac{\varepsilon_0}{2}(\varepsilon\cdot\mathbf{E})\mathbf{E}, \quad \mathbf{P}_f=e_1(\nabla\cdot\mathbf{n})\mathbf{n}-e_3(\mathbf{n}\times\nabla\times\mathbf{n}), \tag{4}$$

wherein γ is the rotational viscosity, $K_{1,2,3}$ are the LC elasticity coefficients, ε is the low-frequency permittivity tensor, the free space dielectric constant $\varepsilon_0 \cong 8.85 \times 10^{-12}$ F/m, $\mathbf{E}$ is the electric field vector, and $\mathbf{P}_f$ is the flexoelectric polarization vector determined by the

flexoelectric coefficients $e_{1,3}$ and the corresponding deformation of the director distrib
In the current work, the simulation results are for $\mathbf{P}_f = 0$. Chirality is determined b
wavenumber q_0, which specifies the natural pitch of the helix $p_0 = 2\pi/q_0$. The param
and the corresponding Lagrange multiplier μ are related to the unit length of the dire
and are taken into account automatically in the numerical solution, when normalizati
is performed at each discrete moment of time. Also, in present simulations, we use p
alignment along the y-axis (see Figure 1) at rigid boundary conditions (infinite ancho

a. Induced orientation transition in the first Grandjean zone

Figure 3a shows the simulation results and experimental photographs of the obs textures in the case of an electric field-induced orientational transition in the first Gran zone. Initially, in the absence of electrical voltage, the axis of the cholesteric helix is ori perpendicular to the xy-plane of the layer and directed along the z-axis. In the first ($m = 1$), only one half-turn of the helix is across the thickness of the simulated LC ($d = 1$ μm). At an electric voltage ($U = 26$ V) in the interelectrode space, a transformat the Grandjean texture, uniform in the xy-plane, into a planar-modulated structure o As can be seen from the spatial distribution of the LC director, a helix director distrib with a pitch of 0.95 μm along the x-axis appears in the center of the layer, as sho Figure 3a. In the interelectrode space between the lines AD and CF, two half-turns helix are laid. That is, the observed transition can be characterized as a transition to a with LH. The transition is local, since above the electrodes, excluding areas at the e the original Grandjean texture is preserved. An important feature of the transition emergence of special localized areas (below, for simplicity, we will call them special p clearly visible in the gaps marked in the figure as AD, BE, and CF. If, for example, we along the line AD, then as it passes through a singular point near the lower surfac y-component of the director changes sign (in Figure 3a, this is reflected by the color cl of the end of the cylinder showing the director from red to blue). That is, as in the or Grandjean texture, we have a rotation of the director by π at the layer thickness alor AD line. Since, almost along the entire length of the segment AD, except the vicinity mentioned singular point, the director of the LC is oriented planarly, we will call s local orientation "quasi-planar". A similar quasi-planar orientation occurs along th and CF lines. However, if the singular point on line BE is located near the upper surf the layer, then on lines AD and CF, it is near the bottom. Singular points marked in drawing in 3D space form lines along the y-axis.

In crossed polarizers of a polarizing microscope, when the y-axis is oriented a to the polarizer axes, these three quasi-planar areas along the y-direction appear i calculated optical image as three bright stripes, as shown in Figure 3b. A completely si optical picture is observed in the experiment, as shown in Figure 3c. In addition, i experiment, one can observe the defects noted in Figure 3c with red circles. A discuss the mechanism of formation of these defects is beyond the scope of this work.

An important feature of the transition in the first Grandjean zone is that when a helix is formed, the Grandjean texture with one half-turn in the z-direction is essen preserved, but at the same time, it is strongly deformed in the z-direction and is mogeneous in the xy-plane. As will be shown below, this feature associated wit preservation of a half-turn of the helix throughout the thickness of the layer funda tally distinguishes the transition in the first zone from transitions in zones with a h m-number. This also leads to a significantly faster relaxation rate of the induced LH the field is turned off compared to the relaxation rate in higher-order Grandjean z which is also discussed below.

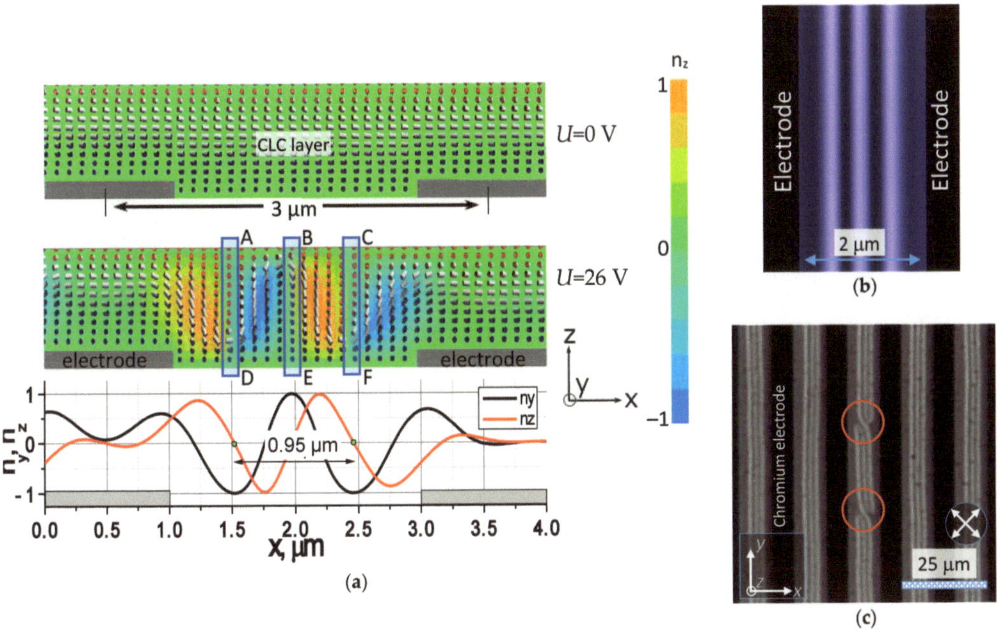

Figure 3. Electric field-induced director distribution in the 1st Grandjean zone ($m = 1$): (**a**) numerical modeling of the director distribution with the electric field turned off ($U = 0$ V) and on ($U = 26$ V); (**b**) numerical modeling of the texture with the electric field turned on, virtually observed in a polarizing microscope in crossed polarizers with axes at 45° in the xy-plane relative to the electrode strips; (**c**) experimentally observed texture in a polarizing microscope with the electric field turned on under the conditions specified for (**b**). The thicknesses of the CLC layers in the simulation and experiment were $d = 1$ and 1.2 µm, respectively. The regions marked as AD, BE, and CF show the areas of local quasi-planar alignment at which the LC director shows π-flips (see the text). The red circles in (**c**) indicate the characteristic defects in the observed texture.

b. Induced orientation transition in the second Grandjean zone

According to numerical simulation data, the transition in the second Grandjean zone, where two half-turns of the helix are across the thickness of the layer, is fundamentally different from the transition considered above in the first zone. Figure 4a shows the results of modeling the transition at an electrical voltage of $U = 40$ V and after turning off the electric field. The difference between the induced texture (Figure 4a) and the texture in the first zone (Figure 3) is the appearance of two local regions AD and CF with a truly planar orientation throughout the entire thickness. Thus, in these regions, the initial 2π-state transforms into an untwisted homogeneous planar state. Such a transition is well known as topologically equivalent, i.e., realized as a result of continuous deformation of the director distribution in the volume of the layer. This transition was observed in bistability effects in planarly aligned CLC cells [35,36]. The twisted 2π-state of the second Grandjean zone is topologically equivalent to the homogeneous planar state of the zeroth Grandjean zone. A continuous (defect-free) transition between these two states is possible through the homeotropic (vertical) state of the director at the center of the layer, which, in this case, is induced by the electric field. To achieve this homeotropic state, a certain electric field is required, which is why the transition is characterized by some threshold voltage. Since topologically equivalent states differ in twist angles that are multiples of 2π, topologically equivalent transitions are possible between states characteristic of Grandjean zones with either exclusively even or odd m-numbers. A defect-free transition between states of even and odd zones turns out to be forbidden under planar boundary conditions [37]. Only in the case of hybrid boundary conditions is a field-induced transition possible with a c in the twist angle by π, i.e., between neighboring Grandjean zones [38]. This informat important for understanding the transitions in the third and fourth Grandjean zones

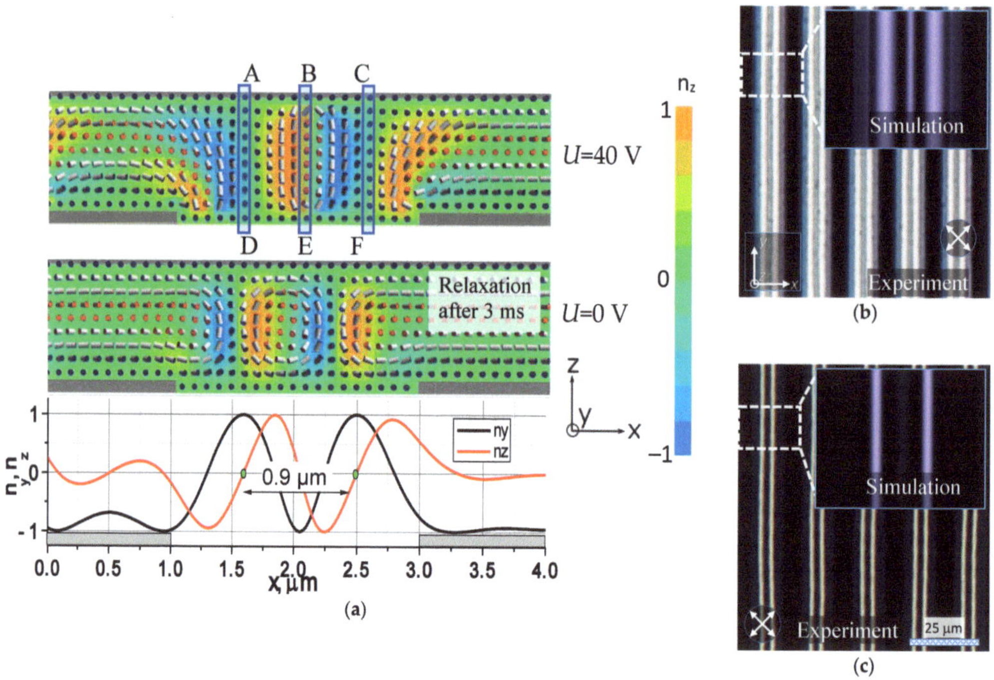

Figure 4. Electric field-induced director distribution and optical textures in the 2nd Grandjea $m = 2$: (**a**) numerical modeling of the director distribution at the electric field switched on (U = and off (after 3 ms); (**b**) texture observed in a polarizing microscope with the electric field turr in crossed polarizers with axes at 45° in the xy-plane relative to the electrode strips (the inset the result of numerical modeling of the type of texture in a polarizing microscope); (**c**) the sam (**b**), but after turning off the electric field (see text). The thickness of the CLC layer in the exper is 2.3 μm. The regions *AD* and *CF* correspond to local planar orientation, while in the *BE* regio LC director has two π-flips near the surfaces (see text).

In an electric field, the *AD* and *CF* areas with planar orientation are created alon y-direction and appear in the simulated and experimentally observed optical picture form of two wide bright stripes, as seen in Figure 4b. In the center between the elect (line *BE*), the transition to the planar state does not occur. It can be seen that when m along *BE*, the y-component of the director changes sign twice, respectively, near the and lower surface. That is, in this case, there is a highly deformed quasi-planar state electric field. The total change in the angle across the layer thickness along *BE* is eq 2π and corresponds to the initial 2π-state characteristic of the second Grandjean zo the numerically simulated optical picture (see inset in Figure 4b), this state appear narrow bright stripe in the center between the electrodes. This bright and narrow is also visible in the experimental image. Since in this narrow spatial region, in fac original 2π-state, although deformed, is preserved, then after turning off the electric this state relaxes very quickly, and the narrow stripe in the image in the center betwee electrodes disappears, as seen in Figure 4c. However, the local planar states *AB* ar separated from the 2π-state by an energy barrier [35], are preserved, and two bright s between the electrodes remain in the model and experimental images, as seen in Figu

Since, in the second Grandjean zone, local planar states are separated by an energy barrier from the lowest-energy ground 2π-state, their relaxation occurs extremely slowly.

Below, we will also call the induced planar state the 0-state, thereby indicating that it can arise in Grandjean zones of a higher order with a total director twist angle that is a multiple of 2π.

c. Induced orientation transition in the third and fourth Grandjean zones

It was shown above that a distinct feature of the transition in the second Grandjean zone, compared to the transition in the first zone, is the emergence of topologically equivalent *0*-states. In the third and fourth zones, the situation is similar. Since topologically equivalent states are characterized by a change in the twist angle across the layer's thickness by values that are multiples of 2π, then for the third zone, local states with a twist angle of $3\pi \pm 2\pi m$ are topologically equivalent, where m is an integer. That is, the initial director distribution with three half-turns at the layer thickness, characteristic of the third Grandjean zone, corresponds to a topologically equivalent π-state, which is the ground one in the first Grandjean zone. For the fourth zone, two topologically equivalent transitions are possible: a transition to the 2π-state and a transition to the *0*-state. It is also important to note that since, in the initial Grandjean texture, the electric field is directed predominantly perpendicular to the helix axis, transitions to states with a less twisted helix are energetically favorable [1], i.e., topologically equivalent transitions occur in the direction with a decrease in the initial number of half-turns over the layer thickness.

What has been said regarding the third zone can be seen in Figure 5a, which shows the simulation results at an electrical voltage U = 50 V. Along the four lines, *AE*, *BF*, *CG*, and *DH*, transitions to local highly deformed topologically equivalent π-states occur. Since these states are characterized by a quasi-planar orientation, they appear in the optical image as the four brightest stripes in the y-direction, as shown in Figure 5b. Since new topologically equivalent π-states are locally formed during the transition, after turning off the electric field, their relaxation is very slow, and the induced state with a lying helix can be observed for quite a long time, as shown in Figure 5c.

The behavior in the fourth Grandjean zone (m = 4), although it looks more complex, also fully fits into the concept of topologically equivalent transitions.

Figure 6a shows that in an electric field in the center between the electrodes, a topologically equivalent transition from the ground 4π-state to the 2π-state is realized. In optics (both in simulation and experiment), this highly deformed 2π-state appears as a single bright stripe in the center between the electrodes, as shown in Figure 6b. After turning off the electric field, relaxation also proceeds through topologically equivalent states, which appear in optical images in the form of many bands, among which two bands correspond to the *0*-state and are therefore the brightest, as shown in Figure 6c. Thus, in the fourth zone, we deal with multiple ways of relaxation to the ground 4π-state from topologically equivalent 2π- and 0-states.

Concluding this section, we generalize the features of orientational transitions to the LH state in four primary Grandjean zones, which can be accompanied by local topologically equivalent transitions. Figure 7 schematically shows the energy diagram of allowed transitions in different Grandjean zones. As can be seen, in the first zone, where the ground state is the π-state, there are no allowed transitions, since the nearest untwisted *0*-state is not topologically equivalent. In the second and third zones, only one transition to a state with the change for two half-turns is allowed, and in the fourth zone, there are two such transitions. The scheme shown in Figure 7 can easily be continued for zones with $m > 4$ taking into account that only transitions between the states with an even number of half-turns are allowed.

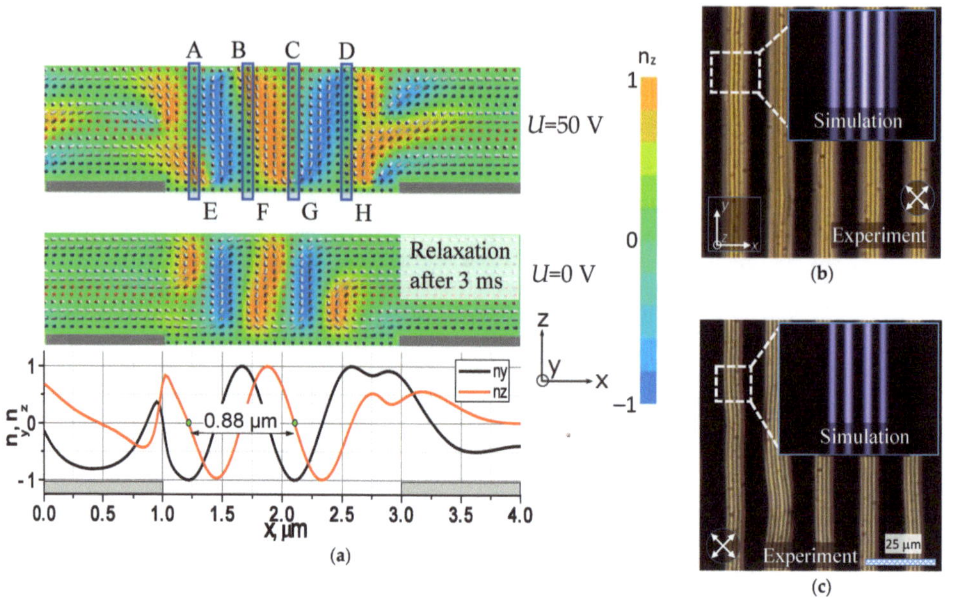

Figure 5. Electric field-induced director distribution in the 3rd ($m = 3$) Grandjean zone: (**a**) num modeling of the director distribution with the electric field on (U = 50 V) and off (after (**b**) texture observed in a polarizing microscope with the electric field turned on in crossed pola with axes at 45° in the xy-plane relative to the electrode strips (the inset shows the result of num modeling of the type of texture in a polarizing microscope); (**c**) the same as in (**b**), but after tu off the electric field (see text). The thickness of the CLC layer in the experiment is ~3.5 μm. T *BF*, *CG*, and *DH* lines indicate the quasi-planar regions at which the LC director has π-flips.

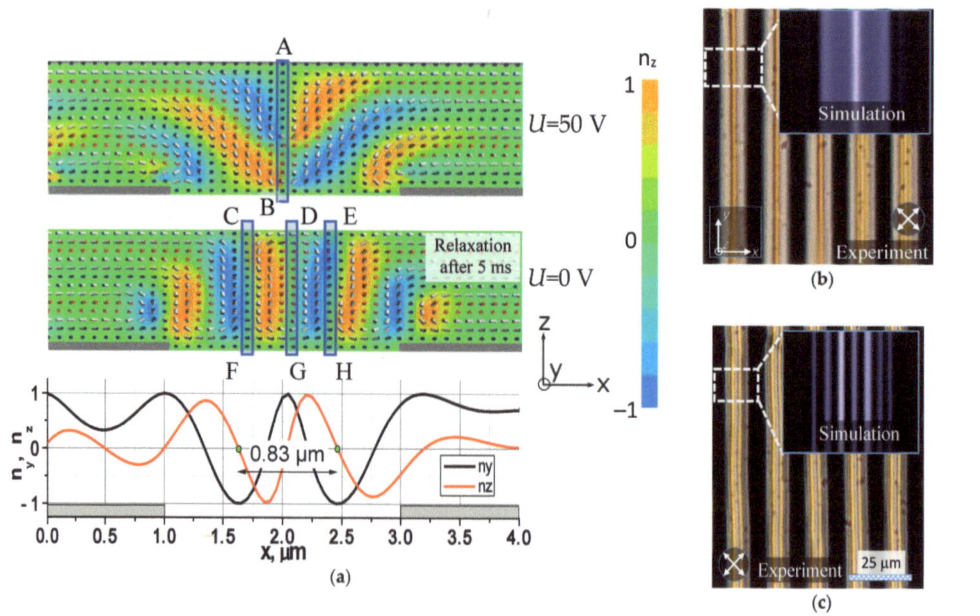

Figure 6. Distribution of the CLC director field in the 4th Grandjean zone ($m = 4$, i.e., 720° twist) a ages U = 50 V and after 5 ms when the voltage is turned off (**a**). The photographs on the right

experimentally observed and calculated (inset) images in a polarizing microscope when the voltage is applied (**b**) and a second after it is turned off (**c**). The thickness of the CLC layer in the experiment is ~4.6 μm. The regions *AB*, *CF*, *DG*, and *EH* are for local quasi-planar and topologically equivalent 2π- and 0-states with 2 and 0 π-flips across the thickness.

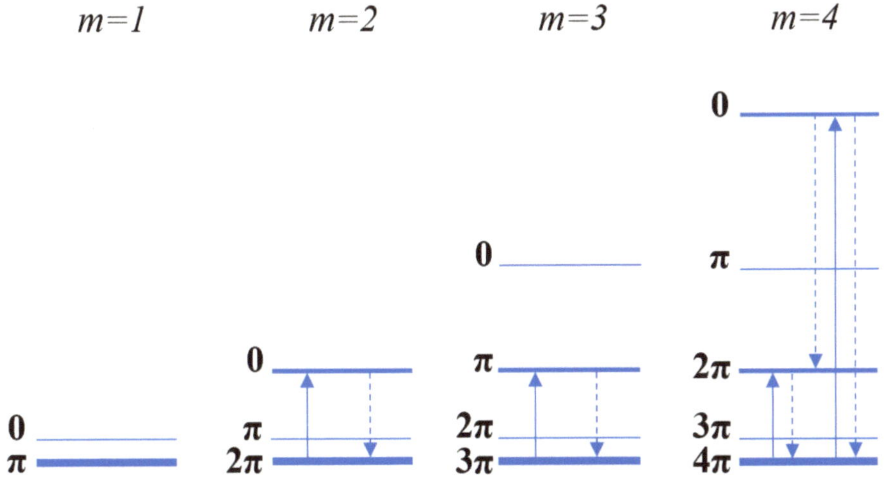

Figure 7. Schematic representation of free energy levels for fixed layer thickness $d=mp_0/2$ and allowed topologically equivalent transitions (shown by arrows) in various Grandjean zones accompanying the orientational transition to the LH state. The allowed relaxation paths are shown by dashed arrows.

d. Dynamics of induced transitions

Figure 8 shows the simulated dynamics of relaxation of induced states for four Grandjean zones (m = 1, 2, 3, 4), obtained for the same thickness of the LC layer d = 1.5 μm. However, as already mentioned, the natural pitch p_0 for each zone was set to obtain a stressed helix at an LC layer thickness $d = mp_0/2 + 0.2p_0$, so we shift from a zone center towards the next zone with a higher number m. In this case, we obtain the best agreement with the experimental observations. Calculations were performed for the following CLC parameters: elasticity coefficients K_1 = 13.9 pN, K_2 = 11.4 pN, K_3 = 15.5 pN, and rotational viscosity γ = 0.1 Pa·s. To illustrate the dynamics of transitions, the value of the director z-component n_z in the center of the layer between the electrodes was chosen as a parameter characterizing the induced state.

In the first Grandjean zone (m = 1), the relaxation rate is characterized by the shortest time τ~5 ms. At the same time, for large values of m, the relaxation rate drops sharply, reaching its lowest value in the fourth zone (t > 250 ms). All this is well explained within the framework of the model of topologically equivalent transitions discussed above, as seen in Figure 7. Indeed, for m = 1, transitions to topologically equivalent states do not occur since there is no corresponding topologically equivalent state with fewer half-turns. Therefore, after turning off the field, the usual viscoelastic relaxation takes place with a characteristic time, which can be estimated from relation (1). By substituting the LC parameters into (1), it is easy to obtain a characteristic time of ~5 ms, which is consistent with numerical modeling and experimental observations. For Grandjean zones with m > 1, topologically equivalent states exist, and when a lying helix is induced, corresponding local transitions occur. Thus, for a zone with m = 2, local transitions to the *0*-state with a uniform planar director distribution take place. Relaxation of these *0*-states into the original 2π-states requires activation of the local homeotropic states in the layer. Such activations are possible through spatial regions adjacent in the x-direction, where across the layer thickness, local

director orientation is homeotropic (see the director distribution to the left and right
AD line in Figure 4a).

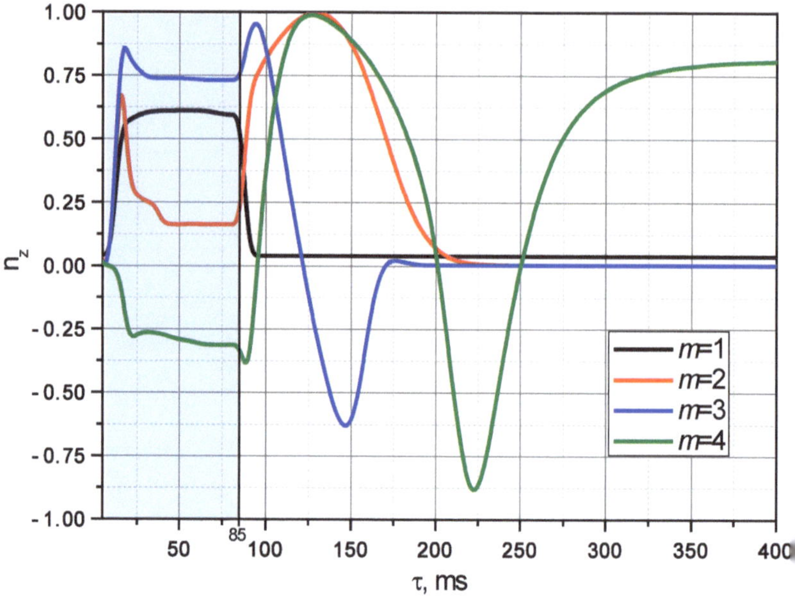

Figure 8. Dynamics of the formation of the LH structure and its relaxation when considering director component for the four Grandjean zones in the interelectrode space. The data were ob at the same thickness of d = 1.5 μm in the center of the LC layer, but with a shift with respect center of zones towards the next zone with a higher m-value (see the text). Values of ±1 corre to the homeotropic orientation of the LC director, and 0 corresponds to the planar orientat rectangular control pulse is applied for 75 ms starting from 10 ms. The blue background indica duration of the electric field applied to the LC cell.

In order to interpret a correlation between the relaxation times in the secon third zones, it is important to take into account the difference in the free energy bet topologically equivalent states. According to Figure 2, the 0- and π-states for the se and third zones, respectively, have the same free energy, and the difference in the rela time, which is shorter for the third zone (85 ms vs 125 ms), looks surprising at first g However, the energy levels in Figure 7 are given for an unstressed helix located i center of Grandjean zones. In general cases, the total free energy of the CLC helix pe area of a layer (integral of the free energy density over the thickness) at rigid bou conditions is expressed as follows [37]:

$$\Phi_{m,k} = \frac{K_2 d}{2}(q_{m,k} - q_{m,0})^2,$$

where index m is for the Grandjean zone number; k is for a number of helix half-tu a given zone m; and $q_{m,k}$ and $q_{m,0}$ are for actual and natural wavenumbers of the respectively. The ground state in the m-th zone is characterized by wavenumber which corresponds to m helix half-turns over the layer thickness. The difference in th energy between the topologically equivalent states with k half-turns and ground stat m half-turns can be derived from (5) as follows:

$$\Delta\Phi_{k,m} = \Phi_{m,k} - \Phi_{m,m} = \frac{K_2 d}{2}(\Delta k q_{1,1})^2\left[1 + \frac{2m(q_{1,0} - q_{1,1})}{\Delta k q_{1,1}}\right],$$

where $q_{m,k} = kq_{1,1}$, $q_{m,m} = mq_{1,1}$, and $\Delta k = m\text{-}k$ are taken into account.

According to (6), in the case of the "expanded" helix that we deal with when $q_{1,0} > q_{1,1}$, the $\Delta\Phi$ increases with the Grandjean zone number m. This fact explains why, for the third zone ($m = 3$), the relaxation is faster compared to the second zone ($m = 2$). However, this seems to work only in simplest cases, when we deal with a single topologically equivalent transition like in the case of the second and third zone, as shown in Figure 7.

In the case of the fourth zone, two topologically equivalent states (0- and 2π-state) can exist, and the relaxation has three possible ways: (i) 0-state $\rightarrow$ 4π-state; (ii) 0-state $\rightarrow$ 2π-state$\rightarrow$$4\pi$-state; and (iii) 2π-state $\rightarrow$ 4π-state. This is why the relaxation process turns out to be more complicated and slowest among the first four Grandjean zones.

4. Conclusions

In a CLC layer with negative dielectric anisotropy, local transitions induced by a periodic electric field from planar Grandjean textures with the helix axis normal to the layer into local states with a lying helix were studied experimentally and numerically. For transitions in the Grandjean zones with the number of half-turns at a layer thickness of more than one, the induced new states with the helix axis lying in the plane of the LC layer are metastable and characterized by slow relaxation from several seconds to tens of minutes, depending on the Grandjean zone and layer thickness.

Numerical modeling has shown that for Grandjean zones with $m > 1$, the formation of a lying helix is associated with multiple local transitions to topologically equivalent states, characterized by a change in the director twist angle by values that are multiples of 2π. An important circumstance is that, in contrast to topologically equivalent transitions observed previously [35,36], there is no need to involve hydrodynamic flows.

Electric field-induced lying helix states are interesting not only for their topological features but also as superperiodic structures for electro-optical and photonic applications. For example, their properties, characteristic of photonic crystals, can manifest themselves in the waveguide mode, and the topologically equivalent transition to an unwound planar state is of interest for new types of bistable switching.

Author Contributions: Conceptualization and supervision, S.P.P.; methodology and experiment, S.P.P. and I.V.S.; investigation, I.V.S.; verification, S.P.P.; writing—original draft preparation, I.V.S., A.R.G., and I.V.K.; writing—review and editing, all authors. All authors have read and agreed to the published version of the manuscript.

Funding: This research work was performed within the state assignment of the National Research Centre "Kurchatovsky Institute".

Data Availability Statement: Data are available upon reasonable request.

Acknowledgments: We thank Alexey Voloshin for administrative support of this publication.

Conflicts of Interest: No potential conflicts of interest were reported by the authors.

rences

De Gennes, P.G.; Prost, J. *The Physics of Liquid Crystals*; Oxford University Press: Oxford, UK, 1993; ISBN 9780198517856.

Oseen, C.W. The theory of liquid crystals. *J. Chem. Soc. Faraday Trans.* **1933**, *29*, 883–899. [CrossRef]

Stebryte, M. Reflective optical components based on chiral liquid crystal for head-up displays. *Liq. Crystals Today* **2021**, *30*, 36–45. [CrossRef]

Nam, S.; Wang, D.; Lee, G.; Choi, S.S. Broadband wavelength tuning of electrically stretchable chiral photonic gel. *Nanophotonics* **2022**, *11*, 2139–2148. [CrossRef]

Wang, D.; Nam, S.; Jung, W.; Yang, H.J.; Choi, S.S. Electrically wavelength-controllable color filters with high optical transmittance using heterogeneous chiral liquid crystals. *Adv. Opt. Mater.* **2023**, *11*, 2202906. [CrossRef]

Mysliwiec, J.; Szukalska, A.; Szukalski, A.; Sznitko, L. Liquid crystal lasers: The last decade and the future. *Nanophotonics* **2021**, *10*, 2309–2346. [CrossRef]

Coles, H.; Morris, S. Liquid-crystal lasers. *Nat. Photon.* **2010**, *4*, 676–685. [CrossRef]

Papič, M.; Mur, U.; Zuhail, K.P.; Ravnik, M.; Muševič, I.; Humar, M. Topological liquid crystal superstructures as structured light lasers. *Proc. Nat. Acad. Sci. USA* **2021**, *118*, e2110839118. [CrossRef]

9. Ortega, J.; Folcia, C.L.; Etxebarria, J. Laser emission at the second-order photonic band gap in an electric-field-distorted chol liquid crystal. *Liq. Cryst.* **2019**, *46*, 2159–2166. [CrossRef]
10. Folcia, C.L.; Ortega, J.; Sierra, T.; Martínez-Bueno, A.; Etxebarria, J. Chiral ferroelectric nematic liquid crystals as materi versatile laser devices. *Giant* **2024**, *19*, 100316. [CrossRef]
11. Dadalyan, T.; Ninoyan, Z.; Nys, I.; Alaverdyan, R.; Beeckman, J.; Neyts, K. Light-induced multi-wavelength lasing in dye-chiral nematic liquid crystals due to strong pumping illumination. *Liq. Cryst.* **2018**, *45*, 1272–1278. [CrossRef]
12. Sarukhanyan, T.M.; Gharagulyan, H.; Rafayelyan, M.S.; Golik, S.S.; Gevorgyan, A.H.; Alaverdyan, R.B. Multimode Robust in a Dye-Doped Polymer Layer Embedded in a Wedge-Shaped Cholesteric. *Molecules* **2021**, *26*, 6089. [CrossRef] [PubMed
13. Inoue, Y.; Yoshida, H.; Inoue, K.; Fujii, A.; Ozaki, M. Improved lasing threshold of cholesteric liquid crystal lasers with in helix alignment. *Appl. Phys. Express* **2010**, *3*, 102702. [CrossRef]
14. Carbone, G.; Salter, P.; Elston, S.J.; Raynes, P.; De Sio, L.; Ferjani, S.; Strangi, G.; Umeton, C.; Bartolino, R. Short pitch chol electro-optical device based on periodic polymer structures. *Appl. Phys. Lett.* **2009**, *95*, 011102. [CrossRef]
15. Palto, S.P.; Shtykov, N.M.; Kasyanova, I.V.; Umanskii, B.A.; Geivandov, A.R.; Rybakov, D.O.; Simdyankin, I.V.; Artemo Gorkunov, M.V. Deformed lying helix transition and lasing effect in cholesteric LC layers at spatially periodic boundary conc *Liq. Cryst.* **2020**, *47*, 384–398. [CrossRef]
16. Komitov, L.; Bryan-Brown, G.P.; Wood, E.L.; Smout, A.B.J. Alignment of cholesteric liquid crystals using periodic anch *J. Appl. Phys.* **1999**, *86*, 3508–3511. [CrossRef]
17. Carbone, G.; Corbett, D.; Elston, S.J.; Raynes, P.; Jesacher, A.; Simmonds, R.; Booth, M. Uniform lying helix alignment on pe surface relief structure generated via laser scanning lithography. *Mol. Cryst. Liq. Cryst.* **2011**, *544*, 37–49. [CrossRef]
18. Yip, W.C.; Welch, C.; Mehl, G.H.; Wilkinson, T.D. A cholesteric liquid crystal device having stable uniform lying helix str *J. Mol. Liq.* **2020**, *299*, 112141. [CrossRef]
19. Yu, C.H.; Wu, P.C.; Lee, W. Polymer Stabilization of Uniform Lying Helix Texture in a Bimesogen-Doped Cholesteric I Crystal for Frequency-Modulated Electro-Optic Responses. *Materials* **2022**, *15*, 771. [CrossRef]
20. Kim, S.H.; Shi, L.; Chien, L.C. Fast flexoelectric switching in a cholesteric liquid crystal cell with surface-localized pc network. *J. Phys. D Appl. Phys.* **2009**, *42*, 195102. [CrossRef]
21. Jia, Z.; Pawale, T.; Guerrero-García, G.I.; Hashemi, S.; Martínez-González, J.A.; Li, X. Engineering the Uniform Lying H Structure in Chiral Nematic Liquid Crystals: From Morphology Transition to Dimension Control. *Crystals* **2021**, *11*, 414. [Cro
22. Zheng, Z.; Li, Y.; Bisoyi, H.K.; Wang, L.; Bunning, T.J.; Li, Q. Three-dimensional control of the helical axis of a chiral ne liquid crystal by light. *Nature* **2016**, *531*, 352–356. [CrossRef] [PubMed]
23. Bisoyi, H.K.; Bunning, T.J.; Li, Q. Stimuli-Driven Control of the Helical Axis of Self-Organized Soft Helical Superstructure *Mater.* **2018**, *30*, 1706512. [CrossRef] [PubMed]
24. Salter, P.S.; Carbone, G.; Jewell, S.A.; Elston, S.J.; Raynes, P. Unwinding of the uniform lying helix structure in cholesteric crystals next to a spatially uniform aligning surface. *Phys. Rev. E* **2009**, *80*, 041707. [CrossRef]
25. Geivandov, A.R.; Simdyankin, I.V.; Barma, D.D.; Shtykov, N.M.; Palto, S.P. High-quality deformed lying helix in chiral surface with periodic alignment prepared by two-step optical treatment. *Liq. Cryst.* **2022**, *49*, 2027–2036. [CrossRef]
26. Palto, S.P.; Barnik, M.I.; Geivandov, A.R.; Kasyanova, I.V.; Palto, V.S. Spectral and polarization structure of field-induced ph bands in cholesteric liquid crystals. *Phys. Rev. E* **2015**, *92*, 032502. [CrossRef] [PubMed]
27. Simdyankin, I.V.; Geivandov, A.R.; Barnik, M.I.; Palto, V.S.; Palto, S.P. Selective reflection zones of cholesteric liquid cr induced in a non-uniform spatially periodic electric field. *Liq. Cryst. Its Pract. Use* **2019**, *19*, 48–56. (In Russian) [CrossRef]
28. Rumi, M.; Tondiglia, V.P.; Natarajan, L.V.; White, T.J.; Bunning, T.J. Non-Uniform Helix Unwinding of Cholesteric Liquid C in Cells with Interdigitated Electrodes. *ChemPhysChem* **2014**, *15*, 1311–1322. [CrossRef]
29. Jewell, S.A.; Sambles, J.R. Optical imaging of the effect of in-plane fields on cholesteric liquid crystals. *Phys. Rev. E* **2008**, *78*, ([CrossRef]
30. Gardiner, D.J.; Morris, S.M.; Castles, F.; Qasim, M.M.; Kim, W.-S.; Choi, S.S.; Park, H.J.; Chung, I.J.; Coles, H.J. Polymer stab chiral nematic liquid crystals for fast switching and high contrast electro-optic devices. *Appl. Phys. Lett.* **2011**, *98*, 2 [CrossRef]
31. Gardiner, D.J.; Morris, S.M.; Hands, P.J.W.; Castles, F.; Qasim, M.M.; Kim, W.-S.; Choi, S.S.; Wilkinson, T.D.; Coles, H.J. Sponta induction of the uniform lying helix alignment in bimesogenic liquid crystals for the flexoelectro-optic effect. *Appl. Phy* **2012**, *100*, 063501. [CrossRef]
32. Joshi, V.; Chang, K.H.; Paterson, D.A.; Storey, J. P-151: Fast Flexoelectro-optic Response of Bimesogen-doped Polymer Stab Cholesteric Liquid Crystals in Vertical Standing Helix Mode. *SID Symp. Dig. Tech. Pap.* **2017**, *48*, 1849–1852. [CrossRef]
33. Choi, S.S.; Castles, F.; Morris, S.M.; Coles, H.J. High contrast chiral nematic liquid crystal device using negative dielectric m *Appl. Phys. Lett.* **2009**, *95*, 193502. [CrossRef]
34. Umanskii, B.A.; Simdyankin, I.V. Circular dichroism in cholesteric liquid crystals. *Crystall. Rep.* **2019**, *64*, 437–442. [Cross
35. Berreman, D.W.; Heffner, W.R. New bistable cholesteric liquid-crystal display. *Appl. Phys. Lett.* **1980**, *37*, 109–111. [CrossR
36. Palto, S.P.; Barnik, M.I. Bistable switching in chiral nematic liquid crystal layers with a 2π-twist ground state. *J. Expert. Phys.* **2005**, *100*, 199–207. [CrossRef]

Palto, S.P. On mechanisms of the helix pitch variation in a thin cholesteric layer confined between two surfaces. *J. Expert. Theor. Phys.* **2002**, *94*, 260–269. [CrossRef]
Palto, S.P.; Rybakov, D.O.; Umanskii, B.A.; Shtykov, N.M. Spiral Pitch Control in Cholesteric Liquid Crystal Layers with Hybrid Boundary Conditions. *Crystals* **2022**, *13*, 10. [CrossRef]

crystals

Article

Effects of Photopatterning Conditions on Azimuthal Surface Anchoring Strength

Nilanthi P. Haputhanthrige [1,2], Mojtaba Rajabi [1] and Oleg D. Lavrentovich [1,2,3,4,*]

[1] Advanced Materials and Liquid Crystal Institute, Kent State University, Kent, OH 44242, USA; hpadmini@kent.edu (N.P.H.); mrajabi@kent.edu (M.R.)
[2] Department of Physics, Kent State University, Kent, OH 44242, USA
[3] Materials Science Graduate Program, Kent State University, Kent, OH 44242, USA
[4] Faculty of Chemistry, University of Warsaw, Zwirki i Wigury 101, 02-089 Warsaw, Poland
* Correspondence: olavrent@kent.edu

Abstract: Spatially varying alignment of liquid crystals is essential for research and applic One widely used method is based on the photopatterning of thin layers of azo-dye molecules as Brilliant Yellow (BY), that serve as an aligning substrate for a liquid crystal. In this stuc examine how photopatterning conditions, such as BY layer thickness (b), light intensity (I), irrad dose, and age affect the alignment quality and the strength of the azimuthal surface anchorin azimuthal surface anchoring coefficient, W, is determined by analyzing the splitting of i disclinations into half-integer disclinations at prepatterned substrates. The strongest ancho achieved for b in the range of 5–8 nm. W increases with the dose, and within the same dc increases with I. Aging of a non-irradiated BY coating above 15 days reduces W. Our stud demonstrates that sealed photopatterned cells filled with a conventional nematic preserve alignment quality for up to four weeks, after which time W decreases. This work sugges optimization pathways for photoalignment of nematic liquid crystals.

Keywords: azimuthal anchoring; surface anchoring; liquid crystals; photopatterning cond alignment stability; Brilliant Yellow

Citation: Haputhanthrige, N.P.; Rajabi, M.; Lavrentovich, O.D. Effects of Photopatterning Conditions on Azimuthal Surface Anchoring Strength. *Crystals* **2024**, *14*, 1058. https://doi.org/10.3390/cryst14121058

Academic Editor: Ingo Dierking

Received: 18 November 2024
Revised: 3 December 2024
Accepted: 3 December 2024
Published: 6 December 2024

1. Introduction

Alignment of liquid crystals (LCs) is crucial for their applications. While th directional alignment can be achieved by mechanical rubbing [1–5], alignment w spatially varying "easy axis" requires a more sophisticated approach. Spatially va alignment becomes exceedingly important in many academic and applied projects, su the fabrication of planar optics elements [6–12], LC elastomer coatings with predes topography [13–19], orientationally ordered environments that control collective and vidual dynamics of microswimmers [20–25], and substrates that align living tissues [2 The most popular approach to achieving the spatially varying alignment of LCs is toalignment [13,30–38]. Photoalignment allows one to design complex director pat with high spatial resolution [34,35,39], controllable surface anchoring [40], dynamic di alignment [41], and the capability to pattern the alignment on flexible and curved strates [42–44]. Photoalignment does not induce impurities, electric charges, or mecha damage to the treated surfaces, unlike conventional rubbing [45].

Brilliant Yellow is one of the azo-dye materials used for photoalignment [46–48 tensive studies by Yaroshchuk et al. [49] brought the conclusion that BY provides "exc photoalignment of nematic, smectic ferroelectric and reactive liquid crystals" and "s extraordinarily high photo and thermal stability". The study also noted that the d C=C bond in the core of BY molecules results in an absorption peak at 432 nm [49]. peak, being in the lower part of the visible spectrum, might result in reduced stability exposed to visible light. Nevertheless, the study by Yaroshchuk et al. [49] demonst that BY has excellent stability when irradiated with ultraviolet light and kept at an elevated temperature of 150 °C for one hour. Since BY allows one to produce high-quality alignment in a relatively easy process, it continues to attract interest in research and applications, see, for example, some recent publications in Refs. [33,49–54].

Azo-dye molecules undergo photoinduced reorientation when exposed to light [50,55,56]. For irradiation with a linearly polarized beam, the probability of absorption is $P \propto \cos^2\beta$ where β is the angle between the long axis of the molecule in its *trans* state and the light polarization direction. The absorption–reorientation process repeats itself until the dye molecule aligns perpendicularly to the light polarization, $\beta = \pi/2$. Studies by Wang et al. [57] and Shi et al. [58] show that alignment quality strongly depends on exposure to humidity and only slightly depends on the type of surface used for depositing BY. Our previous work showed that a longer light exposure produces a stronger in-plane (azimuthal) surface anchoring [40]. However, the effect of other important factors remained unexplored.

This study investigates the effect of photopatterning conditions, such as BY layer thickness (b), light intensity (I), irradiation dose, and age of non-irradiated substrate, on the alignment quality and the strength of azimuthal anchoring expressed by the anchoring coefficient W in the surface potential $\frac{1}{2}W\sin^2\alpha$, where α is the angle between the alignment direction imposed by the patterned BY layer and the actual director specifying the local orientation of the liquid crystal. The coefficient W is determined by analyzing the splitting of integer disclinations into half-integer disclinations at photopatterned substrates [26,40]. We find that b in the range 5–8 nm yields the strongest azimuthal anchoring. W increases with the dose and with I when the dose is fixed. BY-coated substrates, which are photopatterned within 15 days of substrate preparation (after BY spin coating and baking, but before irradiation), show no significant change in W. However, substrates aged for more than 15 days before irradiation exhibit a decline in W. Our study also demonstrates that photopatterned cells filled with a conventional nematic LC and sealed with epoxy glue preserve their alignment strength for about 4 weeks, but further aging of the filled cell leads to a reduction in W. The results facilitate the optimization of BY photoalignment for liquid crystal applications.

2. Materials and Methods

2.1. Cell Preparations

Indium tin oxide (ITO)-coated glass plates are sonicated in water with a small amount of detergent at 60 °C for 15 min. Although the studies do not require an application of an electric field, the choice of ITO-coated plates is justified by the fact that most applications of liquid crystals involve electro-optic effects, thus the presence of the ITO electrodes is often a necessity. The plates are rinsed with isopropanol, dried in an oven at 80 °C for 15 min, and exposed to UV in an ozone chamber for 15 min. The plates are spin-coated with a solution of azo-dye BY in N, N-dimethylformamide (DMF), Figure 1a (both purchased from Sigma Aldrich, St. Louis, MO, USA) at 3000 rpm for 30 s, and baked at 80 °C for 30 min. We use DMF solutions with various BY concentrations, 0.2, 0.4, 0.5, 0.6, 0.8, 1.0, 2.0, and 4.0 wt%, in order to vary the thickness b of the resulting BY layer and to explore its effect on the anchoring strength. All other experiments are performed with substrates coated with a 0.5 wt% BY solution, resulting in $b = 7.6$ nm, which is in the optimal thickness range of 5–8 nm that yields the strongest W. To avoid the detrimental effects of humidity on BY alignment [57], we limit the relative humidity (RH) of the environment to less than 20% during the spin coating and baking, and to 20–35% during substrate storage, cell assembly, and photopatterning. During imaging, the sealed LC filled cells are kept at an RH of less than 50% and a constant temperature of 45 °C.

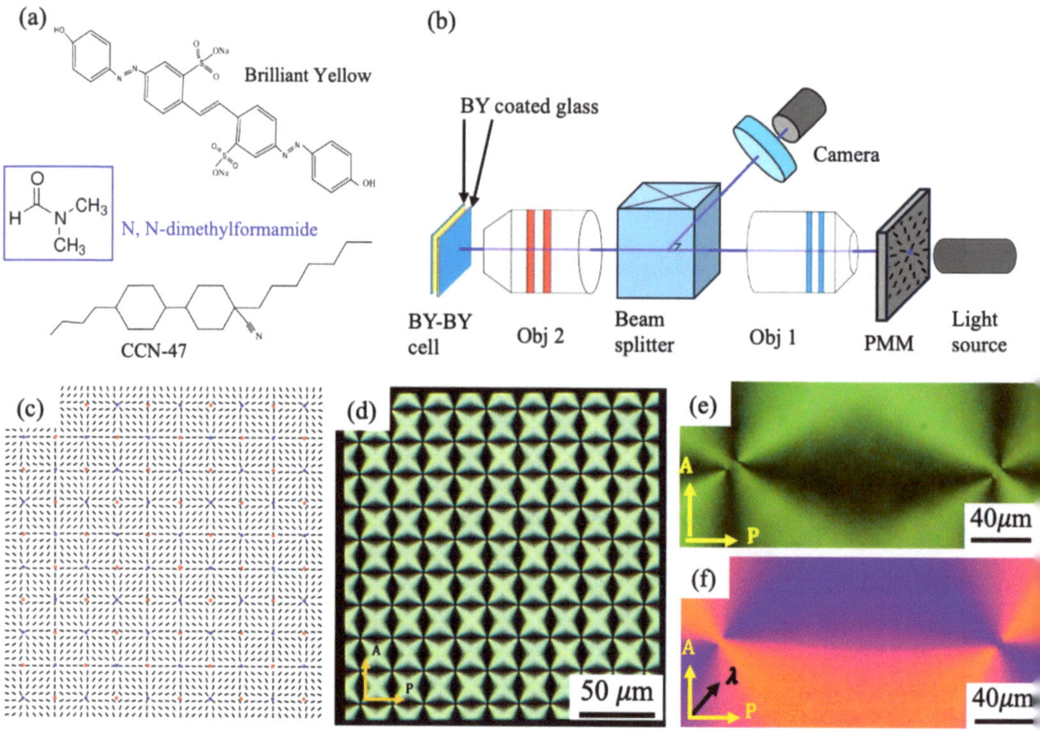

Figure 1. Chemical structures of materials and photopatterned defect array. (**a**) Photorespc azo-dye Brilliant Yellow, liquid crystal CCN-47, and solvent N, N-dimethylformamide. (**b**) Sche of the photopatterning setup. Obj: objective, PMM: plasmonic metamask. (**c**) Director pattern defects array. The blue dots represent the cores of +1 defects, and the red dots represent the cc -1 defects. (**d**) A polarizing optical microscope texture of the plasmonic meta mask with ± 1 defect array. (**e**) Polarized optical microscope images of a portion of a photopatterned cell witl radial defect on the left and a -1 defect on the right. The cell thickness is 1.1 μm. (**f**) The polarizing microscopy with a full-wavelength optical compensator with the slow axis λ. P represent the polarizer and analyzer, respectively.

Cells are assembled from two glass substrates with the BY-coated surfaces f each other; these are called BY-BY cells throughout the text. The gap is fixed using e glue NOA 65 (Norland Products Inc., Jamesburg, NJ, USA), without spacers, to ac a thickness of ~1 μm. The thickness h of the gap between two plates is measure an interferometric technique using a UV/VIS spectrometer Lambda 18 (Perkin E Waltham, MA, USA).

2.2. Photopatterning

We use the plasmonic metamask (PMM) technique introduced by Guo et al. [3 to pattern the substrates with an array of +1 and -1 defects (these defects should n confused with "defects" induced by mishandling of the samples or by dust particles). is an aluminum film with a thickness of 150 nm containing an array of nanoslits, with a length of 220 nm and a width of 100 nm. When an unpolarized light beam p through a nanoslit, the transmitted light becomes polarized along the short axis c nanoslit. The degree of polarization of the transmitted light depends on the wavele Guo et al. [34] demonstrated experimentally and through numerical simulations th polarization contrast ratio exceeds 7 dB for wavelengths ranging from 400 to 800 nm transmitted light beam irradiates the cells, as shown in Figure 1b. We use two types o in this work: BY-BY and BY-PS cells. The BY-PS cells are assembled with one BY-coated substrate and one polystyrene (PS)-coated substrate, with the coated surfaces facing each other. Preparation of the PS-coated substrates is explained later in the text, Section 3.1. During photoalignment, BY-PS cells are positioned so that the BY-coated substrate is closer to the light source. Irradiated light aligns azobenzene molecules perpendicularly to the light polarization direction. As a result, a desired pattern, which replicates the pattern of nanoslits in the PMM, is produced in the azo-dye layer. The BY molecule exhibits an absorption range of 350–500 nm, with a peak at 432 nm [49]. We use a light source EXFO X-Cite (Excelitas Technologies, Pittsburgh, PA, USA) with a wavelength range of (320–750) nm, which fully covers the absorption spectrum of BY. The light beam propagates along the normal to the PMM and the cell, Figure 1b. The light intensity I is measured at the point of cell incidence using a power and energy meter console PM 100D (Thorlabs, Newton, NJ, USA).

A periodic square lattice of defects with strength +1 and −1 is designed by using a superposition rule for the in-plane director, $\hat{\mathbf{n}}_{\mathrm{BY}} = (n_x, n_y, 0) = (\cos\varphi, \sin\varphi, 0)$, where $\varphi = \sum_{i=1}^{p}\sum_{j=1}^{q}(-1)^{i+j}\arctan\left(\frac{y-jb}{x-ia}\right)$, x and y are the Cartesian coordinates, p and q are the numbers of defects in rows and columns, respectively, $p = q = 10$; $a = b = 200$ μm is the distance between the defects along the x and y directions, respectively. The +1 defects in the pattern are of a radial type, so that the director around them experiences mostly splay, Figure 1c.

2.3. Nematic Material

The photopatterned cells are filled with 4-butyl-4-heptyl-bicyclohexyl-4-carbononitrile (CCN-47), as shown in Figure 1a, by capillary action in the isotropic state at the temperature 70 °C. The material exhibits the following phase transitions upon heating: Smectic A 29.9 °C Nematic 58.5 °C Isotropic. After filling the cells, they are kept at 45 °C during the experiments using a Linkam hot stage (Linkam Scientific, Redhill, UK). At 45 °C, the elastic constants K_1 of splay and K_3 of bend for CCN-47 are equal, $K_1 = K_3 = K = 8$ pN [59,60], which allows one to use the superposition rule for the director field and to analyze the elastic properties of the patterns in the so-called one-constant approximation [61].

2.4. Optical Microscopy Characterization

The optical textures of photopatterned LC cells are recorded using an Olympus BX51 polarized optical microscope (Olympus, Tokyo, Japan) equipped with a Basler (acA1920-155um) digital color camera (Basler, Ahrensburg, Germany), Figure 1e,f. A full-wavelength (530 nm) optical compensator is used to reconstruct the director field, Figure 1f. Regions where the director aligns parallel to the slow axis of the compensator exhibit a blue interference color, while areas where the director is perpendicular to the slow axis appear yellow. The left defect in Figure 1e,f is a + 1 radial defect, while the right one is a − 1 defect. The separation distance between half integer defects, d, is measured using the open-source software package Fiji/ImageJ, version 2.140/1.54f.

2.5. Theoretical Background

Defects of strength ±1 tend to split into pairs of ±1/2 to reduce their elastic energy [61]. In the so-called one-constant approximation, the elastic repulsive potential of two +1/2 or two −1/2 defects is weakly dependent on their separation d: $F_{\mathrm{E}} = -\frac{\pi K h}{2}\ln\frac{d}{2r_c}$, where K is the average Frank elastic modulus, h and r_c are the cell thickness and the radius of the disclination core, respectively [26,61]. In a photopatterned cell, the separation of the defects is resisted by surface anchoring that tends to enforce the patterned ±1 defects. The elasticity-anchoring balance determines the equilibrium separation distance d of the semi-integer cores. The surface anchoring energy of a patterned cell can be found by integrating the Rapini–Papoular potential, $F_{\mathrm{S}} = 2\int_0^{2\pi}\int_0^{d}\frac{1}{2}W\left[1-(\hat{\mathbf{n}}_{\mathrm{BY}}\cdot\hat{\mathbf{n}}_{\mathrm{LC}})^2\right]rdrd\varphi$ which yields $F_{\mathrm{S}} = 2\alpha W d^2$; here $\hat{\mathbf{n}}_{\mathrm{LC}}$ is the actual director field of a split defect pair, $\hat{\mathbf{n}}_{\mathrm{BY}}$ is the ideal radial pattern of the easy axis at the substrates, $\alpha \approx 0.184$ is a numerical coefficien the factor 2 reflects anchoring at both plates [26,40,61]. The equilibrium value of d a one to calculate the anchoring strength as $W = \pi Kh/(8\alpha d^2)$. For cells assembled wi photopatterned surface and the second plate providing degenerate in-plane anch $W = \pi Kh/(4\alpha d^2)$.

We calculate W by measuring d and h and using the known values of K and α. Tl olution of the optical microscope is sufficient for precise measurements, as the experi show that the parameter d is higher than 3 μm even for the highest achieved anch The distance between neighboring photopatterned +1 and −1 defects is set to be ∼ 20 which is sufficiently large to prevent any interactions between them. The experi are designed with thin cells to ensure core splitting is the prevailing director stru as opposed to the escape in the third dimension [40,62]. Each data point for d a represents the average value obtained from 50 defects of the same sign within the with the errors calculated as the standard deviation [63].

3. Results

3.1. *Effect of BY Layer Thickness on Photoalignment*

The thickness b of BY coatings affects the anchoring strength. Therefore, BY coa with different thicknesses are produced by spin coating solutions with different B centrations. The thickness b is measured using a digital holographic microscope (L Tec., Chino, CA, USA) in a reflection mode with a vertical resolution better than A portion of the BY coating is removed by wiping the substrate with water, follow isopropanol using a cotton bud, exposing the glass surface to serve as the referen thickness measurement, Figure 2a. After imaging the surface, the height difference bet the coating and the glass substrate is measured using the open-source software pa Fiji/ImageJ, Figure 2b. Thickness is measured at 10 different locations on the coated su and the average of these measurements is used as b. Changing the BY concentration 0.4, 0.5, 0.6, 0.8, 1.0, 2.0, and 4.0 wt%) in DMF results in different values of b, ranging 3.2 to 80.9 nm, Figure 2c. The thickness b was also verified by scratching the BY-c glass with a sharp blade to create a groove; the depth b of the groove is measured digital holographic microscope. The b values obtained by the two methods are in agreement with each other, being within the measurement error.

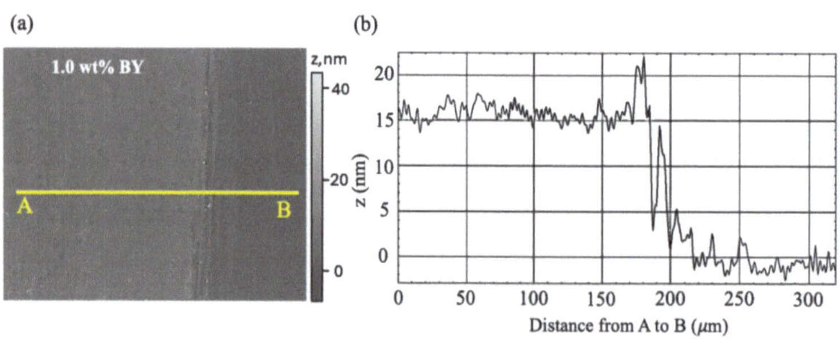

(c)

BY concentration, wt%	BY layer thicknes nm
0.2	3.2 ± 0.1
0.4	5.5± 0.2
0.5	7.6± 0.2
0.6	8.2± 0.2
0.8	12.5± 0.2
1.0	17.1± 0.3
2.0	37.3± 0.9
4.0	80.9± 1.8

Figure 2. Brilliant Yellow layer thickness measurement. (**a**) An image of a 1.0 wt% BY-coated, pa wiped surface taken using a digital holographic microscope. The left side of the image shov BY coating, while the right side shows the glass surface after wiping off the coating. (**b**) The s profile plot along the line AB marked in (**a**). (**c**) BY layer thicknesses for different BY concentr in the spin-coated solutions.

A larger b reduces the intensity of light that reaches the second plate that is fu away from the source due to absorption by the BY layer on the front plate. To ex the effect, we prepare the BY-PS cells with one BY-coated substrate as the front plat a polystyrene (PS)-coated glass plate as the back substrate. During photoirradiatio pattern is focused on the BY-coated front plate. PS-coated plates yield a negligibly weak azimuthal anchoring, $W_{PS} \sim 10^{-10}\text{J}/\text{m}^2 \ll W$ [64]. Thus, it is the patterned substrate that dictates the director orientation and the separation distance between the defects. The anchoring coefficient for these BY-PS cells is calculated as $W = (\pi K h)/(4\alpha d^2)$. The PS-coated plates are prepared by spin coating (3000 rpm, 30 s) a solution of 0.5 wt% PS in chloroform (Sigma-Aldrich, >98%) on clean glass substrates. The plates are kept at 80 °C for 30 min to evaporate the solvent. All cell preparation steps and observations are conducted on the same day. The cells are photopatterned with $I = 5.50 \times 10^2\text{Wm}^{-2}$ for 30 min.

The anchoring coefficient is measured for BY-PS cells with different b, Figure 3a–c. For a thin BY layer of 3.2 nm, W is weak $\sim 0.2 \times 10^{-6}\text{Jm}^{-2}$, Figure 3c. Anchoring increases sharply with b and reaches a maximum of $0.98 \times 10^{-6}\text{Jm}^{-2}$ for b ranging from 5.5 nm to 8.2 nm, Figure 3c. For layers with $b = (12.5 - 37.3)$ nm, W sharply decreases to around 10^{-7}Jm^{-2} and remains constant for thicker BY layers, Figure 3c.

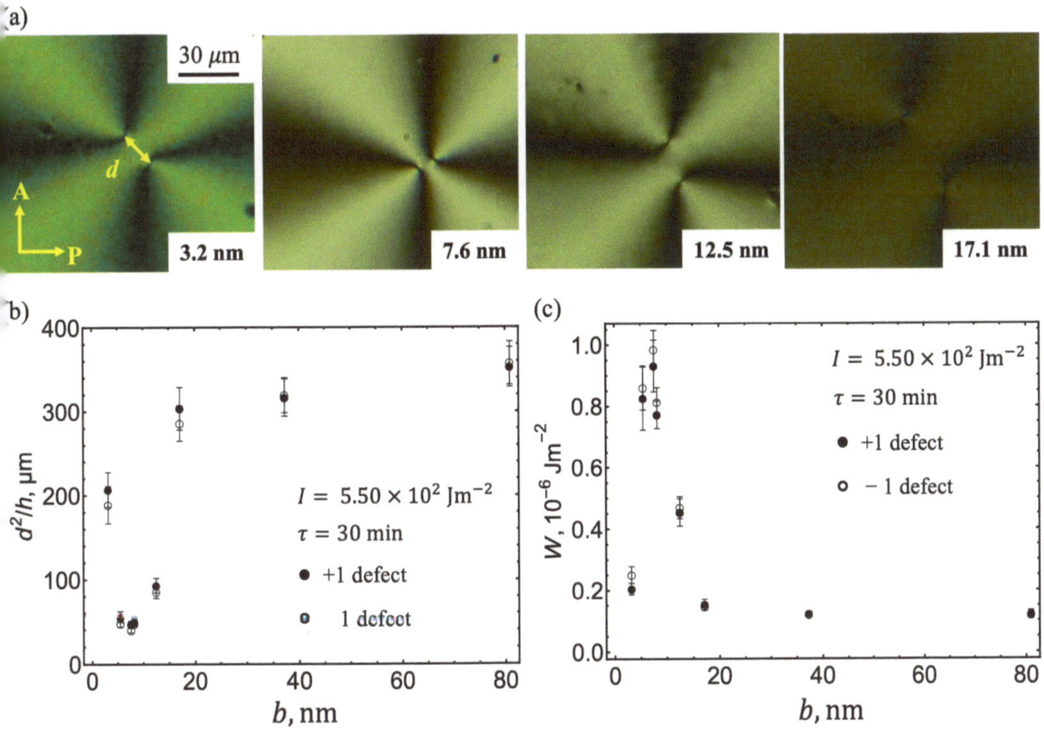

Figure 3. Effect of Brilliant Yellow coating thickness on photoalignment in BY-PS cells. (**a**) Optical microscopy textures of photopatterned cells with different BY layer thicknesses, b. $h = 2.2 \pm 0.2$ μm. Cells are photopatterned with $I = 5.50 \times 10^2$ Wm^{-2} and a dose of 9.90×10^5 Jm^{-2}, corresponding to an irradiation time $\tau = 30$ min. Cell preparation and observation steps are performed on the same day. (**b**) $\frac{d^2}{h}$, and (**c**) W, as functions of b. Each data point represents the average value obtained from 50 defects of the same sign within the array, with the errors calculated as the standard deviation.

The weak W at the thinnest BY coatings ($h = 3.2$ nm) is most likely caused by the surface roughness of the glass, ITO, and the coating itself. As clear from Figure 2b, the thickness 3.2 nm is within the range of the measured variations of the coating's surface, which implies that in some places there might not be enough BY molecules. Another potential reason for weak anchoring at thin coatings is a formation of hydrogen bonds between the BY molecules and the hydrophilic UV/O_3-treated ITO surface. As noted by Wang et al. [57], BY films of 3 nm thickness on a hydrophilic substrate of polyvinyl alcohol

(PVA) demonstrated a low degree of orientational order as compared to thicker films effect was attributed to the formation of hydrogen bonds between the BY molecule the hydroxide OH group of PVA which hinders the *trans-cis* isomerization of BY mol needed for light-induced alignment [57,65].

The weak W at thick BY layers is attributed to weaker light intensity received molecules at the interface with the LC, due to absorption by BY along the light path light transmittance through BY layers of varying thicknesses is measured at wavel 410 nm using a UV/VIS spectrometer Lambda 18 (Perkin Elmer, Waltham, MA, USA data show 90% of incident light is transmitted by the BY layers when $b \leq 6$ nm, b transmission is reduced to 22% for $b = 80.9$ nm, Figure 4b. As a result, the light int received by BY molecules on the surface that will meet the LC is substantially red Figure 4a,b. As the effective intensity decreases, the effective dose also decreases; factors yield a weaker W. Since BY layers with thicknesses less than 8 nm transmit ove of the incident light, and the thickness of 5–8 nm results in stronger azimuthal anch a 7.6 nm thick BY-coating layer (spin coated with a 0.5 wt% BY solution) is used f rest of experiments, presented in Figures 5–7. For a BY-BY cell, each with a 7.6 nm BY layer, the substrate that is further from the light source receives 87% of the inc light intensity during photoalignment. In such a cell, both BY surfaces that will later the LC are aligned with the same light intensity, 87% of the incident light. As s in Figure 3c, this intensity is sufficient to achieve the strongest anchoring. Therefo subsequent experiments are performed using BY-BY cells assembled with two BY-c glass substrates.

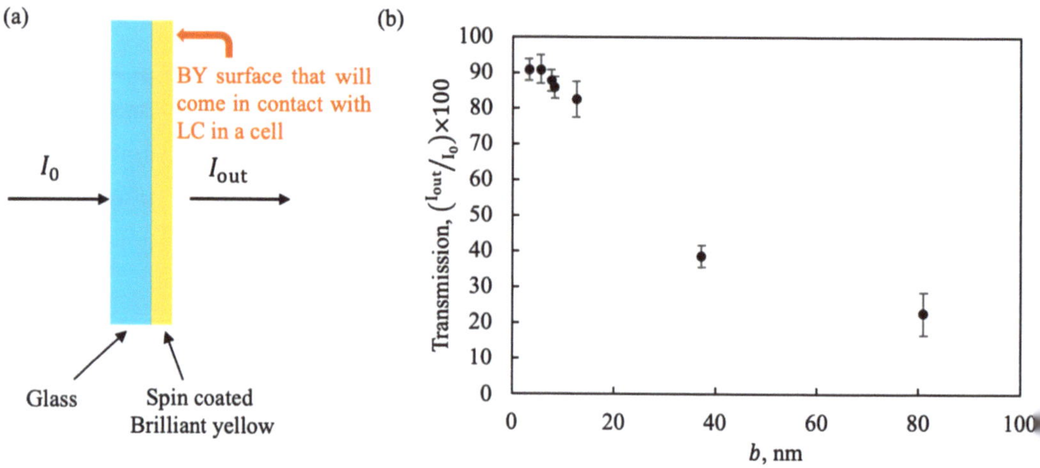

Figure 4. Transmittance measurements of different Brilliant Yellow layer thicknesses. (**a**) Sche illustration of BY-coated glass. Here, I_0 is the incident light intensity and I_{out} is the intensity o transmitted through BY layer. (**b**) Transmission as a function of BY coating thickness. The wave of light is 410 nm.

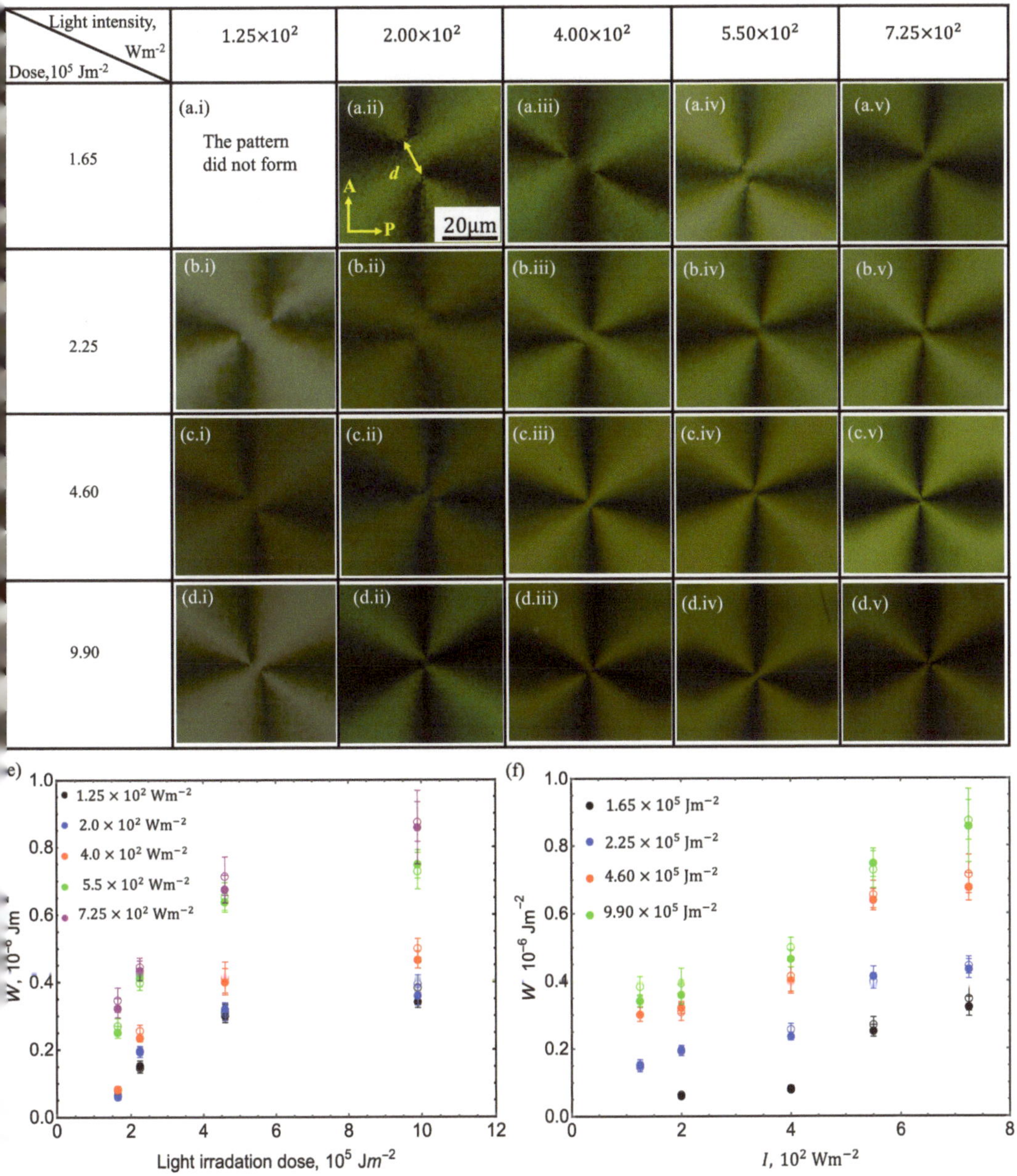

Figure 5. Effect of dose and light intensity on photoalignment in BY-BY cells. (**a–d**) Optical microscopy textures of photopatterned cells at different irradiation doses and light intensities. The images show a pair of +1/2 defects split from a photopatterned +1 defect. $h = 1.3 \pm 0.3$ μm and $b = 7.6$ nm. Cell preparation and characterization are performed within two consecutive days. (**e**) Azimuthal surface anchoring coefficient, W, as a function of dose for various light intensities. (**f**) W as a function of I for different doses. Colors indicate different intensities and doses in (**e**,**f**), respectively. Filled circles represent +1 defects, and open circles represent -1 defects. Each data point for W represents the average value obtained from 50 defects in the array, with the errors calculated as the standard deviation.

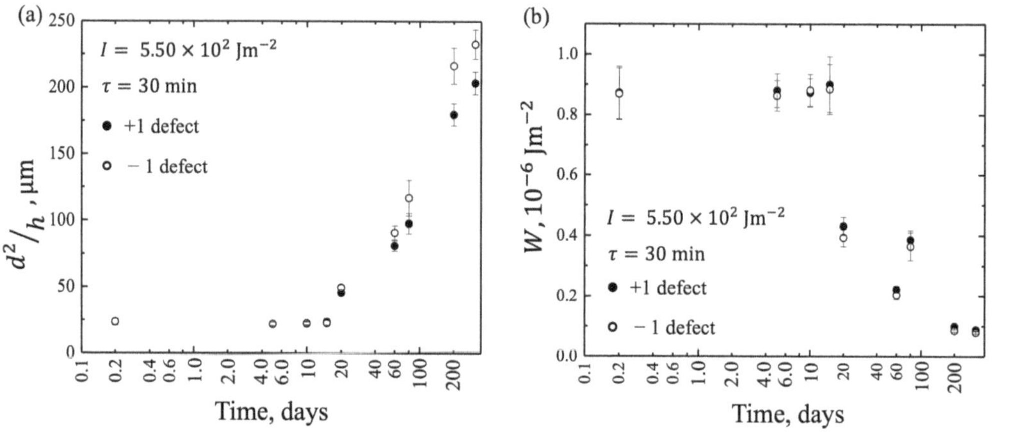

Figure 6. Effect of non-irradiated BY-coated layer aging on photoalignment in BY-BY (**a**) $\frac{d^2}{h}$, and (**b**) W, as functions of non-irradiated BY layer aging time. Each data point represer average value obtained from 50, +1 and 50, −1 defects within the array, with the errors calcu as the standard deviation. $h = 1.1 \pm 0.2$ µm and $b = 7.6$ nm. Cell assembly, photopatternin characterization are performed on the same day. Cells are photopatterned with $I = 5.50 \times 10^2$ and dose of 9.90×10^5 Jm^{-2}.

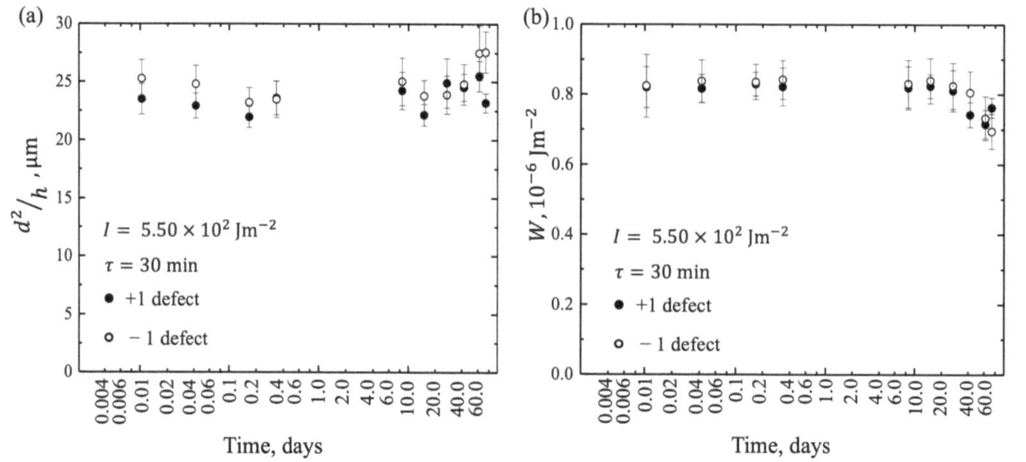

Figure 7. Surface patterning stability of aged LC-filled BY-BY cells. (**a**) d^2/h, and (**b**) W, as functi the age of the LC filled cell. Each data point represents the average value obtained from 50 def the same sign within the array, with the errors calculated as the standard deviation. $h = 1.5 \pm 0.$ and $b = 7.6$ nm. The cell is photopatterned with $I = 5.50 \times 10^2$ Wm^{-2} and a dose of 9.90×10^5 Cell preparation steps are performed on the same day and observation is performed as the cel

3.2. Effect of Light Dose and Intensity on Photoalignment

The BY-BY cells are photopatterned with four different light irradiation doses dose is defined as a product of irradiation time (τ) and the light intensity I. τ is cha to vary the dose while I is kept constant. When studying the effect of I, the dose is constant by varying τ. The cell preparation and characterization are completed wi maximum of two consecutive days to minimize the effects of aging.

Increasing the dose from 1.65×10^5 Jm^{-2} to 9.90×10^5 Jm^{-2} results in an incre W, Figure 5(a–d rows) and Figure 5e; W saturates at doses higher than 4.6×10^5

This behavior is consistent with the previous study on the effect of photopatterning time on W [40].

Increasing I from 1.25×10^2 Wm^{-2} to 7.25×10^2 Wm^{-2} at a fixed dose increases W and creates better alignment, Figure 5(i–v columns) and Figure 5f. It is notable that although dose and I are interrelated, both must be set properly to achieve a high-quality patterning. For instance, photopatterning with a high dose of 9.90×10^5 Jm^{-2} and a low I of 1.25×10^2 Wm^{-2} results in a poor alignment (Figure 5d.i), and a weak $W = 0.36 \pm 0.02 \times 10^{-6}$ Jm^{-2}, Figure 5e,f. However, a moderate dose of 2.25×10^5 Jm^{-2} at $I > 5.50 \times 10^2$ Wm^{-2} results in a better alignment with $W \geq 0.4 \times 10^{-6}$ Jm^{-2}. Using W values, one can establish criteria for setting irradiation conditions to achieve a good alignment. The results also show that W can be tuned by adjusting the dose and I.

At a constant dose, photopatterning with high I for a short τ is more efficient than low intensity irradiation over a prolonged τ (compare columns i and v in Figure 5), Figure 5f. This indicates that the total number of photons is not the single decisive factor defining the anchoring strength. The intensity-dependent behavior of W suggests that photoisomerization occurs as a collective process, where the isomerization of individual BY molecules depends on the isomerization probability of neighboring molecules. Schönhoff et al. [66] observed a similar dependency of molecular photoreorientation on intensity for 4-(4′-N-octadecylamino)phenylazocyanobenzene (amino azobenzene) films irradiated by polarized light with varying intensity but a constant dose and concluded that photoreorientation is a pronounced collective effect.

3.3. *Effect of Aging of Non-Irradiated BY Coatings*

The glass substrates, which were spin-coated with BY and then baked, are stored in a Humidity and Temperature-Controlled Cabinet (SIRUI HC series, Zhongshan, China) for up to 310 days at relative humidity (RH) 25–35%, and 23 °C. Then BY-BY cells are assembled using two aged, non-irradiated BY-coated plates. After a predetermined aging time, the BY coatings are photopatterned with an I of 5.50×10^2 Wm^{-2} for 30 min. The cells are filled with the nematic material, sealed and analyzed under the optical microscope.

Aging of the non-irradiated BY layer impacts photopatterning in two phases. For BY layers irradiated within 15 days after the layers were spin-coated, dried, and baked, W remains constant at $\sim (0.88 \pm 0.01) \times 10^{-6}$ Jm^{-2}, Figure 6. This suggests that BY-coated substrates can be safely stored at controlled humidity conditions (at RH 25–35%) for around two weeks without a noticeable effect on their photopatterning quality. However, for older BY layers, W decreases continuously with age and reduces to 0.08×10^{-6} Jm^{-2} for substrates photopatterned 310 days after preparation, Figure 6. A possible reason could be the water absorption by the BY layers during storage. As already noted, we control the RH of the environment at less than 20% during the spin coating and baking, but during the storage, the RH is at higher levels 20–35%. Absorption of water during the prolonged storage can result in aggregation of BY molecules into J-structures, a process called by Shi et al. [58] "hydrogen-bond-assisted self-assembly of BY molecules with water molecule insertion". The molecules in J-aggregates are less likely to undergo an efficient *trans*-to-*cis* isomerization. Accumulation of water at the substrate–BY interface with a suppression of isomerization through hydrogen bonds might also contribute to the effect of aging.

3.4. *Surface Patterning Stability of Aged LC-Filled Cells*

A photopatterned BY-BY cell is filled with CCN-47, and the edges of the cell are sealed with epoxy glue after the filling. Photopatterning is performed with $I = 5.50 \times 10^2$ Wm^{-2} for 30 min. The LC-filled BY-BY cell is maintained in an environment with RH < 50% at 45 °C for 72 days. While maintaining the temperature at 45 °C to stabilize the nematic phase, optical textures of the cell are recorded over time as the cell ages.

The distance d remains at around 10.4 ± 0.3 µm for 26 days, corresponding to $W = (0.82 \pm 0.01) \times 10^{-6}$ Jm^{-2}, Figure 7a,b. Maintaining the cell for a longer period (around 72 days) results in a slight decrease in W to 0.76×10^{-6} Jm^{-2}, Figure 7b. This result suggests that filling the photopatterned cell with LC and sealing it helps to preserv quality of photopatterning. The observed decline could be due to the gradual disso of LC into the BY coating [67,68] or the dissolving of epoxy glue into the LC over tim

4. Conclusions

We have demonstrated how various photopatterning conditions, including the ness b of the azobenzene alignment layers the intensity I of the light used for pho terning, the irradiation dose, and substrate aging, affect the anchoring strength of a topatterned nematic LC. Our results show that BY layers of thickness b =5–8 nm pro the strongest anchoring coefficient W. Moreover, W increases with both irradiation and I. Also, aging of non-irradiated substrates beyond 15 days significantly reduc However, if the cell is filled with a LC immediately after photopatterning of the BY lay remains constant for up to 4 weeks. This work provides practical strategies for enha the azimuthal strength of the photopatterned anchoring of nematics. The results guidelines for optimizing BY photoalignment parameters and storage.

The method to measure the azimuthal anchoring coefficient described in this is based on the properties of topological defects produced by photopatterning; it not require one to use a second plate, for example, a rubbed polyimide plate, wi anchoring much stronger than the photoalignment anchoring [47,51,70]. It also doe require one to use any external fields [71–73] or to prepare wedge samples of va thickness [74–76]. Within a broader prospectus, our approach to measuring the azim anchoring coefficient can be extended to other photoalignment materials, such as SD- does not feature absorption peaks in the visible spectral range [49] and thus might be suited for applications that require a long-term stability.

Author Contributions: N.P.H. conducted the experiments, analyzed and discussed the dat contributed to the writing, M.R. discussed the data and contributed to the writing, O.D.L. contri to the writing and directed the research. All authors have read and agreed to the published v of the manuscript.

Funding: The work is supported by NSF grant DMR-2215191.

Data Availability Statement: The datasets generated during and/or analyzed during the c study are available from the corresponding author upon reasonable request.

Conflicts of Interest: The authors declare no conflicts of interest.

References

1. Cognard, J. *Alignment of Nematic Liquid Crystals and Their Mixtures*; Molecular Crystals and Liquid Crystals: Supplement Gordon and Breach Science Publishers: London, UK, 1982; Volume 78, p. 1.
2. Lee, K.W.; Paek, S.H.; Lien, A.; Durning, C.; Fukuro, H. Microscopic molecular reorientation of alignment layer polymer su induced by rubbing and its effects on LC pretilt angles. *Macromolecules* **1996**, *29*, 8894–8899. [CrossRef]
3. Chen, J.; Cranton, W.; Fihn, M. *Handbook of Visual Display Technology*; Springer Publishing Company, Incorpo Berlin/Heidelberg, Germany, 2011.
4. Babakhanova, G.; Lavrentovich, O.D. The techniques of surface alignment of liquid crystals. In *Proceedings of the Modern Pr of the Physics of Liquid Systems: Selected Reviews from the 8th International Conference "Physics of Liquid Matter: Modern Prob Kyiv, Ukraine, 18–22 May 2018*; Springer: Cham, Switzerland, 2019; pp. 165–197.
5. Muravsky, A.; Murauski, A. Q&A of liquid crystal alignment: Theory and practice. *Front. Soft Matter* **2024**, *4*, 1382925.
6. Huang, Y.H.; Ko, S.W.; Li, M.S.; Chu, S.C.; Fuh, A.Y.G. Modulation of shape and polarization of beam using a liquid c q-plate that is fabricated via photo-alignment. *Opt. Express* **2013**, *21*, 10954–10961. [CrossRef] [PubMed]
7. Jiang, M.; Yu, H.; Feng, X.; Guo, Y.; Chaganava, I.; Turiv, T.; Lavrentovich, O.D.; Wei, Q.H. Liquid crystal Pancharatnam- micro-optical elements for laser beam shaping. *Adv. Opt. Mater.* **2018**, *6*, 1800961. [CrossRef]
8. Chen, P.; Wei, B.Y.; Hu, W.; Lu, Y.Q. Liquid-crystal-mediated geometric phase: From transmissive to broadband reflective optics. *Adv. Mater.* **2020**, *32*, 1903665. [CrossRef] [PubMed]
9. Tabiryan, N.V.; Roberts, D.E.; Liao, Z.; Hwang, J.Y.; Moran, M.; Ouskova, O.; Pshenichnyi, A.; Sigley, J.; Tabirian, A.; Ve R.; et al. Advances in transparent planar optics: Enabling large aperture, ultrathin lenses. *Adv. Opt. Mater.* **2021**, *9*, 20 [CrossRef]

Tang, D.; Shao, Z.; Zhou, Y.; Lei, Y.; Chen, L.; Xie, J.; Zhang, X.; Xie, X.; Fan, F.; Liao, L.; et al. Simultaneous surface display and holography enabled by flat liquid crystal elements. *Laser Photonics Rev.* **2022**, *16*, 2100491. [CrossRef]

Chen, F.; Zheng, J.; Xing, C.; Sang, J.; Shen, T. Applications of liquid crystal planer optical elements based on photoalignment technology in display and photonic devices. *Displays* **2024**, *82*, 102632. [CrossRef]

Tabiryan, N.V.; Pshenychnyi, A.; Ouskova, O.; Gerosa, K.; Crenshaw, E.; Carter, M.J.; McConney, M.; Urbas, A.; Bunning, T.J.; Slagle, J. Liquid crystalline geometrical phase Alvarez-Lohmann lenses. *Opt. Mater. Express* **2024**, *14*, 2786–2799. [CrossRef]

McConney, M.E.; Martinez, A.; Tondiglia, V.P.; Lee, K.M.; Langley, D.; Smalyukh, I.I.; White, T.J. Topography from topology: Photoinduced surface features generated in liquid crystal polymer networks. *Adv. Mater.* **2013**, *25*, 5880–5885. [CrossRef]

Ware, T.H.; McConney, M.E.; Wie, J.J.; Tondiglia, V.P.; White, T.J. Voxelated liquid crystal elastomers. *Science* **2015**, *347*, 982–984. [CrossRef] [PubMed]

Mostajeran, C.; Warner, M.; Ware, T.H.; White, T.J. Encoding Gaussian curvature in glassy and elastomeric liquid crystal solids. *Proc. R. Soc. A Math. Phys. Eng. Sci.* **2016**, *472*, 20160112. [CrossRef] [PubMed]

Babakhanova, G.; Turiv, T.; Guo, Y.; Hendrikx, M.; Wei, Q.H.; Schenning, A.P.; Broer, D.J.; Lavrentovich, O.D. Liquid crystal elastomer coatings with programmed response of surface profile. *Nat. Commun.* **2018**, *9*, 456. [CrossRef] [PubMed]

Feng, W.; Liu, D.; Broer, D.J. Functional liquid crystal polymer surfaces with switchable topographies. *Small Struct.* **2021**, *2*, 2000107. [CrossRef]

Chen, J.; Akomolafe, O.I.; Jiang, J.; Peng, C. Light-actuated liquid crystal elastomer prepared by projection display. *Materials* **2021**, *14*, 7245. [CrossRef]

Chen, J.; Johnson, A.S.; Weber, J.; Akomolafe, O.I.; Jiang, J.; Peng, C. Programmable light-driven liquid crystal elastomer kirigami with controlled molecular orientations. *Adv. Intell. Syst.* **2022**, *4*, 2100233. [CrossRef]

Peng, C.; Turiv, T.; Guo, Y.; Wei, Q.H.; Lavrentovich, O.D. Command of active matter by topological defects and patterns. *Science* **2016**, *354*, 882–885. [CrossRef]

Straube, A.V.; Pagès, J.M.; Ortiz-Ambriz, A.; Tierno, P.; Ignés-Mullol, J.; Sagués, F. Assembly and transport of nematic colloidal swarms above photo-patterned defects and surfaces. *New J. Phys.* **2018**, *20*, 075006. [CrossRef]

Turiv, T.; Koizumi, R.; Thijssen, K.; Genkin, M.M.; Yu, H.; Peng, C.; Wei, Q.H.; Yeomans, J.M.; Aranson, I.S.; Doostmohammadi, A.; et al. Polar jets of swimming bacteria condensed by a patterned liquid crystal. *Nat. Phys.* **2020**, *16*, 481–487. [CrossRef]

Dhakal, N.P.; Jiang, J.; Guo, Y.; Peng, C. Self-assembly of aqueous soft matter patterned by liquid-crystal polymer networks for controlling the dynamics of bacteria. *ACS Appl. Mater. Interfaces* **2020**, *12*, 13680–13685. [CrossRef]

Koizumi, R.; Turiv, T.; Genkin, M.M.; Lastowski, R.J.; Yu, H.; Chaganava, I.; Wei, Q.H.; Aranson, I.S.; Lavrentovich, O.D. Control of microswimmers by spiral nematic vortices: Transition from individual to collective motion and contraction, expansion, and stable circulation of bacterial swirls. *Phys. Rev. Res.* **2020**, *2*, 033060. [CrossRef]

Ma, L.L.; Liu, C.; Wu, S.B.; Chen, P.; Chen, Q.M.; Qian, J.X.; Ge, S.J.; Wu, Y.H.; Hu, W.; Lu, Y.Q. Programmable self-propelling actuators enabled by a dynamic helical medium. *Sci. Adv.* **2021**, *7*, eabh3505. [CrossRef] [PubMed]

Turiv, T.; Krieger, J.; Babakhanova, G.; Yu, H.; Shiyanovskii, S.V.; Wei, Q.H.; Kim, M.H.; Lavrentovich, O.D. Topology control of human fibroblast cells monolayer by liquid crystal elastomer. *Sci. Adv.* **2020**, *6*, eaaz6485. [CrossRef]

Dhakal, N.P.; Jiang, J.; Guo, Y.; Peng, C. Photopatterning DNA structures with topological defects and arbitrary patterns through multiple length scales. *Phys. Rev. Appl.* **2020**, *13*, 014026. [CrossRef]

Jiang, J.; Dhakal, N.P.; Guo, Y.; Andre, C.; Thompson, L.; Skalli, O.; Peng, C. Controlled dynamics of neural tumor cells by templated liquid crystalline polymer networks. *Adv. Healthc. Mater.* **2020**, *9*, 2000487. [CrossRef]

Zhao, Z.; Li, H.; Yao, Y.; Zhao, Y.; Serra, F.; Kawaguchi, K.; Zhang, H.; Chate, H.; Sano, M. Integer topological defects provide a new way to quantify and classify cell sheets. *bioRxiv* **2024**. [CrossRef]

Culbreath, C.; Glazar, N.; Yokoyama, H. Note: Automated maskless micro-multidomain photoalignment. *Rev. Sci. Instrum.* **2011**, *82*, 126107. [CrossRef]

Shteyner, E.A.; Srivastava, A.K.; Chigrinov, V.G.; Kwok, H.S.; Afanasyev, A.D. Submicron-scale liquid crystal photo-alignment. *Soft Matter* **2013**, *9*, 5160–5165. [CrossRef]

Chigrinov, V. Liquid crystal devices based on photoalignment and photopatterning materials. In Proceedings of the Emerging Liquid Crystal Technologies IX, San Francisco, CA, USA, 1–6 February 2014; pp. 38–47.

Guo, Y.; Jiang, M.; Peng, C.; Sun, K.; Yaroshchuk, O.; Lavrentovich, O.D.; Wei, Q.H. Designs of plasmonic metamasks for photopatterning molecular orientations in liquid crystals. *Crystals* **2016**, *7*, 8. [CrossRef]

Guo, Y.; Jiang, M.; Peng, C.; Sun, K.; Yaroshchuk, O.; Lavrentovich, O.; Wei, Q.H. High Resolution and High Throughput Plasmonic Photopatterning of Complex Molecular Orientations in Liquid Crystals. *Adv. Mater.* **2016**, *28*, 2353–2358. [CrossRef]

Yun, H.; Jiang, S.; Chen, H.; Zhu, Y.; Xu, X.; Li, B.; Xi, P.; Jiang, M.; Wei, Q.H. Ultra-high spatial resolutions in photopatterning molecular orientations. *Opt. Express* **2024**, *32*, 31107–31119. [CrossRef] [PubMed]

Ropač, P.; Hsiao, Y.T.; Berteloot, B.; Ussembayev, Y.; Nys, I.; Ravnik, M.; Neyts, K. Liquid Crystal 3D Optical Waveguides Based on Photoalignment. *Adv. Opt. Mater.* **2024**, 2402174. [CrossRef]

Kudreyko, A.; Chigrinov, V.; Neyts, K.; Chausov, D.; Perestoronina, A. Photonic Devices with Multi-Domain Liquid Crystal Structures. *Crystals* **2024**, *14*, 512. [CrossRef]

Modin, A.; Leheny, R.L.; Serra, F. Spatial Photo-Patterning of Nematic Liquid Crystal Pretilt and its Application in Fabricating Flat Gradient-Index Lenses. *Adv. Mater.* **2024**, *36*, 2310083. [CrossRef]

39. Liu, S.; Nys, I.; Neyts, K. Two-Step Photoalignment with High Resolution for the Alignment of Blue Phase Liquid Crysta *Opt. Mater.* **2022**, *10*, 2200711. [CrossRef]
40. Padmini, H.N.; Rajabi, M.; Shiyanovskii, S.V.; Lavrentovich, O.D. Azimuthal anchoring strength in photopatterned alignme nematic. *Crystals* **2021**, *11*, 675. [CrossRef]
41. Jiang, J.; Wang, X.; Akomolafe, O.I.; Tang, W.; Asilehan, Z.; Ranabhat, K.; Zhang, R.; Peng, C. Collective transpo reconfigurable assembly of nematic colloids by light-driven cooperative molecular reorientations. *Proc. Natl. Acad. Sci. US 120*, e2221718120. [CrossRef]
42. Chigrinov, V.G.; Kozenkov, V.M.; Kwok, H.S. *Photoalignment of Liquid Crystalline Materials: Physics and Applications*; John W Sons: Hoboken, NJ, USA, 2008.
43. Chigrinov, V.; Sun, J.; Wang, X. Photoaligning and photopatterning: New LC technology. *Crystals* **2020**, *10*, 323. [CrossRef
44. Chigrinov, V.; Kudreyko, A.; Sun, J. Flexible Optically Rewritable Electronic Paper. *Crystals* **2023**, *13*, 1283. [CrossRef]
45. Schadt, M.; Schmitt, K.; Kozinkov, V.; Chigrinov, V. Surface-induced parallel alignment of liquid crystals by linearly polym photopolymers. *Jpn. J. Appl. Phys.* **1992**, *31*, 2155. [CrossRef]
46. West, J.L.; Su, L.; Reznikov, Y. Photo-alignment using adsorbed dichroic molecules. *Mol. Cryst. Liq. Cryst. Sci. Technol. Sect. Cryst. Liq. Cryst.* **2001**, *364*, 199–210. [CrossRef]
47. Yaroshchuk, O.; Reznikov, Y. Photoalignment of liquid crystals: Basics and current trends. *J. Mater. Chem.* **2012**, *22*, 28 [CrossRef]
48. McGinty, C.P.; Kołacz, J.; Spillmann, C.M. Large rewritable liquid crystal pretilt angle by in situ photoalignment of brilliant films. *Appl. Phys. Lett.* **2021**, *119*, 141111. [CrossRef]
49. Yaroshchuk, O.; Gurumurthy, H.; Chigrinov, V.G.; Kwok, H.S.; Hasebe, H.; Takatsu, H. Photoalignment properties of br yellow dye. In Proceedings of the 14th International Display Workshops (IDW '07), Sapporo, Japan, 5–7 December pp. 1665–1668.
50. Yin, K.; Xiong, J.; He, Z.; Wu, S.T. Patterning liquid-crystal alignment for ultrathin flat optics. *ACS Omega* **2020**, *5*, 31485– [CrossRef] [PubMed]
51. Folwill, Y.; Zeitouny, Z.; Lall, J.; Zappe, H. A practical guide to versatile photoalignment of azobenzenes. *Liq. Cryst. 48*, 862–872. [CrossRef]
52. Chigrinov, V.; Kudreyko, A.; Guo, Q. Patterned photoalignment in thin films: Physics and applications. *Crystals* **2021**, [CrossRef]
53. Lall, J.; Zappe, H. In situ, spatially variable photoalignment of liquid crystals inside a glass cell using brilliant yell *Proceedings of the Photosensitive Materials and Their Applications II, Proceedings of the SPIE Photonics Europe, Strasbourg, Fr April–23 May 2022*; SPIE: Bellingham, WA, USA, 2022; pp. 91–97. [CrossRef]
54. Lovšin, M.; Petelin, A.; Berteloot, B.; Osterman, N.; Aya, S.; Huang, M.; Drevenšek-Olenik, I.; Mandle, R.; Neyts, K.; Mert et al. Patterning of 2D second harmonic generation active arrays in ferroelectric nematic fluids. *Giant* **2024**, *19*, 100315. [Cro
55. Fang, G.; Shi, Y.; Maclennan, J.E.; Clark, N.A.; Farrow, M.J.; Walba, D.M. Photo-reversible liquid crystal alignment azobenzene-based self-assembled monolayers: Comparison of the bare monolayer and liquid crystal reorientation dyn *Langmuir* **2010**, *26*, 17482–17488. [CrossRef]
56. Fang, G.; Maclennan, J.; Yi, Y.; Glaser, M.; Farrow, M.; Korblova, E.; Walba, D.; Furtak, T.; Clark, N. Athermal photofluidiza glasses. *Nat. Commun.* **2013**, *4*, 1521. [CrossRef]
57. Wang, J.; McGinty, C.; West, J.; Bryant, D.; Finnemeyer, V.; Reich, R.; Berry, S.; Clark, H.; Yaroshchuk, O.; Bos, P. Effects of hu and surface on photoalignment of brilliant yellow. *Liq. Cryst.* **2017**, *44*, 863–872. [CrossRef]
58. Shi, Y.; Zhao, C.; Ho, J.Y.L.; Song, F.; Chigrinov, V.G.; Luo, D.; Kwok, H.S.; Sun, X.W. High photoinduced ordering and contr photostability of hydrophilic azobenzene material based on relative humidity. *Langmuir* **2018**, *34*, 4465–4472. [CrossRef] [Pu
59. Dhara, S.; Madhusudana, N. Physical characterisation of 4′-butyl-4-heptyl-bicyclohexyl-4-carbonitrile. *Phase Transit.* **20** 561–569. [CrossRef]
60. Zawadzki, A.; Walton, H. Measurements of the splay and bend elastic constants of 4′-butyl-4-heptyl-bicyclohexyl-4-carbono CCN47. *Mol. Cryst. Liq. Cryst.* **2012**, *569*, 10–14. [CrossRef]
61. Kleman, M.; Lavrentovich, O.D. *Soft Matter Physics: An Introduction*; Springer: New York, NY, USA, 2003.
62. Chiccoli, C.; Feruli, I.; Lavrentovich, O.; Pasini, P.; Shiyanovskii, S.V.; Zannoni, C. Topological defects in schlieren text biaxial and uniaxial nematics. *Phys. Rev. E* **2002**, *66*, 030701. [CrossRef] [PubMed]
63. Taylor, J. *Introduction to Error Analysis, the Study of Uncertainties in Physical Measurements*, 2nd ed.; University Science Melville, NY, USA, 1997.
64. Ramdane, O.O.; Auroy, P.; Forget, S.; Raspaud, E.; Martinot-Lagarde, P.; Dozov, I. Memory-free conic anchoring of liquid cr on a solid substrate. *Phys. Rev. Lett.* **2000**, *84*, 3871. [CrossRef]
65. Chaplanova, Z.D.; Murauski, A.A.; Rogachev, A.A.; Agabekov, V.E.; Gracheva, E.A. Multi-layered anisotropic films based azo dye brilliant yellow and organic polymers. *J. Appl. Spectrosc.* **2013**, *80*, 658–662. [CrossRef]
66. Schönhoff, M.; Mertesdorf, M.; Lösche, M. Mechanism of photoreorientation of azobenzene dyes in molecular films. *J. Phys.* **1996**, *100*, 7558–7565. [CrossRef]

Perny, S.; Barny, P.L.; Delaire, J.; Buffeteau, T.; Sourisseau, C.; Dozov, I.; Forget, S.; Martinot-Lagarde, P. Photoinduced orientation in poly (vinylcinnamate) and poly (7-methacryloyloxycoumarin) thin films and the consequences on liquid crystal alignment. *Liq. Cryst.* **2000**, *27*, 329–340. [CrossRef]

Vilfan, M.; Olenik, I.D.; Mertelj, A.; Čopič, M. Aging of surface anchoring and surface viscosity of a nematic liquid crystal on photoaligning poly-(vinyl-cinnamate). *Phys. Rev. E* **2001**, *63*, 061709. [CrossRef]

Drzaic, P.S. *Liquid Crystal Dispersions*; World Scientific: Singapore, 1995.

Andrienko, D.; Dyadyusha, A.; Iljin, A.; Kurioz, Y.; Reznikov, Y. Measurement of azimuthal anchoring energy of nematic liquid crystal on photoaligning polymer surface. *Mol. Cryst. Liq. Cryst. Sci. Technol. Sect. A Mol. Cryst. Liq. Cryst.* **1998**, *321*, 271–281. [CrossRef]

Gu, D.-F.; Uran, S.; Rosenblatt, C. A simple and reliable method for measuring the liquid crystal anchoring strength coefficient. *Liq. Cryst.* **1995**, *19*, 427–431. [CrossRef]

Yang, F.; Cheng, H.; Gao, H.; Sambles, J. Technique for characterizing azimuthal anchoring of twisted nematic liquid crystals using half-leaky guided modes. *JOSA B* **2001**, *18*, 994–1002. [CrossRef]

Zhang, B.; Sheng, P.; Kwok, H.S. Optical measurement of azimuthal anchoring strength in nematic liquid crystals. *Phys. Rev. E* **2003**, *67*, 041713. [CrossRef] [PubMed]

Polossat, E.; Dozov, I. New optical method for the measurement of the azimuthal anchoring energy of nematic liquid crystals. *Mol. Cryst. Liq. Cryst. Sci. Technol. Sect. A Mol. Cryst. Liq. Cryst.* **1996**, *282*, 223–233. [CrossRef]

Fonseca, J.G.; Galerne, Y. Simple method for measuring the azimuthal anchoring strength of nematic liquid crystals. *Appl. Phys. Lett.* **2001**, *79*, 2910–2912. [CrossRef]

Vilfan, M.; Čopič, M. Azimuthal and zenithal anchoring of nematic liquid crystals. *Phys. Rev. E* **2003**, *68*, 031704. [CrossRef]

crystals

Article

Structural Study of Nematogenic Compound 5OS5

Aleksandra Deptuch [1,*], Bartosz Sęk [2], Sebastian Lalik [3], Wojciech Zając [1], Mirosława D. Ossowska-Chruściel Janusz Chruściel [4] and Monika Marzec [3]

1 Institute of Nuclear Physics Polish Academy of Sciences, Radzikowskiego 152, PL-31342 Kraków, Pol wojciech.zajac@ifj.edu.pl
2 Faculty of Physics and Applied Computer Science, AGH University of Kraków, Reymonta 19, PL-30059 Kraków, Poland
3 Institute of Physics, Jagiellonian University, Łojasiewicza 11, PL-30348 Kraków, Poland; sebastian.lalik@uj.edu.pl (S.L.); monika.marzec@uj.edu.pl (M.M.)
4 Faculty of Science, University of Siedlce, 3 Maja 54, PL-08110 Siedlce, Poland; dch@uph.edu.pl (M.D. janusz.chrusciel@uph.edu.pl (J.C.)
* Correspondence: aleksandra.deptuch@ifj.edu.pl

Abstract: The S-(4-pentylphenyl) 4-(pentyloxy)benzothioate, forming the nematic phase, is i gated by X-ray diffraction in temperatures between 263 K and 365 K, with the support of diffe scanning calorimetry and polarizing optical microscopy. The microscopic observations show c within the solid state, while X-ray diffraction does not indicate any transitions between the phases. The Rietveld refinement shows that the crystal phase formed from the melt is the monoclinic crystal phase with the $P2_1/c$ space group as reported for a single crystal grown fr ethanol solution. The temperature dependence of the unit cell parameters in the 263–335 K is determined and the coefficients of thermal expansion are obtained. The unit cell expar heating along the longer *ac*-diagonal and *b*-axis while, along the shorter *ac*-diagonal, a very shrinkage occurs. The diffraction patterns of the liquid crystalline nematic phase indicate the tion of dimers via hydrogen bonding. Density functional theory calculations (def2TZVPP ba B3LYP-D3(BJ) correlation-exchange functional) are applied for geometry optimization of an is molecule and selected dimers.

Keywords: nematic liquid crystal; thermal expansion of crystal; X-ray diffraction

Citation: Deptuch, A.; Sęk, B.; Lalik, S.; Zając, W.; Ossowska-Chruściel, M.D.; Chruściel, J.; Marzec, M. Structural Study of Nematogenic Compound 5OS5. *Crystals* **2024**, *14*, 367. https://doi.org/10.3390/cryst14040367

Academic Editor: Vladimir Chigrinov

Received: 25 March 2024
Revised: 9 April 2024
Accepted: 12 April 2024
Published: 13 April 2024

1. Introduction

The nematic phase is the simplest liquid crystalline phase, which possesses on long-range orientational order. The long axes of molecules are oriented on averag common direction, denoted by the director versor $\hat{n}$ (Figure 1a). The nematic phase not show long-range positional order, but short-range positional order is present Compounds forming the nematic phase have an important place in the display nology, which is continuously developed [3–5]. They are also subjects of ongoing studies, as new types of the nematic phase are searched for, e.g., biaxial or twist- nematics [6–8]. S-(4-pentylphenyl) 4-(alkyloxy)benzothioates (Figure 1b) are the h ogous series known from their mesomorphic properties, which depend on the leng the C_nH_{2n+1} chain [9–19]. They were tested as components of mixtures with other l crystals, e.g., [19]. Homologues with n = 4–6 show only the nematic phase and l homologues with n = 7–12 exhibit also the smectic C phase (n = 7) or a few smectic p (n = 8–12) [9–19] with the lamellar order [1,2]. For the homologues with n = 4–7, c structures were solved by single-crystal X-ray diffraction (XRD) [14–17]. While, = 7, the crystal phase and both liquid crystalline phases (nematic and smectic C) investigated by XRD at various temperatures [14], for n = 4–6, as far as we are a only the XRD results at room temperature were published [15–17]. Herein, we pr the structural investigations by XRD for the S-(4-pentylphenyl) 4-(pentyloxy)benzot compound, with n = 5, abbreviated either as $\bar{5}$S5 or 5OS5 (used in this paper). The phase sequence of 5OS5 on heating is crystal (335 K) nematic (353 K) isotropic liquid [18]. These three phases are abbreviated further as Cr, N and Iso, respectively. The crystal structure obtained at 295 K for the 5OS5 single-crystal grown from an ethanol solution is monoclinic ($a \neq b \neq c$, $\beta \neq 90°$, $\alpha = \gamma = 90°$), with the $P2_1/c$ space group. The molecules are arranged parallel to each other, which corresponds well to the N phase observed above the melting temperature. Weak hydrogen bonds of the C-H...O type exist in the Cr phase [17]. Our aim is to examine whether 5OS5 shows polymorphism in the solid state, as well as to investigate the short-range positional order in the N and Iso phases. Although the main experimental method is XRD, the results of differential scanning calorimetry (DSC) and polarizing optical microscopy (POM) are also presented for comparison.

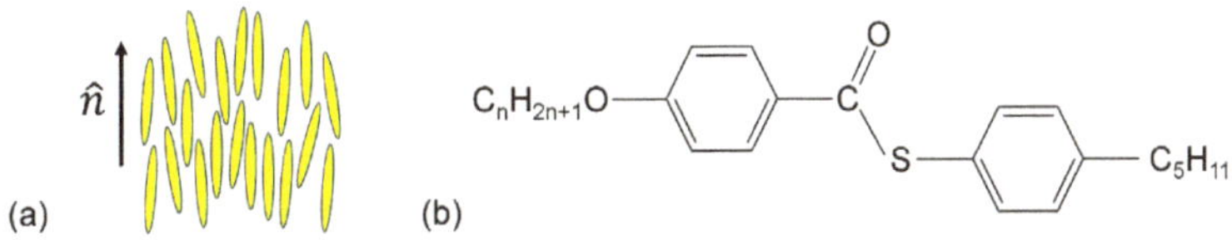

Figure 1. Scheme of the nematic phase (**a**) and the molecular formula of S-(4-pentylphenyl) 4-(alkyloxy)benzothioates (nOS5) (**b**). For the homologue in this study, n = 5.

2. Materials and Methods

Polycrystalline S-(4-pentylphenyl) 4-(pentyloxy)benzothioate was synthesized according to the method described in [13].

DSC measurements were performed with the PerkinElmer DSC 8000 calorimeter in the 263–373 K range in two cycles of heating and cooling with a 5 K/min rate. The results from the second cycle were used in analysis. The calibration was based on the melting points of indium and water. The sample weighting 3.10 mg was sealed within an aluminum pan. Data analysis was carried out using the PerkinElmer software. The phase transition temperatures and enthalpy changes were determined with uncertainties of 0.5 K and 0.08 kJ/mol, respectively.

The POM measurements were carried out with the Leica DM2700 P microscope in the 263–365 K range in three cycles of cooling and heating with a 5 K/min rate. The temperature was controlled by the Linkam attachment. The phase transitions in the vicinity of the clearing temperature were additionally observed during cooling with a 1 K/min rate from 363 K to 343 K. The transition within the solid state was investigated in isothermal conditions for temperatures of 280, 282 and 284 K. The sample was heated to 365 K and cooled down to 263 K at 20 K/min, then heated to a selected temperature, also at 20 K/min, and observed at a constant temperature. The sample was placed between two glass slides not covered by any aligning layer. The average luminance of recorded textures as a function of temperature was calculated using the TOApy program [20,21].

The XRD measurements were performed with the X'Pert PRO diffractometer (PANalytical) using the CuKα radiation with the Bragg–Brentano geometry in the 2θ = 2–30° or 2–8° range. The temperature in the 263–365 K range was controlled by the TTK-450 stage (Anton Paar). The sample in polycrystalline form was placed into a flat sample holder and pressed with a glass slide. The diffraction patterns were collected during different temperature programs:

(a) first heating of a pristine sample which was not melted after synthesis,
(b) heating after direct cooling with a 5 K/min rate from 365 to 263 K,
(c) cooling from 365 to 298 K,
(d) cooling from 365 to 298 K in the low-angle region only.

In programs (a–c), the measurements were carried out in the 2θ = 2–30° range with a rate of 0.078°/s, the 0.033° step and 4 diffraction patterns collected for each temperature. In program (d), the measurements were carried out in the 2θ = 2–8° range with a rate of

0.046°/s, the 0.017° step and 24 diffraction patterns collected for each temperature data analysis was performed in FullProf [22], PASCal [23,24] and OriginPro.

The DFT calculations for an isolated molecule and dimers were carried out in sian, version 16.C.01 [25], with the def2TZVPP basis set [26], B3LYP correlation-exch functional [27,28] and Grimme's D3 dispersion with Becke–Johnson damping [29] optimized models were visualized in Avogadro [30].

3. Results

3.1. DSC Thermograms

The DSC results are presented in Figure 2 and Table 1. During cooling, 5OS5 show N phase at the wide temperature range and crystallization is observed at room temper The large supercooling of the nematic phase is likely to be caused by too low a nucle rate above room temperature, which prevents crystallization [31]. Supercooling is monly reported for liquid crystals [32–34]. No transition between the crystal pha observed during cooling down to 263 K and during heating. However, the absolute va the enthalpy change during crystallization, 14.6 kJ/mol, is much smaller than the ent of melting, 33.2 kJ/mol. This indicates that the recrystallization effect occurs in the sa before melting, but its rate is too slow to be visible as an anomaly in the DSC curve anomaly between the N and Iso phases has an irregular shape both for cooling and he which can be interpreted as two overlapping anomalies. The unknown phase, preser narrow temperature range between N and Iso, is denoted as X. In some chiral compo the blue phase is present between N* and Iso [35], but this is not the case for 5OS5, wh achiral. Another explanation is the presence of two types of the nematic phase, e.g., re N at higher temperatures and twist-bent N_{TB} at lower temperatures [6,7]. On the hand, the N_{TB} phase is formed by molecules with a bent shape, while 5OS5 molecul rather rod-like and are expected to form only a regular N phase.

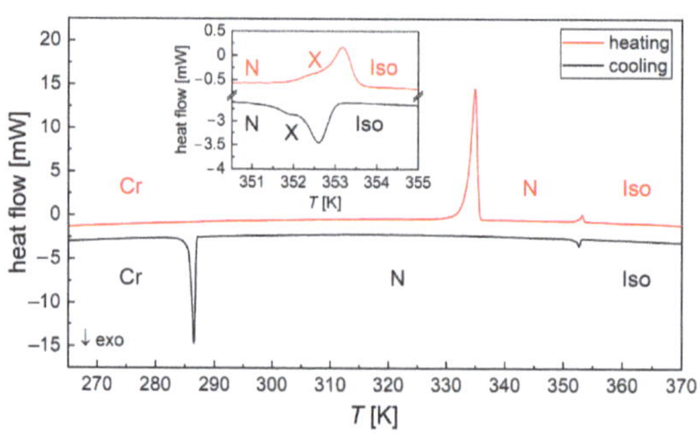

Figure 2. DSC thermogram of 5OS5 registered during the second heating/cooling cycle at a 5 K rate. The inset shows the enlarged region around the clearing temperature.

Table 1. Thermodynamic parameters of the phase transitions of 5OS5 determined by the DSC m (5 K/min): onset temperature, peak temperature, enthalpy change and entropy change.

Transition	T_o (K)	T_p (K)	ΔH (kJ/mol)	ΔS (J/(mol·
Cr → N	333.6	335.1	33.2	99.1
N → X	352.0	352.6	0.2	0.6
X → Iso	352.8	353.3	0.6	1.8
Iso → X	353.1	352.7	−0.7	−1.9
X → N	-	352.1	−0.3	−0.7
N → Cr	287.0	286.6	−14.6	-

3.2. POM Observations

The POM measurements at the 5 K/min rate (Figure 3) confirm the presence of the nematic phase with a schlieren texture during cooling [2]. After cooling to 288–289 K, the transition to the crystal phase denoted as Cr2 is observed. The transition occurs via a single front of crystallization. During heating, the Cr2 phase transforms to another crystal phase, Cr1. The new crystal phase is formed by gradual growth of crystallites, instead of the single front of crystallization. In isothermal conditions at 260–264 K, the Cr2 → Cr1 transition lasts for a few minutes (up to 440 s). After melting of the crystal phase, the texture of N resembles that of Cr1. The phase transition temperatures, obtained as an average over three cooling/heating cycles, are Iso (355 K) N (288 K) Cr2 during cooling and Cr2 (287 K) Cr1 (337 K) N (356 K) Iso during heating. The results from three cycles agree with each other within 1 K, except for the Cr2/Cr1 transition temperature, which is 291, 282, 288 K for the first, second and third cycle, respectively. The melt crystallization and Cr2/Cr1 transition occur at the same temperature region close to room temperature. The schlieren textures collected during slow cooling at 1 K/min in the range of the Iso → X → N transitions (Figure 4) show only changes in color, while observed defects are the same. The four-brush disclination, indicated by an arrow in the top-left corner of a texture collected at 355.4 K, is visible down to 343 K, which is well below the X → N transition. Such a defect indicates uniaxiality of the N phase [8]. Thus, the X → N transition cannot be the transition between the uniaxial and biaxial nematic phases.

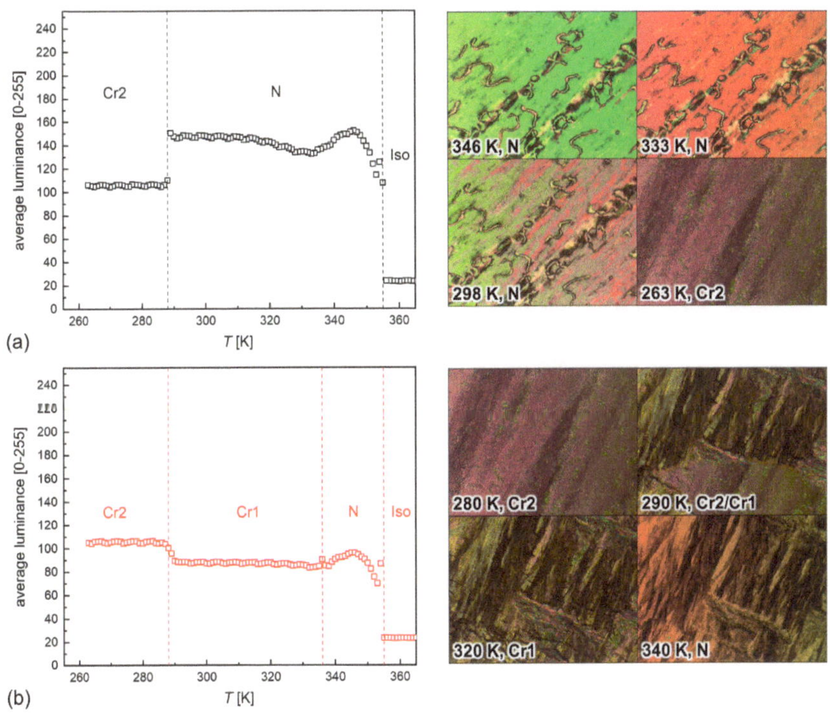

Figure 3. POM textures of 5OS5 collected during cooling (**a**) and heating (**b**) at a 5 K/min rate, with the corresponding results of numerical analysis in the TOApy program. Each texture shows an area of 622 μm × 466 μm.

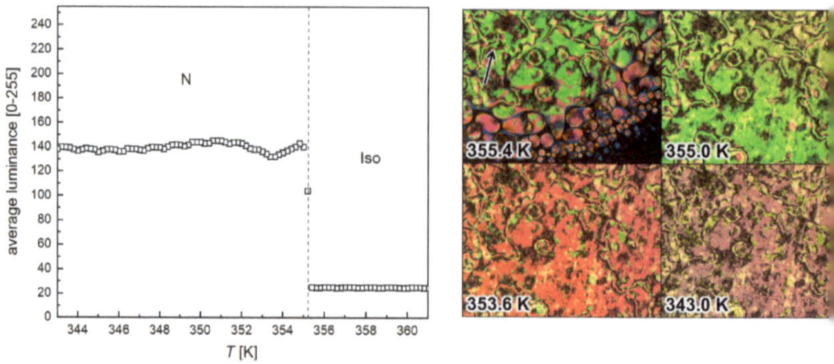

Figure 4. POM textures of 5OS5 collected during cooling at a 1 K/min rate, with the correspc results of numerical analysis in the TOApy program. Each texture shows an area of 622 μm × 4(

3.3. Structural Study by XRD

Only one crystal phase is observed in the XRD patterns of 5OS5 (Figure 5), positions of the diffraction peaks are the same both for the pristine sample and f crystal grown from melt. The differences in the relative intensities of the peaks are c by the effect of the preferred orientation of crystallites. The Rietveld refinemen performed for the patterns of the crystal phase grown from melt (Figure 6), show the observed crystal phase is the same as the monoclinic one reported for a single c in [17]. The coordinates of non-hydrogen atoms and their isotropic thermal coeffi were fixed to values from [17]. The hydrogen's coordinates were generated in the Avo program [30] and their isotropic thermal coefficients were assumed to be the same corresponding carbon atoms. The fitting parameters were the lattice constants, β scale factor, zero of diffractometer (systematic shift in 2θ), half-width of the peak preferred orientation (selection of the [101] direction gave the best fitting results). F pristine sample, the preferred orientation, probably in the [100] direction, was so s that we were unable to perform a satisfactory Rietveld refinement. Instead, the L fitting [37] was applied, where the peak intensities are treated as fitting parameters ar not calculated from the atomic coordinates. The determined parameters of the monc unit cell are presented in Figure 7. Although the values obtained by the Le Bail fitting smaller uncertainty bars, the results of the Rietveld refinement are treated as more re because in the latter method, the peak intensities are based on the experimental c structure. This is supported by the fact that the unit cell parameters reported for a crystal in 295 K in [17] are in better agreement with the results of the Rietveld refine than with the Le Bail fitting (with the exception of the β angle).

The results of the linear fits to the lattice parameters are presented in Table 2. All constants a, b, c increase with increasing temperature; however, the β angle also incr therefore the coefficients of thermal expansion (CTEs) do not have to be positive directions. The PASCal program [23,24] was applied to calculate the relationship bet the principal strain axes x, y, z and the axes of the monoclinic unit cell a, b, c, as well obtain CTEs (Table 3):

$$\alpha_i = \frac{1}{T}\left(\frac{L_i(T)}{L_{i0}} - 1\right),$$

where L_i and L_{i0} are lengths along the i = x, y, z directions at temperatures T respectively. Only the results from the Rietveld refinement for the patterns collect heating from 263 K to 335 K were used. The x principal axis is approximately parallel shorter diagonal of the ac-base ([110] direction) and the corresponding CTE is negativ very close to zero, α_x = −7.5(6.5)·10^{-6}/K. The y axis is oriented along the b crystallogr axis and the CTE along this direction has an intermediate value, α_y = 75(3)·10^{-6}/K largest CTE, α_z = 226(5)·10^{-6}/K, is along the z axis, which is approximately parallel longer diagonal of the ac-base ($[\overline{1}10]$ direction). These results mean that the incre temperature leads mainly to increasing distance between long molecular axes in the 5OS5 crystal, the same as for previously investigated 7OS5 [14].

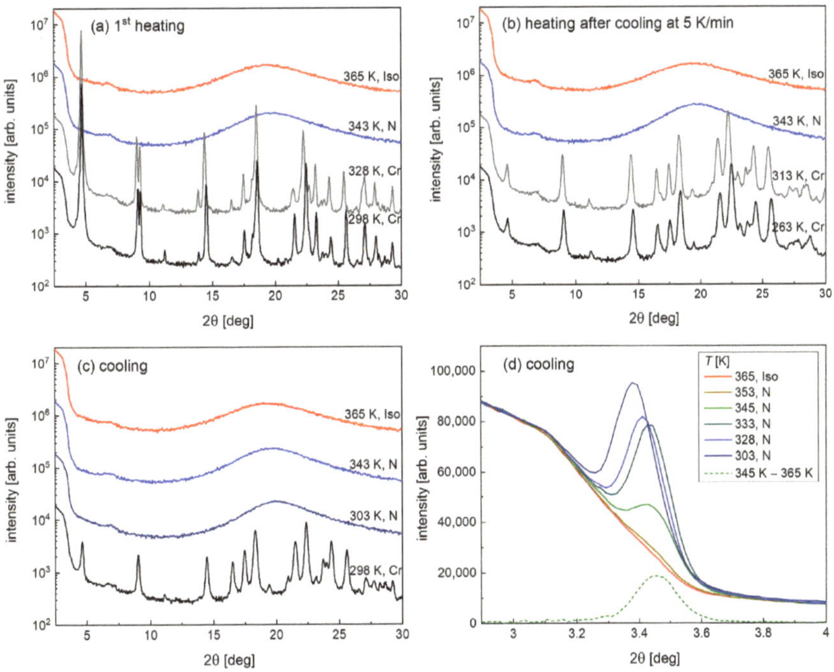

Figure 5. X-ray diffraction patterns of 5OS5 collected during the 1st heating of a pristine sample (**a**), heating after direct cooling at 5 K/min from 365 to 263 K (**b**), cooling from 365 K in the 2θ = 2–30° range (**c**) and slower cooling from 365 K in the 2θ = 2–8° range (**d**). The wide maximum at 2θ ≈ 7° is a background contribution. The 2θ values were corrected by the zero of the diffractometer determined from the Rietveld refinement for the crystal phase.

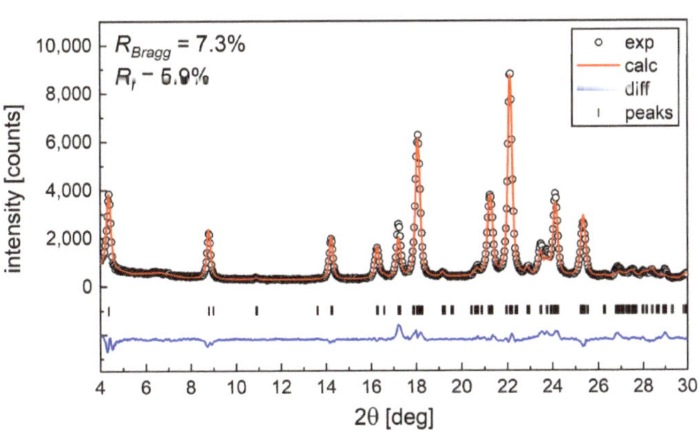

Figure 6. Example of the Rietveld refinement for the diffraction pattern of 5OS5 collected after slow cooling to 298 K.

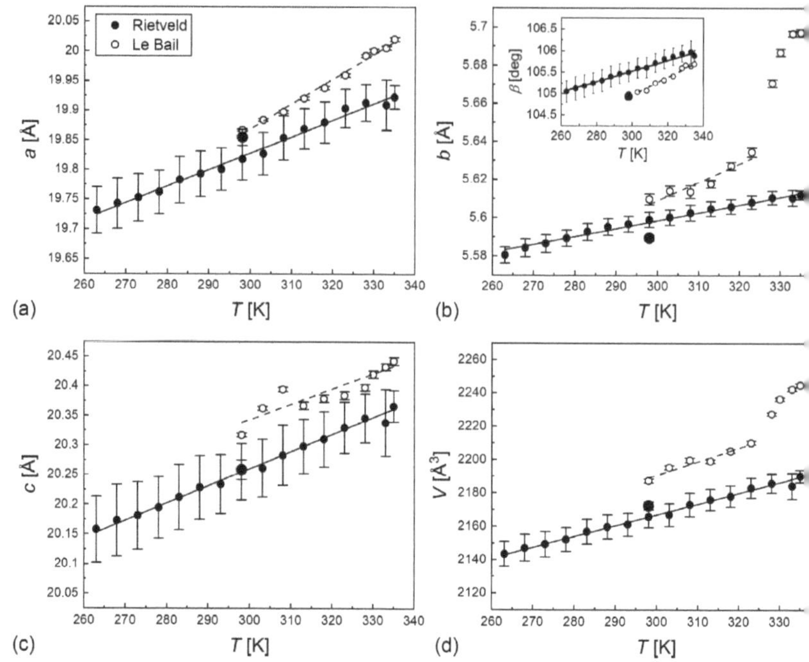

Figure 7. Monoclinic unit cell parameters of 5OS5: lattice constants (**a**–**c**), volume (**d**) and the *β* (inset in (**b**)) as a function of temperature. Open points denote results obtained by the Le Bail for the pristine sample, and solid points denote results obtained by the Rietveld refinement f crystal formed from the melt. The linear fits to the experimental points are shown. The en solid circles refer to the lattice parameters in 298 K after slow cooling and were excluded fro linear fits.

Table 2. Results of linear fits to the monoclinic unit cell parameters of 5OS5 as a function of incr temperature, determined by the Le Bail fitting for the pristine sample and by the Rietveld refin for the crystal grown from melt.

Parameter	Pristine Sample		Crystal Grown from Melt	
	Intercept	Slope	Intercept	Slope
a	18.61(5) Å	0.0042(2) Å/K	18.99(3) Å	0.00279(7) Å
b	5.32(6) Å	0.0010(2) Å/K	5.476(4) Å	0.00041(2) Å
c	19.6(2) Å	0.0026(5) Å/K	19.39(2) Å	0.00290(7) Å
β	98.8(4)°	0.021(2)°/K	102.0(2)°	0.0119(7)°,
V	1947(33) Å^3	0.8(1) Å^3/K	1974(5) Å^3	0.64(2) Å^3,

Table 3. Matrix of transition between the unit cell axes a, b, c and principal axes x, y, z, ar coefficients of thermal expansion along the principal axes calculated in PASCal based on res the Rietveld refinement for the crystal phase of 5OS5.

Principal Axis	a	b	c	α_i [10^{-6}/K
x	0.7298	0	0.6837	−7.5(6.5
y	0	−1	0	75(3)
z	−0.6880	0	0.7257	226(5)
V	-	-	-	295(8)

The short-range positional order in the N and Iso phases appears in XRD patterns as a wide maximum with a center at 2θ ≈ 20°. If the patterns are plotted as a function of the scattering vector q, the relationship with the 2θ angle of which is as follows:

$$q = \frac{4\pi \sin\theta}{\lambda}, \tag{2}$$

then the wide maximum is described by the Lorentz function [1,38]:

$$I(q) = \frac{A}{1 + \xi^2 (q - q_0)^2} + Bq + C. \tag{3}$$

The position of the maximum q_0 is inversely proportional to the average distance w between molecules: $q_0 = 2\pi/w$ and the half-width of the maximum is inversely proportional to the correlation length ξ of the short-range order. The A parameter is the height of the maximum, while B and C are the slope and intercept of the linear background. The w and ξ values obtained for 5OS5 are shown in Figure 8. The w distance both in the N and Iso phases increases slowly with increasing temperature in the 4.39–4.56 Å range with the slope of 0.00252(9) Å/K and it is interpreted as the average distance between the long axes of molecules (or the width of a molecule) [1,38]. There is no significant step in w at the transition between N and Iso, while this is noticeable for the correlation length, which decreases by ca. 0.4 Å after transition to the Iso phase. Both in N and Iso, the ξ values decrease with increasing temperature with slopes equal within uncertainties, −0.013(2) Å/K and −0.009(3) Å/K, respectively. In the N phase, ξ has values in the 3.90–4.81 Å range, while in Iso it is only 3.55–3.72 Å. The small correlation length indicates that there are only next-neighbor positional correlations between molecules in a direction perpendicular to their long axes.

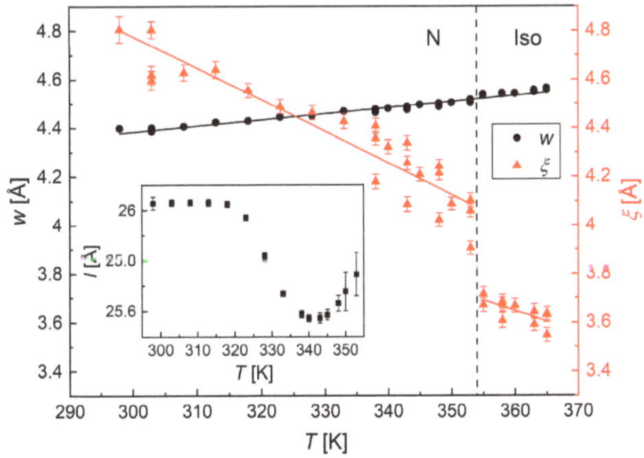

Figure 8. Average distance w between long axes of molecules and correlation length ξ of the short-range order in the N and Iso phases of 5OS5 as a function of temperature. The inset shows the average distance between short axes of molecules in the nematic phase. The solid lines denote the results of linear fits and the dashed line indicates the Iso → N transition temperature.

In the nematic phase, the positional short-range order in the direction parallel to the director results in a diffuse maximum in the low-angle region. The position of this peak is related to the length of molecules and the half-width is inversely proportional to the parallel correlation length, which is much larger than the previously discussed correlation length in a direction perpendicular to the director [1]. In the XRD patterns of 5OS5 collected in the 2θ = 2–30° range (Figure 5a–c), this maximum was in general not visible. Only in 303 K one could notice an additional intensity over the low-angle background after comparison with the patterns of the Iso phase. This is why the slower XRD measurement in the low-region, described as procedure (d) in Section 2, was necessary. The representative pa obtained in this measurement are shown in Figure 5d. The low-angle peak in the N is much stronger than in the previous results because the sample had more time to The high background at low angles was removed by subtracting the pattern collec the Iso phase, as it is presented for 345 K. The diffuse maxima from the short-range are described by the Lorentz function; however, the fitting of Equation (3) to the low-maximum for 5OS5 did not give proper agreement. This indicates that the resolution diffractometer and the short-range order give comparable contribution to the peak's o shape. Because of this, the low-angle peak was fitted with the pseudo-Voigt function, is a sum of the Lorentz and Gaussian peaks, and the correlation length in the dire parallel to the director was not determined. The parameter which was obtained fro peak's position is the characteristic distance l (inset in Figure 8). The l distance evolves temperature: just below the Iso/N transition, it decreases on cooling and has a minim 25.6 Å at 340–343 K, then increases with decreasing temperature and has a constant va 26.0 Å below 323 K. The orientational order parameter in the N phase is expressed as

$$S = \frac{1}{2}\left\langle 3\cos^2\varphi - 1\right\rangle,$$

where φ is the angle between the long molecular axis of each molecule and the di and the averaging is over all molecules. As the orientational order parameter in phase is lower than 1, the l value is expected to be smaller than the molecular le Based on the crystal structure, the length of the 5OS5 molecule is 25.22 Å [17]. Mean the experimental l distance is equal to 25.6–26.0 Å, which exceeds slightly the mole length. The isolated molecule optimized by the DFT method (Figure 9) is slightly le 27.0 Å, where the molecular length is calculated as the maximal H–H distance plu non-bonded H–H contact distance 2.2 Å from [40]. Still, it leads to unusually high val the order parameter in the N phase, S = 0.85–0.89. The possible explanation is the pre of molecular associations [41]. For 5OS5, these can be dimers formed via weak hyd bonds, which was tested using molecular models optimized by the DFT method. examples of dimers are presented in Figure 9. The starting model was the 5OS5 mol with atomic coordinates determined from the single-crystal X-ray diffraction result The numbering of atoms in Figure 9a is the same as in [17]. The C(26)-H(26)... hydrogen bonds present in the crystal phase involve neighbors in the b-direction, bet the aromatic ring from one molecule and the C=O group from another molecule. Thi the basis for the optimization of the head-to-head dimer (Figure 9b). During optimiz the C(26)-H(26)...O(1) bond was preserved and another C(8)-H(8)...O(2) bond was fo between molecules. The length of the head-to-head dimer is 28.5 Å. This correspon the φ angle equal to 23.8–26.0° and the orientational order parameter of 0.71–0.76, wh within the range of S = 0.3–0.8 obtained experimentally for other compounds in the ne phase [1,42,43]. The increased l distance in 348-353 K indicates that other types of d may be present at higher temperatures. It also suggests that the X phase may be a ne phase, only formed by other dimers than in lower temperatures. A head-to-tail (Figure 9c) was prepared based on the hydrogen bonds present in the crystal phase 7OS5 homologue [14]. This dimer is formed by two C(27)-H(27)...O(1) bonds bet the alkyl chain and the C=O group and after optimization it has a length of 34.5 Å, v corresponds to φ = 41.0–42.1° and S = 0.33–0.36, close to the bottom value observe other nematogenic compounds [1,42,43]. The energy of a dimer in Figure 9c is larg 10.2 kJ/mol (mol of molecules, not dimers) than the energy of a dimer in Figure 9b presented dimers are only examples; however, based on the XRD results, it can be ass that the fraction of longer dimers decreases with decreasing temperature. It explai initial decrease of l just below the clearing temperature. The further increase of l b 340 K is caused by an increasing order parameter.

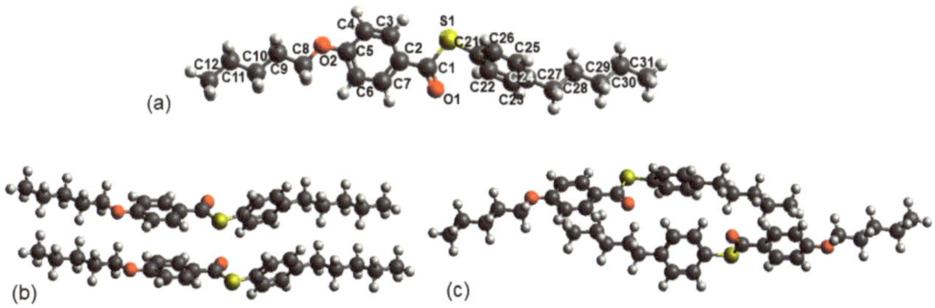

Figure 9. Isolated 5OS5 molecule (**a**) and examples of possible head-to-head (**b**) and head-to-tail (**c**) dimers formed via hydrogen bonds. The geometry optimizations were carried out with the DFT/B3LYP-D3(BJ)-def2TZVPP method.

4. Summary and Conclusions

The phase transitions of liquid crystalline S-(4-pentylphenyl) 4-(pentyloxy)benzothioate (5OS5) were investigated by X-ray diffraction and complementary methods. The additional X phase was detected in a narrow temperature range between the isotropic liquid and nematic phases. The achirality of molecules and their rod-like molecular shape make the presence of the blue phase or the transition between the regular and twist-bend nematic phases very unlikely. The texture observation excluded the transition between the uniaxial and biaxial nematic. For the temperature dependence of the low-angle diffuse maximum in the nematic phase, let us hypothesize that the X $\rightarrow$ N transition is related to the change in a building block, i.e., the types of dimers present in the nematic phase. The X phase is likely to be a nematic formed mainly by longer dimers, while the N phase is a nematic formed mainly by shorter dimers. The XRD patterns confirmed the presence of one crystal phase, for which the coefficients of thermal expansion were determined. In the POM observations, the transition between the crystal phases was clearly visible during heating, which was not observed by any other method, the DSC thermograms indicating only some recrystallization effects. Apparently, in the thin sample placed between two glass slides, used in texture observations, the melt crystallization leads to the metastable crystal phase, and transition to the stable crystal phase investigated by XRD occurs via cold crystallization during heating.

Author Contributions: Conceptualization, A.D.; formal analysis, A.D., B.S. and S.L.; investigation, A.D., B.S., S.L. and W.Z.; resources, M.D.O.-C., J.C. and M.M.; writing—original draft preparation, A.D.; writing—review and editing, B.S., S.L., W.Z., M.D.O.-C., J.C. and M.M.; funding acquisition, M.M. All authors have read and agreed to the published version of the manuscript.

Funding: The PerkinElmer DSC 8000 calorimeter was purchased thanks to the financial support of the European Regional Development Fund in the framework of the Polish Innovation Economy Operational Program (contract no. POIG.02.01.00-12-023/08). This research was supported in part by the Excellence Initiative—Research University Program at the Jagiellonian University in Kraków.

Data Availability Statement: Dataset available on request from the authors.

Acknowledgments: We gratefully acknowledge Poland's high-performance Infrastructure PLGrid Academic Computer Centre Cyfronet AGH for providing computer facilities and support within computational grant no. plgmolkryst8.

Conflicts of Interest: The authors declare no conflicts of interest.

rences

Demus, D.; Goodby, J.; Gray, G.W.; Spiess, H.-W.; Vill, V. (Eds.) *Handbook of Liquid Crystals*; WILEY-VCH Verlag GmbH: Weinheim, Germany, 1998.
Vertogen, G.; de Jeu, W.H. *Thermotropic Liquid Crystals: Fundamentals*; Springer: Berlin/Heidelberg, Germany, 1988.
Schadt, M. Nematic liquid crystals and twisted-nematic LCDs. *Liq. Cryst.* **2015**, *42*, 646–652. [CrossRef]

4. Shivaraja, S.J.; Gupta, R.K.; Kumar, S.; Manjuladevi, V. Enhanced electro-optical response of nematic liquid crystal dope functionalised silver nanoparticles in twisted nematic configuration. *Liq. Cryst.* **2020**, *47*, 1678–1690.
5. Ma, Z.; Gao, Y.; Cao, H. The Effect of Chemically Modified Multi-Walled Carbon Nanotubes on the Electro-Optical Properti Twisted Nematic Liquid Crystal Display Mode. *Crystals* **2022**, *12*, 1482. [CrossRef]
6. Stevenson, W.D.; Zhou, H.; Zeng, X.; Welch, C.; Ungar, G.; Mehl, G.H. Dynamic calorimetry and XRD studies of the ne and twist-bend nematic phase transitions in a series of dimers with increasing spacer length. *Phys. Chem. Chem. Phys.* **20** 25268–25274. [CrossRef]
7. Tomczyk, W.; Longa, L. Role of molecular bend angle and biaxiality in the stabilization of the twist-bend nematic phas *Matter* **2020**, *16*, 4350–4357. [CrossRef]
8. Tschierske, C.; Photinos, D.J. Biaxial nematic phases. *J. Mater. Chem.* **2010**, *20*, 4263–4294. [CrossRef]
9. Brisbin, D.; DeHoff, R.; Lockhart, T.E.; Johnson, D.L. Specific Heat near the Nematic-Smectic-A Tricritical Point. *Phys. Re* **1979**, *43*, 1171–1174. [CrossRef]
10. Gane, P.A.C.; Leadbetter, A.J. Modulated crystal B phases and B-to-G phase transitions in two types of liquid crystalline comp *J. Phys. C Solid State Phys.* **1983**, *16*, 2059–2067. [CrossRef]
11. Leadbetter, A.J.; Tucker, P.A.; Gray, G.W.; Tajbakhsh, A.R. The Phase Behaviour of 4-n-Hexylphenyl 4-n-Tetra-Decyloxybenzth (14S6) and 4-n-Pentylphenyl 4-n-Decyloxybenzthiolate (10S5). *Mol. Cryst. Liq. Cryst. Lett.* **1985**, *1*, 19–24. [CrossRef]
12. Ocko, B.M.; Birgeneau, R.J.; Litster, J.D. Crossover to tricritical behavior at the nematic to smectic A transition: An x-ray scat study. *Z. Phys. B Condens. Matter* **1986**, *62*, 487–497. [CrossRef]
13. Chruściel, J.; Wróbel, S.; Kresse, H.; Urban, S.; Otowski, W. Dielectric Studies of 4-n-Pentylphenyl-4-Octyloxythiobenzoat *Cryst. Liq. Cryst.* **1985**, *127*, 57–65. [CrossRef]
14. Deptuch, A.; Jaworska-Gołąb, T.; Kusz, J.; Książek, M.; Nagao, K.; Matsumoto, T.; Yamano, A.; Ossowska-Chruściel, Chruściel, J.; Marzec, M. Single crystal X-ray structure determination and temperature-dependent structural studies smectogenic compound 7OS5. *Acta Cryst. B* **2020**, *76*, 1128–1135. [CrossRef]
15. Ossowska-Chruściel, M.D.; Karczmarzyk, Z.; Chruściel, J. The Polymorphism Of 4-N-Pentylphenyl-4″-N-Butyloxythio-Ber (4OS5) In The Crystalline State. *Mol. Cryst. Liq. Cryst.* **2002**, *382*, 37–52. [CrossRef]
16. Karczmarzyk, Z.; Ossowska-Chruściel, M.D.; Chruściel, J. The Crystal and Molecular Structure of 4-n-Pentylpheny Hexyloxythiobenzoate (6OS5). *Mol. Cryst. Liq. Cryst.* **2001**, *357*, 117–125. [CrossRef]
17. Chruściel, J.; Pniewska, B.; Ossowska-Chruściel, M.D. The Crystal and Molecular Structure of 4-Pentylphenyl-4′-Pentioxythiobe (5S5). *Mol. Cryst. Liq. Cryst.* **1995**, *258*, 325–331. [CrossRef]
18. Chruściel, J.; Wróbel, S.; Kresse, H.; Urban, S.; Otowski, W. Odd-Even Effect in the Homologous Series of Thioesters. *Mol. Liq. Cryst.* **1990**, *192*, 107–112. [CrossRef]
19. Rudzki, A. Liquid crystal mixtures with the chiral benzoate dopant. *Phase Transit.* **2023**, *96*, 139–148. [CrossRef]
20. Osiecka, N.; Galewski, Z.; Massalska-Arodź, M. TOApy program for the thermooptical analysis of phase transitions. *Therm Acta* **2017**, *655*, 106–111. [CrossRef]
21. Osiecka-Drewniak, N.; Galewski, Z.; Juszyńska-Gałązka, E. Distinguishing the Focal-Conic Fan Texture of Smectic A fro Focal-Conic Fan Texture of Smectic B. *Crystals* **2023**, *13*, 1187. [CrossRef]
22. Rodríguez-Carvajal, J. Recent advances in magnetic structure determination by neutron powder diffraction. *Phys. B Co Matter* **1993**, *192*, 55–69. [CrossRef]
23. Cliffe, M.J.; Goodwin, A.L. PASCal: A principal axis strain calculator for thermal expansion and compressibility determina *Appl. Cryst.* **2012**, *45*, 1321–1329. [CrossRef]
24. Lertkiattrakul, M.; Ewans, M.L.; Cliffe, M.J. PASCal Python: A Principal Axis Strain Calculator. *J. Open Source Softw.* **2023**, *8* [CrossRef]
25. Frisch, M.J.; Trucks, G.W.; Schlegel, H.B.; Scuseria, G.E.; Robb, M.A.; Cheeseman, J.R.; Scalmani, G.; Barone, V.; Petersson Nakatsuji, H.; et al. *Gaussian 16, Revision C.01*, Gaussian, Inc.: Wallingford, CT, USA, 2019.
26. Weigend, F.; Ahlrichs, R. Balanced basis sets of split valence, triple zeta valence and quadruple zeta valence quality for H Design and assessment of accuracy. *Phys. Chem. Chem. Phys.* **2005**, *7*, 3297–3305. [CrossRef]
27. Lee, C.; Yang, W.; Parr, R.G. Development of the Colle-Salvetti correlation-energy formula into a functional of the electron d *Phys. Rev. B* **1988**, *37*, 785–789. [CrossRef]
28. Becke, A.D. Density-functional thermochemistry. III. The role of exact exchange. *J. Chem. Phys.* **1993**, *98*, 5648–5652. [Cros
29. Grimme, S.; Ehrlich, S.; Goerigk, L. Effect of the damping function in dispersion corrected density functional theory. *J. C Chem.* **2011**, *32*, 1456–1465. [CrossRef]
30. Hanwell, M.D.; Curtis, D.E.; Lonie, D.C.; Vandermeersch, T.; Zurek, E.; Hutchison, G.R. Avogadro: An advanced sen chemical editor, visualization, and analysis platform. *J. Cheminform.* **2012**, *4*, 17. [CrossRef]
31. Blaabjerg, L.I.; Lindenberg, E.; Löbmann, K.; Grohganz, H.; Rades, T. Glass Forming Ability of Amorphous Drugs Investiga Continuous Cooling and Isothermal Transformation. *Mol. Pharm.* **2016**, *13*, 3318–3325. [CrossRef]
32. Jasiurkowska-Delaporte, M.; Rozwadowski, T.; Juszyńska-Gałązka, E. Kinetics of Non-Isothermal and Isothermal Crystalli in a Liquid Crystal with Highly Ordered Smectic Phase as Reflected by Differential Scanning Calorimetry, Polarized O Microscopy and Broadband Dielectric Spectroscopy. *Crystals* **2019**, *9*, 205. [CrossRef]

Piwowarczyk, M.; Osiecka-Drewniak, N.; Gałązka, M.; Galewski, Z. Synthesis, mesogenic and photoisomerization studies of (E)-4-[(4-pentyloxyphenyl)diazenyl]phenyl alkanoates. *Phase Transit.* **2019**, *92*, 1066–1076. [CrossRef]
Nakum, K.J.; Katarija, K.D.; Savani, C.J.; Jadeja, R.N. The influence of molecular flexibility on the mesogenic behavior of a new homologous series based on azo-azomethine: Synthesis, characterization, photoisomerization and DFT study. *J. Mol. Struct.* **2022**, *1249*, 131586. [CrossRef]
Bagchi, K.; Emeršič, T.; Martínez-González, J.A.; de Pablo, J.J.; Nealey, P.F. Functional soft materials from blue phase liquid crystals. *Science* **2023**, *9*, eadh9393. [CrossRef] [PubMed]
Rietveld, H.M. A profile refinement method for nuclear and magnetic structures. *J. Appl. Cryst.* **1969**, *2*, 65–71. [CrossRef]
Le Bail, A.; Duroy, H.; Fourquet, J.L. Ab-initio structure determination of $LiSbWO_6$ by X-ray powder diffraction. *Mater. Res. Bull.* **1988**, *23*, 447–452. [CrossRef]
Budai, J.; Pindak, R.; Davey, S.C.; Goodby, J.W. A structural investigation of the liquid crystal phases of 4-(2′-methylbutyl)phenyl 4′-n-octylbiphenyl-4-carboxylate. *J. Phys. Lett.* **1984**, *45*, 1053–1062. [CrossRef]
de Vries, A. The description of the smectic A and C phases and the smectic A-C phase transition of TCOOB with a diffuse-cone model. *J. Chem. Phys.* **1979**, *71*, 25–31. [CrossRef]
Rowland, R.S.; Taylor, R. Intermolecular Nonbonded Contact Distances in Organic Crystal Structures: Comparison with Distances Expected from van der Waals Radii. *J. Phys. Chem.* **1996**, *100*, 7384–7391. [CrossRef]
Sinha, D.; Mandal, P.K.; Dąbrowski, R. High birefringence laterally fluorinated terphenyl isothiocyanates: Structural, optical and dynamical properties. *Phys. B Condens. Matter* **2014**, *441*, 100–106. [CrossRef]
Bielejewska, N.; Chrzumnicka, E.; Mykowska, E.; Przybylski, R.; Szybowicz, M. Comparative Study of Orientational Order of Some Liquid Crystals from Various Homologous Series. *Acta Phys. Pol. A* **2006**, *110*, 777–793. [CrossRef]
Bauman, D.; Zięba, A.; Mykowska, E. Oriental behaviour of some homologues of 4-n-pentyl-phenylthio-4′-n-alkoxybenzoate doped with dichroic dye. *Opto-Electron. Rev.* **2008**, *16*, 244–250. [CrossRef]

crystals

Article

Liquid Crystalline Structures Formed by Sphere–Rod Amphiphilic Molecules in Solvents

Nilanthi P. Haputhanthrige [1,2], Yifan Zhou [3], Jingfan Wei [3], Min Gao [1], Tianbo Liu [3] and Oleg D. Lavrentovich [1,2,*]

1 Advanced Materials and Liquid Crystal Institute, Kent State University, Kent, OH 44242, USA; hpadmini@kent.edu (N.P.H.); mgao@kent.edu (M.G.)
2 Department of Physics, Kent State University, Kent, OH 44242, USA
3 School of Polymer Science and Polymer Engineering, The University of Akron, Akron, OH 44325, US yz99@uakron.edu (Y.Z.); jw357@uakron.edu (J.W.); tliu@uakron.edu (T.L.)
* Correspondence: olavrent@kent.edu

Abstract: Self-assembly of amphiphilic molecules is an important phenomenon attrac broad range of research. In this work, we study the self-assembly of $KTOF_4$ spher amphiphilic molecules in mixed water–dioxane solvents. The molecules are of a T-sh geometry, comprised of a hydrophilic spherical Keggin-type cluster attached by a fl bridge to the center of a hydrophobic rod-like oligodialkylfluorene (OF), which consi four OF units. Transmission electron microscopy (TEM) uncovers self-assembled sph structures of $KTOF_4$ in dilute solutions. These spheres are filled with smectic-like l of $KTOF_4$ separated by layers of the solution. There are two types of layer pac (i) concentric spheres and (ii) flat layers. The concentric spheres form when the di volume fraction in the solution is 35–50 vol%. The flat layers are formed when the di volume fraction is either below (20 and 30 vol%.) or above (55 and 60 vol%.) the indi range. The layered structures show no in-plane orientational order and thus rese thermotropic smectic A liquid crystals and their lyotropic analogs. The layered pac reveal edge and screw dislocations. Evaporation of the solvent produces a bulk birefri liquid crystal phase with textures resembling the ones of uniaxial nematic liquid cr These findings demonstrate that sphere–rod molecules produce a variety of self-asser structures that are controlled by the solvent properties.

Keywords: amphiphilic molecules; self-assembly; TEM; smectic liquid crystal; disloca

Academic Editor: Ingo Dierking

Received: 1 February 2025
Revised: 10 February 2025
Accepted: 11 February 2025
Published: 13 February 2025

Citation: Haputhanthrige, N.P.; Zhou, Y.; Wei, J.; Gao, M.; Liu, T.; Lavrentovich, O.D. Liquid Crystalline Structures Formed by Sphere–Rod Amphiphilic Molecules in Solvents. *Crystals* **2025**, *15*, 177. https://doi.org/10.3390/cryst15020177

1. Introduction

Self-assembly of amphiphilic molecules in aqueous solutions draws significa tention in many fields, such as molecular chemistry, biochemistry, and biomedicine The self-assembled structures are diverse, ranging from isolated spherical micelle vesicles to helices, fibrils, and extended liquid crystals (LCs) such as lyotropic L_α, L_1 L_3 phases [7–22]. Lyotropic LCs, formed by the self-assembly of amphiphilic mole have received considerable attention due to diverse research areas such as drug ery [6,23–27] and molecular engineering [28–30]. Self-assembly of amphiphilic surfa and flexible block copolymers have been well studied with the rule of "packing pa ters", which predicts the formation of micelles, wormlike micelles, vesicles, bilayer However, when some components, especially the solvophobic ones, are rigid and c comply with the conformational changes during the assembly, the common rules wi be applicable. An interesting example was recently introduced by Luo et al. [13], studied amphiphilic $KTOF_4$ molecules in the shape of a letter T, in which a hydrophilic spherical Keggin-type cluster connects by a flexible link to a center of a hydrophobic rod-like oligodialkylfluorene (OF), Figure 1a [13]. In the formula of $KTOF_4$, "4" indicates the number of connected OFs, and "T" represents the shape of the sphere–rod molecule, Figure 1. The chemical formula of Keggin is $(XM_{12}O_{40})^{4-}$ where X is pentavalent phosphorus (P^V), M is tungsten (W) metal, and O is oxygen. The OF is a combination of carbons and hydrogens in the shape of a rod. The Keggin clusters and OFs are rigid, while the chain that connects them is flexible, thus allowing the Keggin and OF segments to move with respect to each other. The studies by Luo et al. [13] revealed that in water and acetonitrile (MeCN) mixed solutions, $KTOF_4$ molecules self-assemble into spherical inclusions with a system of concentric layers in their interior. The concentric spherical layers fill the entire volume of the inclusions with identical interlayer distance. The number of $KTOF_4$ bilayers in the concentric spherical inclusion, and thus the size of the inclusion, depends on the solvent polarity, temperature, and concentration [13]. In a solution with high polarity, 85 vol% water, the inclusion radius is about 19 nm, including three concentric spherical bilayers. In a solution with lower polarity, 50 vol% water, the inclusions grow to a radius of 65 nm with 12 bilayers.

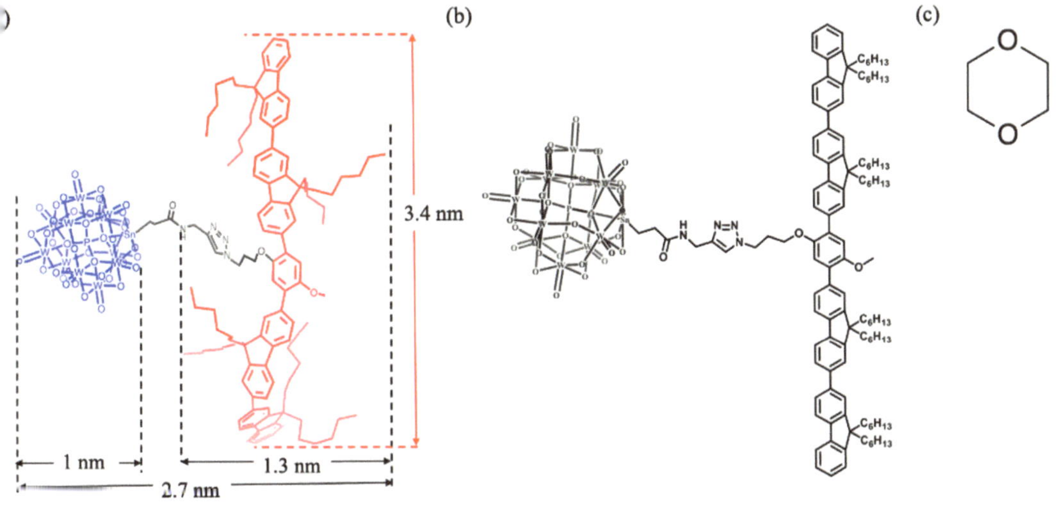

Figure 1. Chemical structures of materials. (**a**) 3D, (**b**) 2D representation of sphere–rod shaped $KTOF_4$ with the hydrophilic Keggin cluster attached to the center of OFs, and (**c**) dioxane. The width and the length of the rod-shaped OF are approximately 1.3 and 3.4 nm, respectively, and the diameter of the Keggin sphere is 1.0 nm. The Keggin cluster is indicated in blue, and the OF units are shown in red in (**a**).

Spherical concentric packings of layers are frequently observed in droplets of thermotropic LCs such as smectics A (SmA) [31], smectics C (SmC) [32], and cholesterics [33–37]. Depending on the type of the order parameter, these spherical concentric packings of layers show two distinct geometries. In SmA, the orientational order is uniaxial, with the molecules oriented along the director $\hat{\mathbf{n}}$ which is along the normal $\hat{\upsilon}$ to the layers. A spherical packing is then of a pure radial type, with $\hat{\mathbf{n}}$ along the radial directions, forming an isolated point defect hedgehog in the center of the droplet. In SmC, the molecular tilt within the layers produces an additional degree of orientational order, described by a polar vector $\mathbf{c} = \hat{\mathbf{n}} - \hat{\upsilon}(\hat{\mathbf{n}}\cdot\hat{\upsilon})$ tangential to the layers. Here, $\hat{\mathbf{n}}$ is tilted with respect to $\hat{\upsilon}$. As a result, the spherical packings still carry a radial point defect in the normal direction to the layers, but now this point defect is not isolated: it is connected to one or two disclir lines in the vector field **c** [32]. Similar "Dirac monopole" concentric packings with disclinations can be observed in cholesteric droplets [33–36,38–42].

Although the spheres of $KTOF_4$ inclusions show concentric packings, it is not whether the structures are of an SmA type, i.e., with no orientational order, o SmC/cholesteric type, i.e., with an orientational order within the layers. Such a entational order might be expected thanks to the presence of the rod-like OFs, which prefer to align parallel to each other and form a two-dimensional nematic ordering v the layers, described by a nonpolar version of the vector **c**. If this is the case, the spherical $KTOF_4$ inclusions would be expected to show radial disclination lines **c** field.

The goal of this work is to explore whether the self-assembled layers of T-sh $KTOF_4$ molecules have an isotropic or orientationally ordered in-plane structure, l alyzing the nanoscale-resolved transmission electron microscopy (TEM) images of spherical inclusions in a mixture of water and 1,4-Dioxane (dioxane), Figure 1c. In solu with 0.2 mg/mL $KTOF_4$, these spheres exhibit two main inner structures: (i) conc spherical layers and (ii) flat parallel layers. In the latter case, we do not observe disclinations, which indicates that there is no orientational order within the layers, i.e $KTOF_4$ molecules show an isotropic in-plane ordering, similar to the structure of an The layered SmA structure shows edge and screw dislocations. The inner geome spherical inclusions is determined by the dioxane volume fraction in the solution. Na the solutions with a scarcity or excess of dioxane form flat layers, in which the conc tion of dioxane in the range 35–50 vol% yields concentric spheres. The flat layers in solvents might be explained by their increased electrostatic repulsion between nega charged Keggin clusters and by the hydrophobic attraction of the rod-like OFs. T crease in the concentration of the $KTOF_4$ molecules through vaporization of the solve produce a homogeneous nematic-like bulk phase with high birefringence of 0.2.

2. Materials and Methods

2.1. The Material Synthesis

The alkyne-functionalized Keggin, $[(PW_{11}O_{39})(SnCH_2CH_2CONHCH_2CCH)]^{4-}$ 4 tetrabutyl ammonium counterions (TBA^+) is synthesized based on a reported me in the literature [43], and the azide-containing OFs are synthesized through the by-step addition of fluorene repeating units [44,45]. Two components are couple ing azide–alkyne Huisgen cyclo-addition to form sphere–rod shaped hybrids K Figure 1a, as explained in the ref. [13]. Different 0.2 mg/mL $KTOF_4$ samples are pre by changing the dioxane volume fraction from 20 to 60 vol% in the water and dic mixed solvents.

To estimate the dimensions of the $KTOF_4$ molecule, the molecular structure is d using ChemDraw software (version: 23.1.2). The 3D structure of an isolated molec a vacuum is obtained using the MM2 Energy minimization engine of Chem3D soft (Perkin Elmer, version: 20.0.0.41). Then, the dimensions of energy-minimized molecu measured in chem 3D and presented in Figure 1a.

2.2. Transmission Electron Microscopy

A small volume (~3 μL) of the sample solution is placed on a carbon-coated coppe and allowed to dry. The dried TEM grids are used for TEM analysis. Regular TEM in are captured using an FEI Tecnai F20 microscope (200 kV) (FEI Company, Hillsborc USA). The basic experimental setup and the procedure are detailed by Gao M. et al

The visibility of the layered structures in the TEM images is improved by enhancing the contrast and brightness using Fiji/ImageJ, version 2.140/1.54f.

2.3. Polarized Optical Microscopy Study of High Concentrated $KTOF_4$

We separately study highly concentrated solutions of $KTOF_4$. A 2.0 mg/mL $KTOF_4$ in 50 vol% dioxane is placed in a capillary tube to allow the mixture to evaporate slowly and to increase the $KTOF_4$ concentration further. A new material with birefringent colors, abbreviated LC_{KTOF4}, forms at the capillary tube's meniscus. The textures of LC_{KTOF4} sessile droplets at glass plates are explored under an Olympus BX51 polarized optical microscope (POM) (Olympus, Tokyo, Japan).

The birefringence of the LC_{KTOF4} is measured in a wedge cell with two PI2555 polyimide-coated, rubbed glass plates as described in ref. [47,48]. The temperature of the cell is increased until LC_{KTOF4} reaches the isotropic phase. The material is then cooled to the nematic phase (N) while observing under a POM with crossed polarizers. The temperature of the cell is controlled using a Linkam hot stage (Linkam Scientific, Redhill, UK).

3. Results

We prepare different samples of 0.2 mg/mL $KTOF_4$ by changing the dioxane to water volume fraction. Throughout the text, "X vol% dioxane" refers to a solution with a dioxane: water ratio of X:(100−X). Unless stated otherwise, a "$KTOF_4$ sample" refers to a sample containing 0.2 mg/mL $KTOF_4$. $KTOF_4$ solutions are studied using bright-field TEM textures. Self-assembled inclusions are observed across the dioxane content of 20–60 vol%. Precipitation occurs when the dioxane content is below 20 vol%, and the self-assembly is not observed when dioxane exceeds 70 vol%.

$KTOF_4$ molecules aggregate into spherical inclusions with radii 25–90 nm, Figure 2a–i. The inclusion size does not correlate with the dioxane volume fraction through the TEM results. As revealed in the TEM images, the assembled inclusions are filled with periodic structures exhibiting alternating dark and bright layers. Two distinct layered structures are observed in the TEM textures: one consists of concentric spherical layers, and the other consists of flat layers arranged inside a spherical shape. In the solutions with 20 and 30 vol% dioxane, the spherical structure with flat layers is dominant. As the dioxane volume fraction increases to 35–50 vol%, the concentric spherical layer structure becomes dominant. However, at higher dioxane volume fractions, 55 and 60 vol%, the flat layers are dominant again, Figure 2 and Table 1. The repeated interlayer distance remains constant at 4.8 nm and does not vary with the dioxane volume fraction. We are not aware of the re-entrant packing scenario with a sequence flat–spherical–flat as a function of composition in any other lyotropic layered liquid crystal system.

The composition of the surface layer of the spherical inclusions varies with the dioxane volume fraction. In solutions with 20, 55, and 60 vol% dioxane, which form flat layers, some inclusions exhibit a dark interfacial layer with a thickness of 0.8–1.1 nm, which can be associated with the Keggin clusters, while others have a bright interfacial layer with a thickness of 1.4–1.9 nm, which might correspond to the combined extension of the flexible bridge and OF rods along the horizontal direction in Figure 1, which amounts to about 1.7 nm. A thick, bright layer, with a thickness of 3.8–6.5 nm greater than that of a $KTOF_4$ molecule, is observed around some inclusions in compositions with 30, 35, and 55 vol% dioxane, Figure 3. These thick, amorphous layers might be a combination of $KTOF_4$ and TBA^+ ions that accumulate around the self-assembled spheres during drying. The layer beneath this thick, bright periphery in 30, 35, and 55 vol% dioxane compositions is dark and thus corresponds to Keggin clusters, Figure 3. When the dioxane volume fractions are 40, 45, and 50 vol%, the surface layer is a bright, thin layer made of OF rods.

Note that the concentrations 35, 40, 45, and 50 vol% dioxane correspond to the sph
packing of layers. Since in these cases the interfacial layer is comprised fully of eithe
(for 40, 45, and 50 vol% dioxane) or Keggin clusters (35 vol% dioxane), we conclud
the spherical packing can be facilitated by the surface tension anisotropy, which p
one component of the molecule to be at the interface. For flat layers, the prefere
apparently weaker.

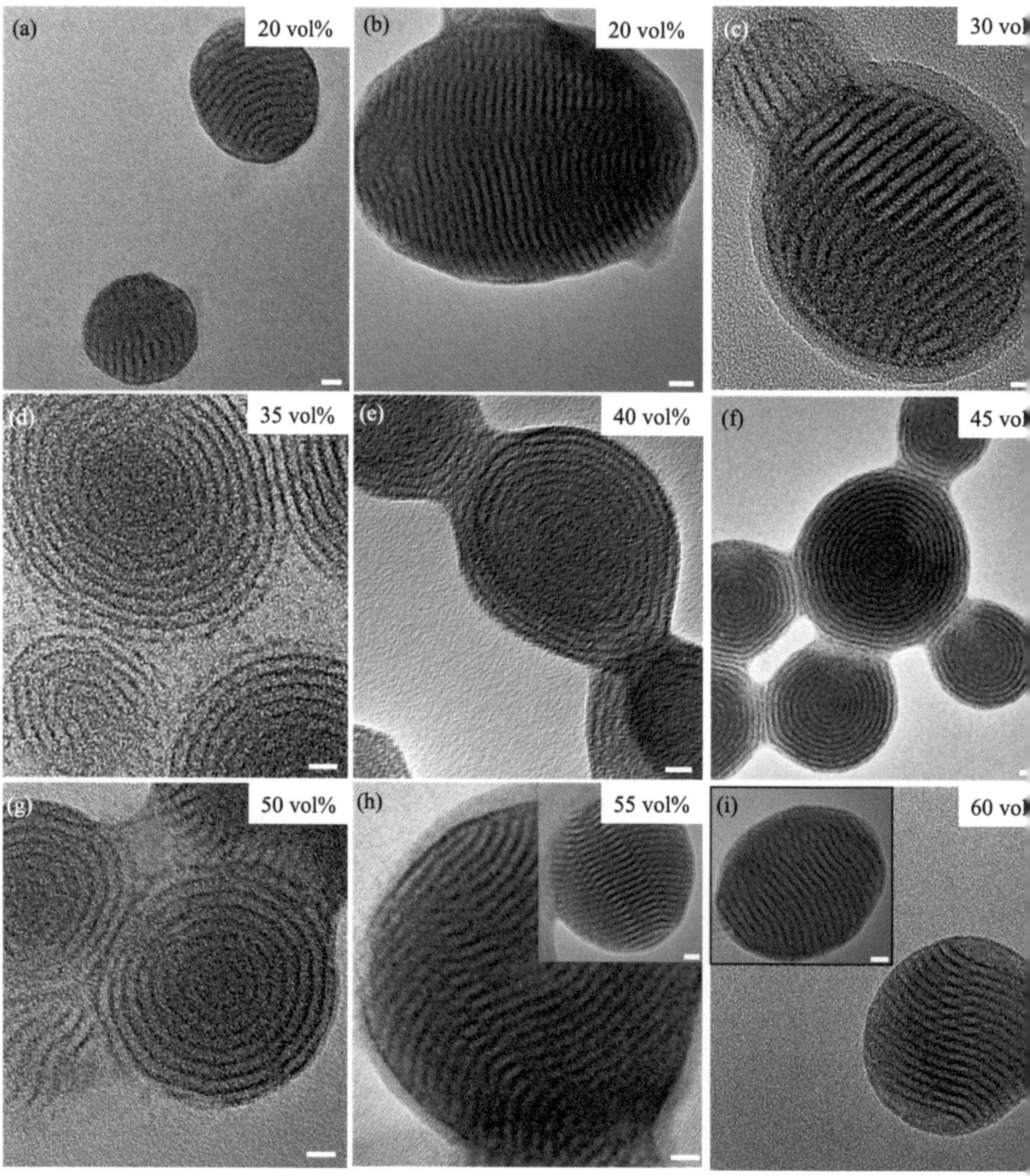

Figure 2. TEM textures of 0.2 mg/mL $KTOF_4$ in a dioxane and water mixture. The solutior
(**a**,**b**) 20, (**c**) 30, (**d**) 35, (**e**) 40, (**f**) 45, (**g**) 50, (**h**) 55, and (**i**) 60 vol% dioxane. The scale bar is 1
X vol% dioxane is a solution with a dioxane: water ratio of X:(100−X).

Table 1. Summary of layer type and surface layer composition of inclusions in solutions with different dioxane volume fractions. (X vol% dioxane; dioxane: water = X:(100−X)).

Dioxane vol%	Concentric Layers	Flat Layers	Surface Composition
20	Not observed	Frequently observed	Keggin/OF
30	Not observed	Frequently observed	Keggin
35	Frequently observed	Not observed	Keggin
40	Frequently observed	Not observed	OF
45	Frequently observed	Not observed	OF
50	Frequently observed	Observed (but rare)	OF
55	Not observed	Frequently observed	Keggin/OF
60	Not observed	Frequently observed	Keggin/OF

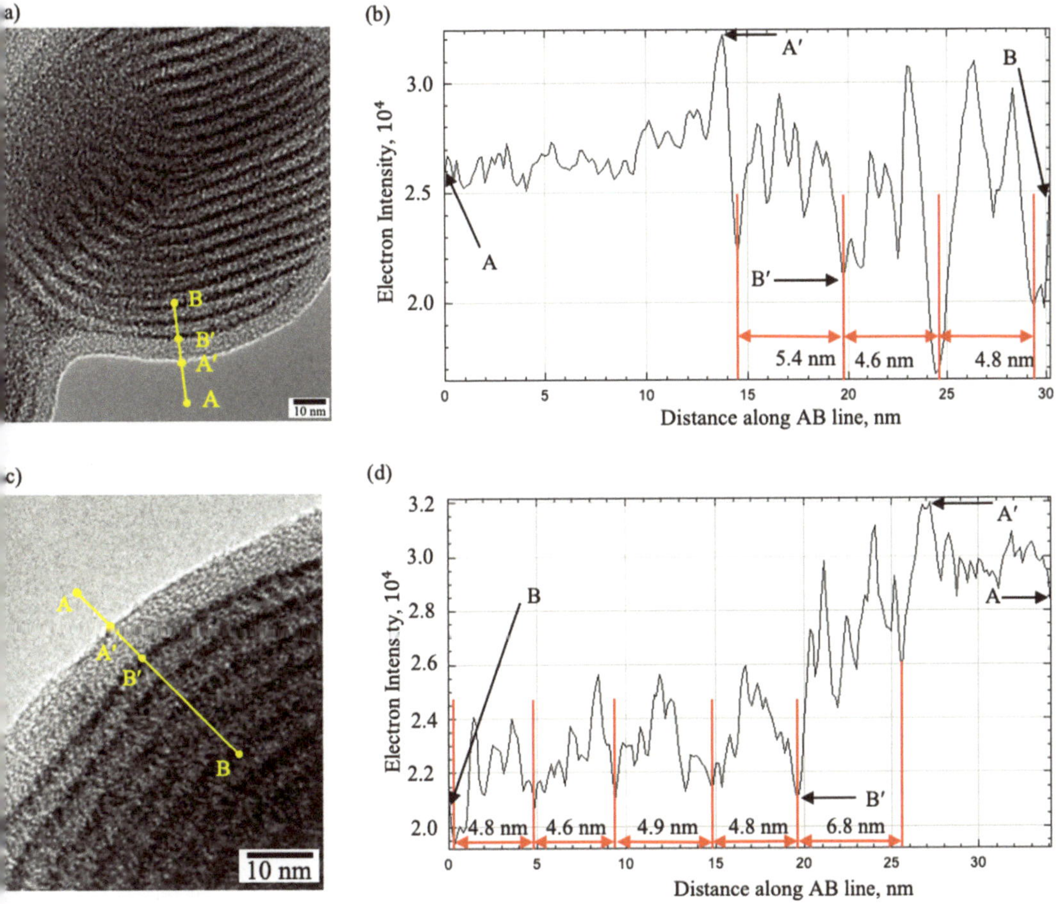

Figure 3. Self-assembled inclusions with thick, bright peripheral layer. TEM textures of 0.2 mg/mL $KTOF_4$ in (**a**) 30 vol%, and (**c**) 35 vol% dioxane. (**b**,**d**) Transmitted electron intensity profile along the line drawn in (**a**,**c**), respectively. X vol% dioxane is a solution with a dioxane: water ratio of X:(100−X).

A zoomed-in TEM texture of the 30 vol% dioxane sample is used to study the assembled structure in detail. The bright-field TEM imaging condition used in this study produces textures mainly governed by mass-thickness contrast. A higher atomic number area generates more scattered electrons at a higher angle, resulting in a reduced amount of

transmitted electron density compared to molecules containing atoms with lower a
numbers [49,50]. For a sample containing $KTOF_4$ molecules, the dark stripes in the TE
tures (marked with white arrows in Figure 4a) correspond to Kegging clusters (molar
of Keggin ~ 2900 g/mol), while the narrow dark stripes (marked with yellow arro
Figure 3a) correspond to the OF (molar mass of four OF ~ 660 g/mol). The bright
represent the flexible chains and the background, Figure 4a.

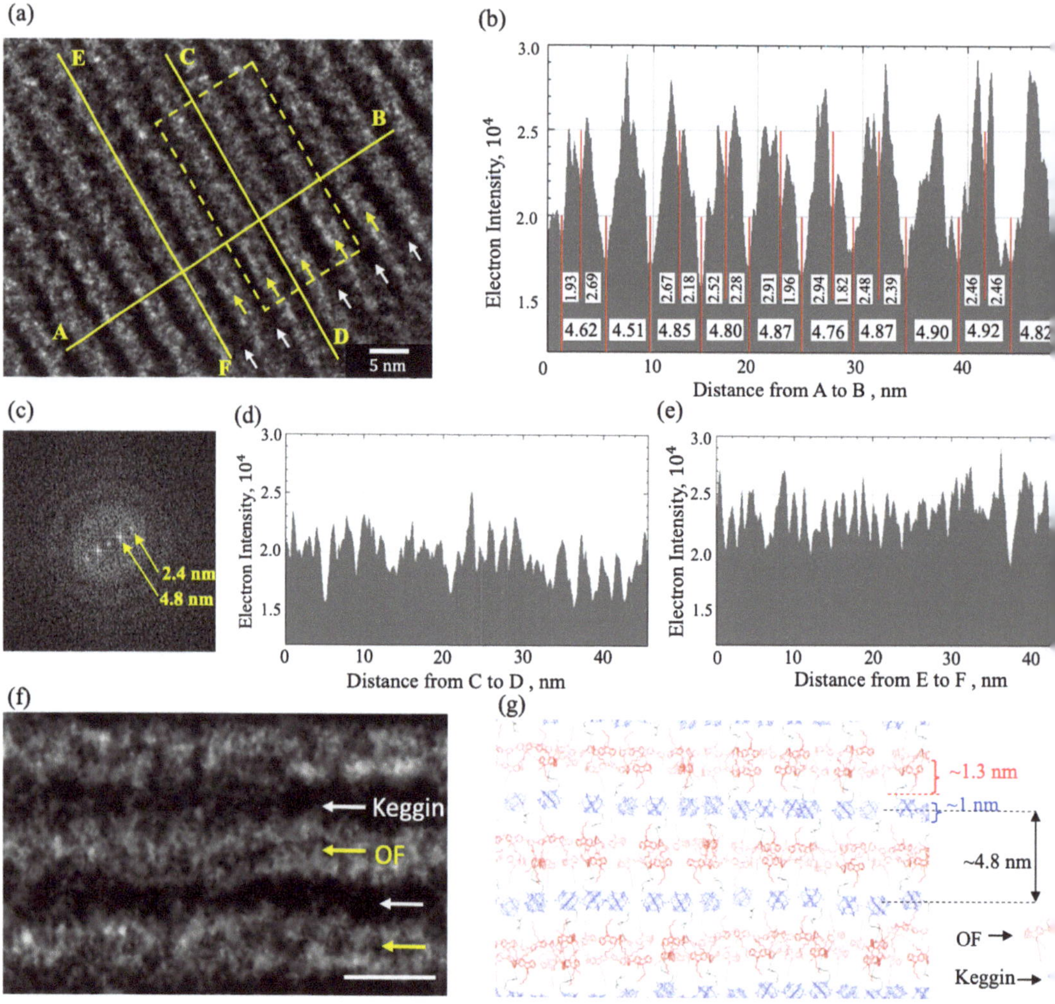

Figure 4. (**a**) Zoomed-in TEM texture of 0.2 mg/mL $KTOF_4$ in 30 vol% dioxane. White a
marking the dark stripes correspond to Kegging clusters, while yellow arrows marking the n
dark stripes correspond to the OFs. (**b**,**d**,**e**) The transmitted electron intensity profile along th
AB, CD, and EF, respectively, (**c**) fast Fourier transform pattern obtained from the texture
(**f**) a zoomed-in section (marked in dashed yellow in (**a**)) of a structure. The scale bar is 5 nm
(**g**) proposed molecular packing within the layers.

The TEM textures and the transmitted electron intensity profile along the layer n
(AB line in Figure 4a) exhibit periodicity between 4.5–4.9 nm, Figure 4b. The fast Fo
transform (FFT) pattern obtained from the texture in Figure 4a is shown in Figure 4c
FFT pattern shows peaks at 4.8 nm, and 2.4 nm along the layer normal. The per

peak at 4.8 nm corresponds to the repeated layer distance observed in TEM textures and the transmitted electron intensity profile. The peak at 2.4 nm is the second-order peak, indicating a well-defined modulation.

The transmitted electron intensity profiles along a dark layer (CD line in Figure 4a) and a bright layer (EF line in Figure 4a) do not show any well-defined periodic intensity peaks, Figure 4d,e. The absence of a periodic peak along the layer in the FFT pattern confirms the absence of periodicity within the layers. Thus, the $KTOF_4$ self-assembled structures exhibit isotropic in-plane ordering, which is similar to periodic layers of an SmA. If there is a long-range ordering within the layers, as observed in cholesteric and SmC phases [33,35,36], one would expect the formation of disclination lines connecting the center of the concentric spherical structure to the surface of the spherical inclusion. However, the absence of radial disclination lines in the concentric spherical packing of $KTOF_4$, Figure 2d–g, supports the conclusion that the lamellar layers correspond to the SmA type of ordering. A proposed molecular packing within the periodic layers in Figure 4f is schematized in Figure 4g.

Periodically modulated LC phases such as SmA exhibit defects of the layered structure such as edge and screw dislocations [51–55]. The dislocations are characterized by Burgers vector, $\boldsymbol{b} = nd\mathbf{v}$ where n is an integer, d is the repeat layer distance, and $\mathbf{v}$ is the unit vector normal to the layers. According to de Gennes [56] and Kleman and Williams [57], the energy per unit length of an edge dislocation is $W_e = K_1 b^2/2\lambda r_c + w_c$, where K_1 is the splay elastic constant, $b = nd$, $\lambda = (K_1/B)^{1/2}$ is the penetration length of material, B is the compressibility modulus, and w_c and r_c are the energy and the radius of the core, respectively. Since the energy varies as the square of the Burgers vector, the dislocations of a small Burgers vector ($n = 1$) are common [58]. Allain and Kleman [59] and Kleman [60] showed the core extension of edge dislocation along the Burgers vector is of the order of $\xi_z = d^2/\lambda$. In our case, ξ_z is of the order of d, and the dislocation core has a similar extension along the Burgers vector and along the direction perpendicular to it; such a core is often called an isotropic core [60]. The isotropic cores of edge dislocations indicate that for the $KTOF_4$ $\lambda \sim d$. Figure 5a illustrates the TEM textures of self-assembled $KTOF_4$ inclusions with edge dislocations indicated by dashed yellow lines. The edge dislocations with small Burgers vectors, $b = d$, are the most frequently observed dislocations in $KTOF_4$ structures. The additional layer of the edge dislocations is always the dark layer in the TEM textures, which corresponds to the hydrophilic part of the $KTOF_4$ molecule (the Keggin cluster); the covalently connected OF rods make a U-turn around this extra layer of Keggin clusters. The core thus can be represented as a pair of disclinations in the OF sublayers, of a strength $+1/2$ around the Keggin clusters and of a strength $-1/2$ in the neighboring region.

In a screw dislocation, the layers are arranged as a spiral staircase [61]. The Burgers vector, $\boldsymbol{b}$, is parallel to the dislocation line. Kleman [58] calculated the energy of screw dislocations is $W_s = Bb^4/128\left(r_c^{-2} - R^{-2}\right) + w_c$, where R is the external radius of the sample. The b^4 term in energy ensures the stability of screw dislocations with small Burgers vectors. Figure 5b illustrates self-assembled $KTOF_4$ inclusions, with screw dislocations indicated by dashed yellow lines. The layers on the two sides of a dislocation line are shifted with respect to each other, by $d/2$, which implies $b = d$. Both edge and screw dislocations are observed in the flat layer structure (formed in 20, 30, and 50–60 vol% dioxane) while dislocations are rarely observed in the concentric spherical layer structure (formed in 35–50 vol% of dioxane).

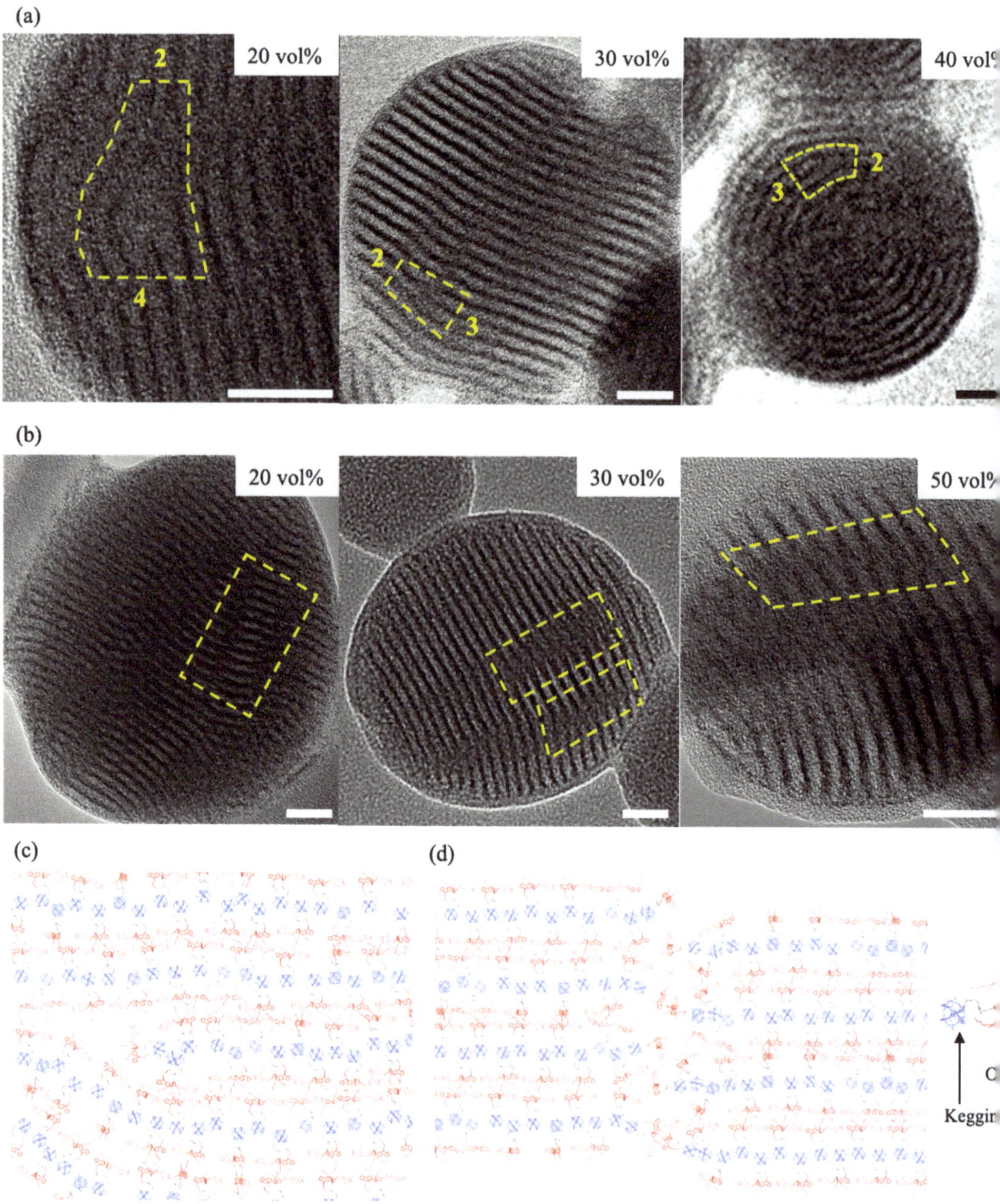

Figure 5. Dislocations in self-assembled $KTOF_4$ structures. TEM textures showing (**a**) ed locations, and (**b**) screw dislocations within the self-assembled layers of 0.2 mg/mL KTOF dioxane content is indicated in the top-right corner of the respective texture. The scale bar is for all textures in (**a**,**b**). X vol% dioxane refers to a solution with a dioxane: water ratio of X:(1 Proposed molecular packing within the layers at the defect site of (**c**) an edge dislocation, an screw dislocation.

The explored $KTOF_4$ 2.0 mg/mL dispersion in 50 vol% dioxane shows an intri mesomorphic behavior when the solvent is slowly evaporated. Evaporation produ birefringent fluid residual with birefringence colors and POM textures with extinction brushes resembling those of uniaxial nematic liquid crystals, Figure 6a. The nematic-like LC_{KTOF4} material aligns perfectly on a rubbed PI2555 layer on a glass substrate, Figure 6b. An apparent reason for the alignment is the orientation of rod-like OF parts along the grooves of the polyimide substrate. The LC_{KTOF4} material shows a biphasic region similar to that of other nematics. Namely, upon cooling, the nematic nuclei appear at 34.5 °C, while at 33.0 °C, one observes a homogeneous nematic bulk structure, Figure 6c. The birefringence of LC_{KTOF4} is measured using a wedge cell method as explained in ref. [47,48], and the measured birefringence of the material is high, 0.2.

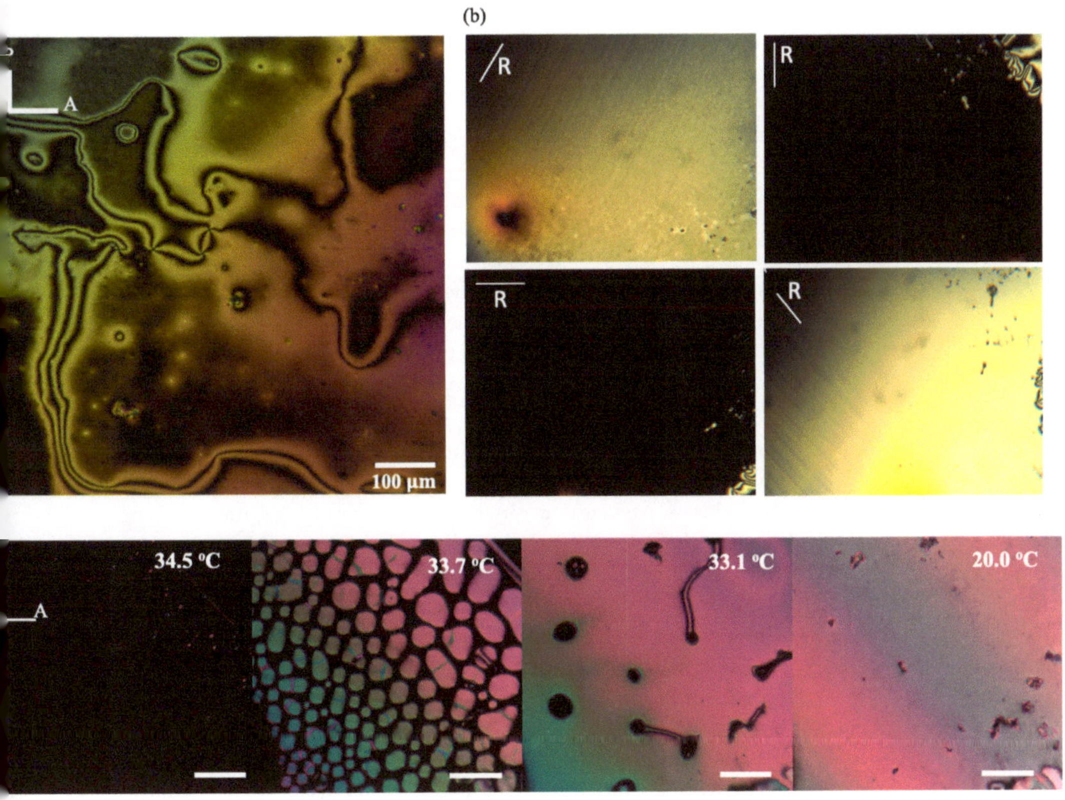

Figure 6. Polarizing optical microscopy (POM) study of LC_{KTOF4}. LC_{KTOF4} is obtained by allowing 2.0 mg/mL $KTOF_4$ with 50 vol% dioxane to slowly evaporate. POM image of a drop of LC_{KTOF4} (**a**) on a glass plate, and (**b**) on the PI2555 coated, rubbed glass plate. The sample is rotated to have different alignments of the rubbing direction, R, with respect to P and A. (**c**) LC_{KTOF4} progression of phase transition from Isotropic to Nematic while cooling. P and A indicate the polarizer and the analyzer directions and stay the same for all the images. The scale bar is 100 μm.

4. Discussion

The study demonstrates that $KTOF_4$ molecules in mixed dioxane–water solutions self-assemble into spherical inclusions with periodic inner structures. FFT and transmitted electron intensity profiles reveal a well-defined periodicity along the layer normal, Figure 4. The measured period of 4.8 nm in FFT represents the bilayer spacing, Figure 4. The spherical inclusions reveal two types of packings: spherical concentric arrangements of layers and flat layers. The observed textures of spherical concentric packings show no radial disclinations, which signals that there is no long-range orientational order within the layers. In words, $KTOF_4$ forms SmA-type layered structures in mixed dioxane–water solution

The inner structure of LC droplets suspended in an isotropic fluid is controll a balance of bulk elasticity, isotropic and anisotropic interfacial interactions [62–6 describe the structure, we introduce an apolar unit vector $\hat{\mathbf{n}} \equiv -\hat{\mathbf{n}}$ which is norm the layers, and a unit vector $\hat{\boldsymbol{\upsilon}}$, which is normal to the spherical surface of the cond inclusions. The interfacial surface energy can be written as $F_{int} = 4\pi\sigma_0 R^2$ for a sph inclusion of radius R, where σ_0 is the interfacial tension [63]. The bulk energy of the s can be written as the elastic energy of the spherically bent layers, $F_{ela} = 8\pi KR$, whe the splay elastic constant for the deformations of $\hat{\mathbf{n}}$ [62,63]. For the tangential align of layers at the interface, i.e., when $\hat{\mathbf{n}} \parallel \hat{\boldsymbol{\upsilon}}$, surface anchoring energy at the bou of the inclusion is, $F_{anch} = \frac{4}{3}\pi W R^2$, where W is the surface anchoring coefficien The balance between the F_{int} and F_{ela} determines the shape of the anisotropic drop isotropic surroundings. Typical values for liquid crystalline cyanobiphenyls in gly are $\sigma_0 \sim (10^{-3} - 10^{-2})\ \mathrm{Jm}^{-2}$, $W \sim (10^{-6} - 10^{-5})\ \mathrm{Jm}^{-2}$ and $K = 10^{-11}$ N [66]. nematic-isotropic interface of 5CB, $\sigma_0 \approx 2 \times 10^{-5}\ \mathrm{Jm}^{-2}$ and $W \approx 5 \times 10^{-7}\ \mathrm{Jm}^{-2}$ [67] result, any macroscopic LC droplet larger than $K/\sigma_0 \approx (1 - 10)$ nm should be spheri our study, we observe that the self-assembled inclusions of $KTOF_4$ in 20–60 vol% di are spherical with radii of 25–90 nm, Figure 2. This observation suggests that in the exp system, $2K/\sigma_0 < 25$ nm. If one assumes $K = 10^{-11}$ N, then $\sigma_0 > 8 \times 10^{-4}\ \mathrm{Jm}^{-2}$.

Anisotropic surface interaction plays a critical role in determining the layer stru inside the LC droplets. For an LC droplet with $R \ll 6K/W$, spatial variations director $\hat{\mathbf{n}}$ are avoided, and $\hat{\mathbf{n}}(\mathrm{r})$ = constant is maintained at the expense of vio boundary conditions. Conversely, droplets with $R \gg 6K/W$ satisfy boundary cond by aligning molecules along the easy axis [62,63]. For $KTOF_4$ inclusions, both fla spherical layer packings are observed in spheres of 25–90 nm. The type of layer do depend on the size of the inclusion, which signals that the type of packing is affected subtle balance of bulk elasticity and surface anchoring. Stiff layers would prefer to for packings, while soft, flexible layers would form concentric spheres. This feature can ex the observation for solutions with low and medium dioxane content, as discussed b

The spherical packings in the medium dioxane content can be explained simp the different affinities of the polar Keggin clusters and nonpolar OF rods to the solv the solvent prefers to interface with one but not the other, the resulting packing w spherical. If the concentration of dioxane decreases and the polarity of the solvent incr then the hydrophobic attractions between OF rods enhance. To minimize the area of c with a polar solvent, the rods would tend to be parallel to each other, Figure 6. The would become stiffer and thus less prone to follow the curvature of the confining vo This is in line with the observation of flat layers at low dioxane concentrations.

The observation of flat layers at the high content of dioxane is less clear. Tenta the non-polar solvent allows the OF rods to form strongly curved U-turn regions ar the layers of Keggin clusters, Figure 7a, as in the cores of edge dislocations, Figure 5c. U-turns are not very costly energetically since the hydrophobic rods are not aligned strictly parallel to each other to minimize the contact with the solvent. These U-tern re might form at the interface of the layers with the non-polar solvent, thus shielding the Keggin clusters from contact with it. Of course, this scenario is likely to be complicat the electrostatic forces that likely change with the solvent polarity; a more detailed an requires further experiments.

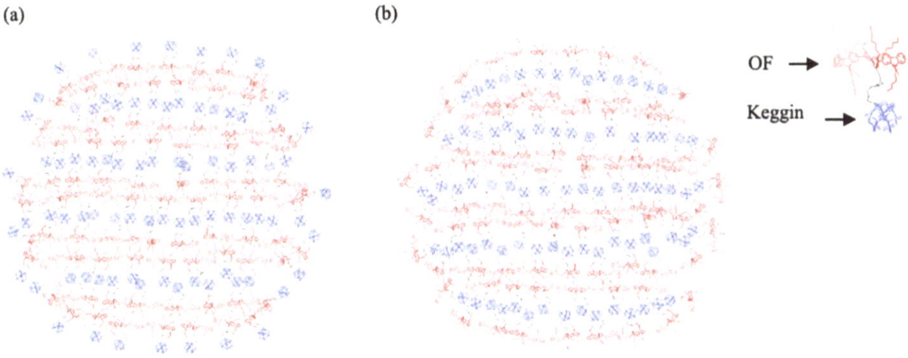

Figure 7. Proposed $KTOF_4$ molecular packing within the flat layers with (**a**) Keggin, and (**b**) OF as surface layer composition.

Finally, we observe an unexpected phenomenon of formation of a nematic-like structure in $KTOF_4$ dispersions when the solvent is slowly evaporated. The residual material is fluid, highly birefringent, and exhibits typical nematic textures, Figure 6a. Unfortunately, attempts to prepare a similar nematic phase by producing highly concentrated $KTOF_4$ dispersions did not produce a nematic phase; apparently, the process of self-organization is complex and requires unusual steps such as slow evaporation of the solvent. The solvent-free $KTOF_4$ material, a solid in powder form, undergoes a transition to an unspecified weekly birefringent phase at 200 °C upon heating. Upon cooling, the material crystallizes at 160 °C. Further studies are needed to explore the strongly concentrated solutions and solvent-free samples of $KTOF_4$.

5. Conclusions

In this work, we study the self-assembled structures of $KTOF_4$ amphiphilic molecules in dioxane and water-mixed solutions. 0.2 mg/mL $KTOF_4$ molecules form self-assembled layered structures of a smectic A type with sublayers of the Keggin and OF of the molecules that segregate away from each other. The self-assembled spherical inclusions show either concentric packing of the layers, when the concentration of dioxane is in the range 35–50 vol% or flat layers when the dioxane concentration is higher or lower. The electron microscopy textures demonstrate elementary dislocations of the edge and screw type. The organization of molecules within the layers shows no long-range orientational order as we do not observe radial disclinations in the spherical packings. In other words, the self-assembled lyotropic phases are of the SmA type. Furthermore, increasing the $KTOF_4$ concentration through slow evaporation leads to the formation of a birefringent bulk phase of the nematic type. This study provides an insight into a rich morphogenesis of self-assembly of $KTOF_4$ amphiphilic molecules of nontrivial shape in mixed dioxane–water solutions.

Author Contributions: N.P.H. conducted the experiments, analyzed and discussed the data, and contributed to the writing, Y.Z. and J.W. synthesized the material ($KTOF_4$) and prepared samples. M.G. guided the TEM imaging and image analysis. T.L. directed research for the portion at the University of Akron. O.D.L. contributed to the writing and directed the research. All authors have read and agreed to the published version of the manuscript.

Funding: O.D.L. acknowledges support from the NSF DMR-2215191. T.L. acknowledges support from the NSF DMR2215190 and the University of Akron.

Data Availability Statement: The datasets generated during and/or analyzed during the c
study are available from the corresponding author upon reasonable request.

Acknowledgments: The TEM imaging was carried out at AMLCI Characterization Facility a
State University.

Conflicts of Interest: The authors declare no conflicts of interest.

References

1. Walde, P.; Ichikawa, S. Enzymes inside lipid vesicles: Preparation, reactivity and applications. *Biomol. Eng.* **2001**, *18*, 14 [CrossRef]
2. Percec, V.; Wilson, D.A.; Leowanawat, P.; Wilson, C.J.; Hughes, A.D.; Kaucher, M.S.; Hammer, D.A.; Levine, D.H.; Kir Bates, F.S. Self-assembly of Janus dendrimers into uniform dendrimersomes and other complex architectures. *Scienc* *328*, 1009–1014. [CrossRef] [PubMed]
3. Kim, B.-S.; Park, S.W.; Hammond, P.T. Hydrogen-bonding layer-by-layer-assembled biodegradable polymeric micelles a delivery vehicles from surfaces. *ACS Nano* **2008**, *2*, 386–392. [CrossRef] [PubMed]
4. Gu, F.; Zhang, L.; Teply, B.A.; Mann, N.; Wang, A.; Radovic-Moreno, A.F.; Langer, R.; Farokhzad, O.C. Precise enginee targeted nanoparticles by using self-assembled biointegrated block copolymers. *Proc. Natl. Acad. Sci. USA* **2008**, *105*, 258 [CrossRef] [PubMed]
5. Guo, X.; Szoka, F.C. Chemical approaches to triggerable lipid vesicles for drug and gene delivery. *Acc. Chem. Res.* **2003**, *36*, 3 [CrossRef]
6. Mulet, X.; Boyd, B.J.; Drummond, C.J. Advances in drug delivery and medical imaging using colloidal lyotropic liquid crys dispersions. *J. Colloid Interface Sci.* **2013**, *393*, 1–20. [CrossRef] [PubMed]
7. Gulik-Krzywicki, T.; Dedieu, J.; Roux, D.; Degert, C.; Laversanne, R. Freeze—Fracture Electron Microscopy of Sheared La Phase. *Langmuir* **1996**, *12*, 4668–4671. [CrossRef]
8. Kato, T. Self-assembly of phase-segregated liquid crystal structures. *Science* **2002**, *295*, 2414–2418. [CrossRef]
9. Dong, R.; Zhong, Z.; Hao, J. Self-assembly of onion-like vesicles induced by charge and rheological properties in anionic–nc surfactant solutions. *Soft Matter* **2012**, *8*, 7812–7821. [CrossRef]
10. Wang, C.; Wang, Z.; Zhang, X. Amphiphilic building blocks for self-assembly: From amphiphiles to supra-amphiphile *Chem. Res.* **2012**, *45*, 608–618. [CrossRef] [PubMed]
11. Miravet, J.F.; Escuder, B.; Segarra-Maset, M.D.; Tena-Solsona, M.; Hamley, I.W.; Dehsorkhi, A.; Castelletto, V. Self-assemb peptide amphiphile: Transition from nanotape fibrils to micelles. *Soft Matter* **2013**, *9*, 3558–3564. [CrossRef]
12. Lombardo, D.; Kiselev, M.A.; Magazù, S.; Calandra, P. Amphiphiles self-assembly: Basic concepts and future perspecti supramolecular approaches. *Adv. Condens. Matter Phys.* **2015**, *2015*, 151683. [CrossRef]
13. Luo, J.; Liu, T.; Qian, K.; Wei, B.; Hu, Y.; Gao, M.; Sun, X.; Lin, Z.; Chen, J.; Bera, M.K. Continuous Curvature Chang Controllable and Responsive Onion-like Vesicles by Rigid Sphere–Rod Amphiphiles. *ACS Nano* **2020**, *14*, 1811–1822. [Cro [PubMed]
14. Gillissen, M.A.; Koenigs, M.M.; Spiering, J.J.; Vekemans, J.A.; Palmans, A.R.; Voets, I.K.; Meijer, E. Triple helix format amphiphilic discotics: Demystifying solvent effects in supramolecular self-assembly. *J. Am. Chem. Soc.* **2014**, *136*, 33 [CrossRef] [PubMed]
15. Percec, V.; Dulcey, A.E.; Balagurusamy, V.S.; Miura, Y.; Smidrkal, J.; Peterca, M.; Nummelin, S.; Edlund, U.; Hudson, S.D.; H P.A. Self-assembly of amphiphilic dendritic dipeptides into helical pores. *Nature* **2004**, *430*, 764–768. [CrossRef] [PubMed]
16. Xing, P.; Sun, T.; Hao, A. Vesicles from supramolecular amphiphiles. *RSC Adv.* **2013**, *3*, 24776–24793. [CrossRef]
17. Bisoyi, H.K.; Kumar, S. Liquid-crystal nanoscience: An emerging avenue of soft self-assembly. *Chem. Soc. Rev.* **2011**, *40*, 30 [CrossRef] [PubMed]
18. Kato, T.; Mizoshita, N. Self-assembly and phase segregation in functional liquid crystals. *Curr. Opin. Solid State Mater. Sci* *6*, 579–587. [CrossRef]
19. Boltenhagen, P.; Kleman, M.; Lavrentovich, O. Freeze-fracture observations in the Lα phase of a swollen surfactant in the v of the L3 and the L1 phase transitions. *J. Phys. II* **1994**, *4*, 1439–1448.
20. Lu, Y.; Lin, J.; Wang, L.; Zhang, L.; Cai, C. Self-assembly of copolymer micelles: Higher-level assembly for constructing hiera structure. *Chem. Rev.* **2020**, *120*, 4111–4140. [CrossRef]
21. Ghosh, S.; Ray, A.; Pramanik, N. Self-assembly of surfactants: An overview on general aspects of amphiphiles. *Biophys.* **2020**, *265*, 106429. [CrossRef]
22. Singh, S. Lyotropic Liquid Crystals. In *Handbook of Liquid Crystals—Volume II: Advanced Aspects and Applications*; Springer: Switzerland, 2024; pp. 203–243.

Guo, C.; Wang, J.; Cao, F.; Lee, R.J.; Zhai, G. Lyotropic liquid crystal systems in drug delivery. *Drug Discov. Today* **2010**, *15*, 1032–1040. [CrossRef] [PubMed]
Chavda, V.P.; Dyawanapelly, S.; Dawre, S.; Ferreira-Faria, I.; Bezbaruah, R.; Gogoi, N.R.; Kolimi, P.; Dave, D.J.; Paiva-Santos, A.C.; Vora, L.K. Lyotropic liquid crystalline phases: Drug delivery and biomedical applications. *Int. J. Pharm.* **2023**, *647*, 123546. [CrossRef] [PubMed]
Mohammady, S.Z.; Pouzot, M.; Mezzenga, R. Oleoylethanolamide-based lyotropic liquid crystals as vehicles for delivery of amino acids in aqueous environment. *Biophys. J.* **2009**, *96*, 1537–1546. [CrossRef] [PubMed]
Mezzenga, R.; Seddon, J.M.; Drummond, C.J.; Boyd, B.J.; Schröder-Turk, G.E.; Sagalowicz, L. Nature-Inspired design and application of lipidic lyotropic liquid crystals. *Adv. Mater.* **2019**, *31*, 1900818. [CrossRef] [PubMed]
Blanco-Fernández, G.; Blanco-Fernandez, B.; Fernández-Ferreiro, A.; Otero-Espinar, F.J. Lipidic lyotropic liquid crystals: Insights on biomedical applications. *Adv. Colloid Interface Sci.* **2023**, *313*, 102867. [CrossRef] [PubMed]
Fong, C.; Le, T.; Drummond, C.J. Lyotropic liquid crystal engineering–ordered nanostructured small molecule amphiphile self-assembly materials by design. *Chem. Soc. Rev.* **2012**, *41*, 1297–1322. [CrossRef] [PubMed]
Gong, X.; Moghaddam, M.J.; Sagnella, S.M.; Conn, C.E.; Danon, S.J.; Waddington, L.J.; Drummond, C.J. Lyotropic liquid crystalline self-assembly material behavior and nanoparticulate dispersions of a phytanyl pro-drug analogue of capecitabine—A chemotherapy agent. *ACS Appl. Mater. Interfaces* **2011**, *3*, 1552–1561. [CrossRef] [PubMed]
Van't Hag, L.; Gras, S.L.; Conn, C.E.; Drummond, C.J. Lyotropic liquid crystal engineering moving beyond binary compositional space–ordered nanostructured amphiphile self-assembly materials by design. *Chem. Soc. Rev.* **2017**, *46*, 2705–2731. [CrossRef]
Lavrentovich, O.D. Hierarchy of defect structures in space filling by flexible smectic-A layers. *Sov. Phys. JETP* **1986**, *64*, 1666–1676.
Kurik, M.; Lavrentovich, O. Monopole structures and shape of drops of smectic C. *Zh. Eksp. Teor. Fiz.* **1983**, *85*, 11–526.
Robinson, C. Liquid-crystalline structures in solutions of a polypeptide. *Trans. Faraday Soc.* **1956**, *52*, 571–592. [CrossRef]
Li, Y.; Jun-Yan Suen, J.; Prince, E.; Larin, E.M.; Klinkova, A.; Thérien-Aubin, H.; Zhu, S.; Yang, B.; Helmy, A.S.; Lavrentovich, O.D.; et al. Colloidal cholesteric liquid crystal in spherical confinement. *Nat. Commun.* **2016**, *7*, 12520. [CrossRef] [PubMed]
Li, Y.; Prince, E.; Cho, S.; Salari, A.; Mosaddeghian Golestani, Y.; Lavrentovich, O.D.; Kumacheva, E. Periodic assembly of nanoparticle arrays in disclinations of cholesteric liquid crystals. *Proc. Natl. Acad. Sci. USA* **2017**, *114*, 2137–2142. [CrossRef] [PubMed]
Kurik, M.; Lavrentovich, O. Topological defects of cholesteric liquid crystals for volumes with spherical shape. *Mol. Cryst. Liq. Cryst.* **1982**, *72*, 239–246. [CrossRef]
Hyun-Gyu, L.; Sundas, M.; Soo-Young, P. Cholesteric Liquid Crystal Droplets for Biosensors. *ACS Appl. Mater. Interfaces* **2016**, *8*, 26407–26417.
Geng, Y.; Seč, D.; Almeida, P.L.; Lavrentovich, O.D.; Žumer, S.; Godinho, M.H. Liquid crystal necklaces: Cholesteric drops threaded by thin cellulose fibres. *Soft Matter* **2013**, *9*, 7928–7933. [CrossRef]
Fialho, A.R.; Bernardino, N.R.; Silvestre, N.M.; Telo da Gama, M.M. Effect of curvature on cholesteric liquid crystals in toroidal geometries. *Phys. Rev. E* **2017**, *95*, 012702. [CrossRef] [PubMed]
Seč, D.; Porenta, T.; Ravnik, M.; Žumer, S. Geometrical frustration of chiral ordering in cholesteric droplets. *Soft Matter* **2012**, *8*, 11982–11988. [CrossRef]
Lopez-Leon, T.; Fernandez-Nieves, A. Drops and shells of liquid crystal. *Colloid Polym. Sci.* **2011**, *289*, 345–359. [CrossRef]
Zhou, Y.; Bukusoglu, E.; Martínez-González, J.A.; Rahimi, M.; Roberts, T.F.; Zhang, R.; Wang, X.; Abbott, N.L.; De Pablo, J.J. Structural transitions in cholesteric liquid crystal droplets. *ACS Nano* **2016**, *10*, 6484–6490. [CrossRef]
Micoine, K.; Hasenknopf, B.; Thorimbert, S.; Lacôte, E.; Malacria, M. A general strategy for ligation of organic and biological molecules to Dawson and Keggin polyoxotungstates. *Org. Lett.* **2007**, *9*, 3981–3984. [CrossRef]
Lin, Z.; Yang, X.; Xu, H.; Sakurai, T.; Matsuda, W.; Seki, S.; Zhou, Y.; Sun, J.; Wu, K.-Y.; Yan, X.-Y. Topologically directed assemblies of semiconducting sphere–rod conjugates. *J. Am. Chem. Soc.* **2017**, *139*, 18616–18622. [CrossRef] [PubMed]
Zhang, R.; Feng, X.; Zhang, R.; Shan, W.; Su, Z.; Mao, J.; Wesdemiotis, C.; Huang, J.; Yan, X.Y.; Liu, T. Breaking parallel orientation of rods via a dendritic architecture toward diverse supramolecular structures. *Angew. Chem. Int. Ed.* **2019**, *58*, 11879–11885. [CrossRef] [PubMed]
Gao, M.; Kim, Y.K.; Zhang, C.; Borshch, V.; Zhou, S.; Park, H.S.; Jákli, A.; Lavrentovich, O.D.; Tamba, M.G.; Kohlmeier, A. Direct observation of liquid crystals using cryo—TEM: Specimen preparation and low-dose imaging. *Microsc. Res. Tech.* **2014**, *77*, 754–772. [CrossRef]
Kędzierski, J.; Raszewski, Z.; Kojdecki, M.; Kruszelnicki-Nowinowski, E.; Perkowski, P.; Piecek, W.; Miszczyk, E.; Zieliński, J.; Morawiak, P.; Ogrodnik, K. Determination of ordinary and extraordinary refractive indices of nematic liquid crystals by using wedge cells. *Opto-Electron. Rev.* **2010**, *18*, 214–218. [CrossRef]
Kowiorski, K.; Kędzierski, J.; Raszewski, Z.; Kojdecki, M.; Herman, J.; Miszczyk, E.; Piecek, W. Application of modified interference wedge method in measurements of indices of refraction and birefringence of nematic liquid crystals. *Acta Phys. Pol. A* **2013**, *124*, 946–948. [CrossRef]

49. Mourdikoudis, S.; Pallares, R.M.; Thanh, N.T.K. Characterization techniques for nanoparticles: Comparison and compleme upon studying nanoparticle properties. *Nanoscale* **2018**, *10*, 12871–12934. [CrossRef] [PubMed]
50. Chen, J. Advanced electron microscopy of nanophased synthetic polymers and soft complexes for energy and medicine a tions. *Nanomaterials* **2021**, *11*, 2405. [CrossRef] [PubMed]
51. Zhang, C.; Sadashiva, B.; Lavrentovich, O.; Jákli, A. Cryo-TEM studies of two smectic phases of an asymmetric bent-core m *Liq. Cryst.* **2013**, *40*, 1636–1645. [CrossRef]
52. Repula, A.; Grelet, E. Elementary edge and screw dislocations visualized at the lattice periodicity level in the smectic pl colloidal rods. *Phys. Rev. Lett.* **2018**, *121*, 097801. [CrossRef] [PubMed]
53. Pershan, P.S. Dislocation effects in smectic—A liquid crystals. *J. Appl. Phys.* **1974**, *45*, 1590–1604. [CrossRef]
54. Abukhdeir, N.M.; Rey, A.D. Edge dislocation core structure in lamellar smectic-A liquid crystals. *Soft Matter* **2010**, *6*, 111 [CrossRef]
55. Lelidis, I.; Blanc, C.; Kléman, M. Optical and confocal microscopy observations of screw dislocations in smectic-A liquid c *Phys. Rev. E—Stat. Nonlinear Soft Matter Phys.* **2006**, *74*, 051710. [CrossRef]
56. de Gennes, P.G. An analogy between superconductors and smectics A. *Solid State Commun.* **1972**, *10*, 753–756. [CrossRef]
57. Kleman, M.; Williams, C.E. Interaction between parallel edge dislocation lines in a smectic A liquid crystal. *J. Phys. Lett* *35*, 49–51. [CrossRef]
58. Kleman, M. Defects in liquid crystals. *Rep. Prog. Phys.* **1989**, *52*, 555. [CrossRef]
59. Allain, M.; Kleman, M. Thermodynamic defects, instabilities and mobility processes in the lamellar phase of a non-ionic sur *J. Phys.* **1987**, *48*, 1799–1807. [CrossRef]
60. Kleman, M. Surfactant-Water Systems with Small Layer Rigidity Phase Stability, Defects Models and Defects Mobi Polyoxyethylene Surfactant with Water. *Liq. Cryst.* **1988**, *3*, 1355–1367. [CrossRef]
61. Kralj, S.; Sluckin, T. Core structure of a screw disclination in smectic-A liquid crystals. *Phys. Rev. E* **1993**, *48*, R3244. [Cro [PubMed]
62. Kleman, M.; Lavrentovich, O.D. *Soft Matter Physics: An Introduction*; Springer: New York, NY, USA, 2003.
63. Lavrentovich, O.D. Topological defects in dispersed liquid crystals, or words and worlds around liquid crystal drops. *Liq.* **1998**, *24*, 117–126. [CrossRef]
64. Volovik, G.E.; Lavrentovich, O.D. Topological dynamics of defects: Boojums in nematic drops. *Zh. Eksp. Teor. Fiz.* *85*, 1997–2010.
65. Kim, W.; Kim, D.S.; Yoon, D.K. Surface-induced orientation of liquid crystal phases. *Giant* **2024**, 100324. [CrossRef]
66. Lavrentovich, O. Defects in liquid crystals: Surface and interfacial anchoring effects. In *Patterns of Symmetry Breaking*; Sp Dordrecht, The Netherlands, 2003; pp. 161–195.
67. Faetti, S. Anchoring at the interface between a nematic liquid crystal and an isotropic substrate. *Mol. Cryst. Liq. Cryst* *179*, 217–231. [CrossRef]

MDPI AG
Grosspeteranlage 5
4052 Basel
Switzerland
Tel.: +41 61 683 77 34

Crystals Editorial Office
E-mail: crystals@mdpi.com
www.mdpi.com/journal/crystals

www.ingramcontent.com/pod-product-compliance
Lightning Source LLC
Chambersburg PA
CBHW041359210926
51850CB00013B/3127
9783725838738